AGRI-FOOD GLOBALIZATION IN PERSPECTIVE

Agri-food Globalization in Perspective

International Restructuring in the Processing Tomato Industry

BILL PRITCHARD
University of Sydney, Australia

DAVID BURCH
Griffith University, Australia

ASHGATE

Published by
Ashgate Publishing Limited
Gower House
Croft Road
Aldershot
Hants GU11 3HR
England

Ashgate Publishing Company
Suite 420
101 Cherry Street
Burlington, VT 05401-4405
USA

Ashgate website: http://www.ashgate.com

British Library Cataloguing in Publication Data
Pritchard, Bill
Agri-food globalization in perspective : international restructuring in the processing tomato industry
1.Tomato industry - Economic aspects 2.Tomatoes - Processing
I.Title II.Burch, David, 1942 Jan. 16-
338.1'75642

Library of Congress Cataloging-in-Publication Data
Pritchard, Bill.
Agri-food globalization in perspective : international restructuring in the processing tomato industry / William Pritchard and David Burch.
p. cm.
Includes bibliographical references and index.
ISBN 0-7546-1508-1
1. Tomato industry. 2. Tomato industry--Technological innovations. 3. Canned foods industry--Technological innovations. 4. Canned tomatoes. 5. Tomato products. 6. Tomato sauces. 7. Globalization--Economic aspects--Case studies. I. Burch, David, 1942 Jan. 16- II. Title.

HD9330.T72P75 2003
338.4'7664805642--dc21

2002043972

ISBN 0 7546 1508 1

Printed in Great Britain by Antony Rowe Ltd, Chippenham, Wiltshire

Contents

List of Figures

List of Tables

List of Plates

About the Authors

Bill Pritchard is Senior Lecturer in Economic Geography at the University of Sydney. He has published widely in the area of agricultural restructuring, food company strategy and food retailing, and has additional research interests in the field of regional development and change. He is the co-author of *Developing Australia's Regions* (UNSW Press, 2003) and co-editor of *Land of Discontent* (UNSW Press, 2000), as well as being a contributor to *Restructuring Global and Regional Agricultures* (Ashgate, 1999) and *Globalization and Agri-Food Restructuring* (Avebury, 1996). Prior to his current position, he was employed in various policy and research positions in the private, public and community sectors.

David Burch is Professor in the School of Science and director of the Science Policy Research Centre, Griffith University, Brisbane. He is the convenor of the Agri-Food Research Network, and co-edited *Restructuring Global and Regional Agricultures* (Ashgate, 1999), *Globalization and Agri-Food Restructuring* (Avebury, 1996), and *Farming in a Globalised Economy* (Monash University, 1998). His most recent book is the co-authored textbook *Science, Technology and Society: An Introduction* (Cambridge, 1998). His research interests focus on agri-food restructuring in southeast Asia and resource use in Australian agriculture.

Preface

This book reflects a labor of love with the tomato, perhaps not surprising for a fruit known historically as the 'love apple', or *pomme d'amour*. In 1999, during an interview with an Australian tomato grower, we were told how difficult it was to 'make a dollar' in this industry. Then why do you keep growing tomatoes, we asked? 'Tomatoes get under your skin', we were told. 'It doesn't give me much money but I like growing tomatoes'.

The completion of this book was not dissimilar. This book was not funded by a dedicated grant to research the tomato industry, but mainly through our ability to deploy discretionary research monies to the project. The tasks of gathering information and translating this into the written text of this book occurred through short bursts of activity, over a three-year period, in between our funded research and teaching obligations. We initially 'sowed the seeds' of the idea for this book in late 1999, and brought it to harvest in late 2002.

As this book evolved, we were drawn progressively into wider debates about the nature of food and globalization in the modern world. Accordingly, the subject matter of this book addresses both the specific details of recent changes in the processing tomato industry, and the wider insights these narratives suggest for an understanding of global restructuring in agriculture and food industries.

In this regard, the key issue here is to challenge generalized accounts of agri-food globalization. It is easy to glibly refer to 'a globalized food system', imagining a seamless and vertically coordinated structure dominated by profit maximizing and rationally behaving transnational corporations. Anyone observing the global march of the 'Golden Arches' over the past two decades could easily conclude that, in certain key ways, this formulation does have merit as a way of thinking about the world's food industries.

However, we argue that for the most part this model does not provide an accurate description of food production and consumption in the contemporary world. What passes for 'the global food system' consists of a set of heterogenous and fragmented processes, bounded in multiple ways by the separations of geography, culture, capital and knowledge. At times these separations are compressed, via corporate takeovers, global branding or the emergence of dramatic new trade relations linking hitherto esparate arenas of production and consumption. But these compressions are neither as smooth, seamless, or permanent as some analysts suggest. The food industry is characterized by low returns to capital; significant exposures to risk; the specificities of culturally and spatially embedded consumption; and sharp variations in the capture of competitive advantage. Global agri-food restructuring needs to be understood as an intricate set of processes operating at many scales, and on many levels, rather than as a unilateral shift towards a single global marketplace.

These arguments are exemplified in the restructuring experiences of the processing tomato industry. This is a particularly interesting commodity system to study, because it is closely attuned to the impulses of globalized competition and change. Tomato products are consumed throughout the world. They are key ingredients in many foods which are typical of contemporary global consumption trends (e.g., pizzas), and some of the world's largest transnational food companies are heavily involved in this production sector. Yet, as the detail of this book reveals, there is rich and unique texture to this sector's experiences of restructuring. The world's processing tomato industries are increasingly being shaped by international flows of products, technologies, ideas and personnel, but the outcomes of these processes do not present as a rational and stabilized, global market place. Rather, what has emerged so far, is an increasingly volatile landscape of winners and losers. Contemporary restructuring of the processing tomato sector is multi-dimensional, exhibits spatial difference, and occurs at a number of different scales. Apparently similar developments in the industry (say, the introduction of mechanized harvesting) are played out in unique ways, in different socio-spatial environments.

Observing the empirical detail of this book, some analysts might conclude that because restructuring is 'messy', indeterminate and varies across space, there is no single logic that accounts for the global transformation of this sector. It could be argued that it is futile to seek a 'grand explanatory narrative' to account for change, and a better course would be to document the various paths of restructuring and rest modestly in meso-scale appreciation of these differences.

We reject such propositions. The processes that create surface manifestations of difference in the processing tomato sector are linked ultimately to a particular, historically specific set of economic conditions. The vital element of recent change involves the elevation of global financial mobility as an agent of restructuring. This has accelerated the pace of restructuring, heightening instability at a global scale, and creating a disjuncture between the price of commodities and their historical cost structures. In this environment, economic actors become increasingly exposed to potentially alternative profit structures, and correspondingly adopt a variety of strategies to extract value and profit in different geographical sites.

Because of this, the pressures to restructure this industry affect some activities more than others. Some value chain components (e.g., the bulk production of industrial grade tomato paste) are especially vulnerable to change, as competitiveness moves from one site to another and geographically flexible operators emerge to take advantage of these shifts. Other components of these chains, particularly those rich in intellectual property (e.g., hybrid seed development), are exposed to a different set of restructuring forces. Consequently, although the manifestations of global restructuring are diverse, they are underpinned by economic processes that identify, separate and exploit profit opportunities more intimately.

On the basis of this analysis, this book argues that the processing tomato industry of the twenty-first century will consist of an uneven landscape involving heightened volatility as various actors (retail outlets, processing firms, machinery manufacturers, grower associations, etc) adjust their strategies in line with shifts in international competition. Some regions and players will be more vulnerable than others. There is considerable truth in the axiom, relayed by many players in this sector, that this is a 'regional industry', meaning that the sector's fundamental geographical logic will result in it remaining wedded to continental 'blocs' (North America, Europe, etc). Actors and processes located at the margin of these 'blocs', and involved in relatively low-value activities (for example, the production of bulk industrial tomato paste) will bear the brunt of global restructuring.

Before readers embark on the chapters that follow, two cautionary notes are necessary. First, this book has been written with two clear audiences in mind; professionals in the processing tomato industry, and academics interested in agri-food globalization. Certain sections of the book therefore, will appeal to some readers more than others. To use a phrase from the processing tomato industry, this is a 'co-branded' product with which we hope to reach two markets on the basis of a single input. Second, our focus here is on the *processing*, and not the *fresh* tomato industry. Both of us have lost count of the number of times people have asked us 'why don't supermarket tomatoes have any flavor these days?'. The answer to this question is not found here.

Throughout the course of preparing this manuscript we discussed the processing tomato industry with an enormous range of industry participants. This research could not have been completed without that assistance. A list of all those persons to whom we owe thanks is provided below. A number of people, however, stand out as deserving particular thanks.

First, the origins of this book owe much to a series of meetings in 1994 and 1995 with the then Industry Development Officer of the Australian Processing Tomato Research Council, Lauren Thompson. Lauren helped us gain an initial understanding of the dynamics of this industry, and urged us to investigate these in greater depth. A few years later, our interest in this industry was rekindled through the support of key participants in the Australian processing tomato industry, notably Louis and Geraldine Chirnside, Rob and Cheryl Hoskins, Peter and Jenny Gray, Liz Mann, Tim and Ann Maree Dyer, Jamie McMaster, and Barry and Lesley Horn. These individuals provided us with a sounding board for our ideas and a source of support within the local industry (although, we emphasize, this book is independent of any interests within the Australian processing tomato industry). We also owe thanks to Michael Kay, of FMC, who we first met in early 2001, when he was employed at the Morning Star Packing Co. Michael's insights into 'the way things come together' provided us with a quantum leap in our understandings of this industry. Finally, we acknowledge the contribution of Bill Friedland, who not only pioneered research in this and other commodities, but who also made available his files and data on the early Californian industry. Of course none of the above persons, nor anyone mentioned below, is responsible for the arguments in this book. The positions we take, and any errors or omissions herein, are entirely our own.

Preparation of this manuscript also could not have occurred without the research assistance at key points of Rebecca Curtis and Jasper Goss, the cartographic skills of Cathi Greve, and the editorial assistance of Nicole Hall. The ongoing support of colleagues at the University of Sydney and Griffith University has helped us throughout the development of this book, as have our links with fellow members of the Australian and New Zealand Agri-Food Research Network, and RC-40 (the International Sociological Association's Research Committee on Agriculture and Food). Our research on the Chinese and Italian tomato industries was assisted through the Australian Research Council (grant A00104175).

In addition to those key informants listed above, we also wish to thank the following individuals for their time and cooperation. In Australia: Ian Bryce, Jim Geltch, Stephen Little, Bryce Merrett, Simon Mills, Nol Neessen, and Rob Rendell. In France: Bernard Bieche, Olivier Vallat, Michel Brousse, Robert Giovinazzo and Patrick Mathiot. In Italy: Dr Carrara Ermenegildo, Marco Crotti, Luciano Bertoni, Fabio Bonvini, Silvestro Pieracci, Giovanni Losa, Armando Gandolfi, Sergio Rizzi, Alfred Giannantonio, Vincenzo Russo, Franco De Martino, Andrea Boscolo, and Antonio Ferraioli. In Spain: Ruth Rama, Alicia Langreo Navarro, and Alfonso Martin. In Turkey: Duncan Blake and Seref Alptekin. In the United States: Mark Evans, Mike Murray, John Welty, Kebede Daniel Gashaw, Dan Sumner, and Brad Rickard. In China: Tania He, Huang Mu, John S. Linn, Jeffrey Wong, Feng Xin Ming, Yu Feng Zhang, Zhang Guoxi, Fuguang Zhong, Peng Zi Lei, Zhou Ai Jiang, and Gordon Wan. In Canada: Dennis Chartrand, Dave Epp, Jane Graham, Peter Hastings, Henry Iacobelli, Jean Laprise, Scott Makey, Larry Martin, John Mumford, Ron Pitblado, Jim Richards, Dale Smith, 'Bub' Stevens, and Mike Trant. In Thailand: Yao Chin-Tsai, Professor Thaworn Kowithayakorn, Fareeda Manichapong, Suhate Mangkang, Watcharuwat Sawamis and Santi Suwannus.

And of course, this book could not have been completed without the support (and tolerance) of Kerry and Arizona Hart, and Anna and Sam Burch, who persevered through our extensive absences and long hours in spent in finalizing this study. We thank them for their help.

List of Abbreviations

ABARE	Australian Bureau of Agricultural and Resource Economics
ACCC	Australian Consumer and Competition Commission
ACS	Australian Customs Service
ADA	Anti-Dumping Authority (Australia)
ADB	Asian Development Bank
AFPA	*Agricultural Fair Practices Act* (US)
AMITOM	*Association Méditerranéenne Internationale de la Tomate Transformée* (French acronym for Mediterranean International Association for the Processing Tomato)
AU	Australia
BIL	Brierley Investments Ltd (New Zealand)
BILD	Board of Industrial Leadership and Development (Ontario)
BOI	Board of Investment (Thailand)
BSE	Bovine Spongiform Encephalopathy
CA	Canada
CAP	Common Agricultural Policy (EU)
CDC	Commonwealth Development Corporation (Thai investor)
CEO	Chief Executive Officer
CFIS	Canned Food Information Service (Australia)
CIQ	Chinese Import/Export Inspection Agency (Chinese acronym)
CITT	Canadian International Trade Tribunal
CNCA	National Catholic Agricultural Confederation (Spain)
CNT	National Labor Confederation (Spain)
CTGA	California Tomato Growers Association
CUNA	*Cooperative d'Utilisation Nationale Agricole* (France)
CUSTA	Canada-US Trade Agreement
ECU	European Currency Units (pre-2002)
EU	European Union
FMC	Food Machinery Company Ltd (US)
FOB	free-on-board
FTC	Federal Trade Commission (US)
GATT	General Agreement on Tariffs and Trade
GDP	Gross Domestic Product
GFS	General Finance and Securities Ltd (Thailand)
GM	genetically modified
ha	hectare
IMF	International Monetary Fund
ISO	International Standards Organization
KFC	Kentucky Fried Chicken

KKR	Kohlberg Kravis Roberts Ltd (US)
lb	pound (Imperial weight measure)
MFA	multifunctionality of agriculture
MOT	material other than tomatoes
NACO	Northeast Agricultural Company Ltd (Thailand)
NACs	new agricultural countries
NAFTA	North American Free Trade Agreement
NAICO	Northeast Agricultural Investment Company Ltd (Thailand)
NGO	Non-government organization
OECD	Organization for Economic Cooperation and Development
OPVG	Ontario Processing Vegetable Growers Inc.
OVGMB	Ontario Vegetable Growers' Marketing Board
oz	ounce (Imperial weight measure)
PACA	Provence Alpes and Cote d'Azur (French region)
PDF	Pacific Dunlop Foods Ltd (Australia)
PLA	Peoples' Liberation Army (China)
PPT	post-productivist transition (in agriculture)
PTAB	Processing Tomato Advisory Board (California)
SPC	Shepparton Preserving Company Ltd (Australia)
TINC	Tomato Industry Negotiating Committee (Australia)
TNC	transnational corporation
TPINC	Tomato Processing Industry Negotiating Committee (Australia)
UCD	University of California at Davis
UCSC	University of California at Santa Cruz
UFC	Universal Food Company Ltd (Thailand)
UFW	United Farm Workers (California labor union)
UK	United Kingdom
UNCTAD	United Nations Conference on Trade and Development
US	United States
USDA	United States Department of Agriculture
WPTC	World Processing Tomato Council
WTO	World Trade Organization
WTOAA	World Trade Organization Agreement on Agriculture
XPCC	Xinjiang Production and Construction Corps
XPCG	Xinjiang Production and Construction Group

Plate 1 **Drums of aseptic-packaged tomato paste, Parma, Italy, 2001**

Photo: Bill Pritchard.

Chapter 1

Introduction

Across the world, recent decades have witnessed profound changes in terms of what people eat, where their food comes from, and who controls and manages the systems by which agri-food products travel from farm to consumer. All too often, the concept of 'agri-food globalization' is enrolled to describe and account for these processes of transformation.

Yet for the thousands of articles and books written about agri-food globalization in recent years, its implications remain nebulous. In broad terms, the parameters of the concept are reasonably well-understood: agri-food globalization is generally associated with reductions in barriers to trade and investment among nations, so that commerce increasingly takes place within an international arena of production and competition. But how advanced are these developments, are they inevitable, and what do they mean for farmers, companies and consumers? Researchers in the area of agri-food studies are divided on these questions, and the issue of how to answer them. The contribution of this book is to use a detailed investigation of one international commodity system, the processing tomato industry, to identify the dimensions and character of 'agri-food globalization'.

The humble tomato exemplifies key attributes in the contemporary transformation of world food production and consumption. This commodity was internationalized four centuries ago, when Spanish conquistadors brought to Europe some small, vine-like plants bearing strange, red fruit. Yet although the tomato plant has for centuries been propagated across continents, its products have nearly always been consumed near to where they were grown. Not until the late nineteenth century, thanks largely to the insights and innovations introduced by the pioneers of modern food processing such as Henry Heinz, were significant quantities of tomatoes grown for industrial food production. Yet even with these innovations, the economic geography of this commodity remained embedded within regional and national production systems. Until relatively recently, the volume of international trade in processing tomato products (for example, canned tomatoes and tomato paste) was relatively small and occurred within stable supply systems orchestrated between neighboring countries (such as the exporting countries of southern Europe and the importing countries of northern Europe).

In recent years, significant changes have been wrought to the global economic geography of the processing tomato system. Spurred on by significant growth in international demand for processing tomato products (in part due to the rise of pizza consumption), processing tomatoes are being grown, canned, pasted and traded in record volumes. Alongside these developments has come unparalleled uncertainty and change. New production and trade systems have emerged, forcing existing industry players along the path of continuous restructuring and adaptation. Government regulations and support regimes for this industry have been re-written, creating new arenas for competitive advantage. This book documents these recent developments, and asks what light they shed on our understanding of contemporary global agri-food restructuring.

Given the dramatic scope of recent change in many elements of the global economy and society, including the emergence of the Internet, shifts in global geopolitics and ideology following the end of the Cold War, and massive growth in cross-border investment, it is tempting to interpret contemporary global agri-food restructuring within a revolutionary framework. For example, the significant changes in the global processing tomato industry over recent years could be seen as an element in a shift towards a 'globalized agri-food system', implying the emergence of a globally integrated industrial system operating within an overall logic of profit and characterized by international flexibility in product sourcing. This book, however, argues against such an interpretation. While elements of these directions of change certainly exist, the central contention here is that the model of a 'globalized agri-food system' over-states and misinterprets the role of globalization in the contemporary global agri-food system.

Rather than viewing contemporary agri-food restructuring as an unproblematic 'project' of accelerated globalization, this book adopts an approach that interprets contemporary change as a contested historical process, featuring a complex interplay of developments at varying scales and in different geographical territories. This is occurring because the conditions for profit accumulation in agri-food industries, and their relationship to national agricultural systems, are shaped by varied and contradictory forces. The individual chapters of this book explicate these processes with respect to a range of different themes, including agricultural technology and restructuring, contract bargaining, farm subsidies, and competition from lower cost, third world producers. The common thread that unites these chapters is that, notwithstanding the scope and 'newness' of recent changes in the processing tomato industry, it is a mistake to interpret restructuring trajectories in the industry within a simplifying, all-embracing, framework that assumes a dominant shift towards global uniformity in production systems, product types, and culinary cultures.

Why tomatoes?

The significance of this book derives from the fact that, in seeking insights into the question of agri-food globalization, we investigate the specific details of a single commodity system, namely, the processing tomato industry. The term *processing tomatoes* refers to those tomatoes grown specifically for conversion to paste, or to be canned, juiced or dried. For readers not familiar with this industry, our selection of this commodity might appear quixotic. Superficially (at least to some readers outside the industry), an investigation into the 'global processing tomato system' may not resonate with the apparent significance that an analysis of some other agri-food commodity systems might hold. We argue, however, that the processing tomato system is precisely the kind of commodity sector likely to reveal major insights into global agri-food restructuring at the current time.

At one level, the processing tomato sector is surprisingly large in size and scope. Outsiders to the industry rarely appreciate the international dimensions of this commodity system. Around the start of the twenty-first century, approximately 25 million tonnes of processing tomatoes were grown annually. The commodity has a significant presence on every inhabited continent, and at the *IV World Processing Tomato Congress*, held in Sacramento, California in June 2000, 435 delegates attended from 34 countries.[1]

At another level, this book examines an industrial food commodity which is at the forefront of current trends in food consumption. Processing tomato products are key ingredients in many prepared, packaged and 'fast' foods, as well as being in the forefront of emergent trends such as functional foods and organic diets. Tomato-based products are key inputs in specialized regional food complexes and diets, such as Mediterranean and Mexican cuisines, which are themselves expanding globally. Moreover, the history of the processing tomato industry has been closely associated with new developments at the interface of science, technology and food consumption. Tomato consumption was restructured and adapted in the nineteenth and early twentieth centuries with the emergence of the new food processing industry. These developments coincided with and reinforced continuous and extensive plant variety research and, ultimately in the 1960s, the development of mechanized harvesting. In the 1990s the commodity was again at the forefront of research and public debate, with the rise of genetically modified tomatoes. Thus, the processing tomato industry provides powerful insights into the articulations of food restructuring, science and public opinion. In other words, we argue that it is possible to see the wider outline of global food relations, within the specific story of the processing tomato.

However, our principal interest in this book lies in the fact that the industry is engaged currently in extensive, global-scale restructuring. These processes are challenging existing geographical structures and can be held to account for the ongoing reorganization of the industry at an international level. The question of how we understand and analyze these processes, is the central issue addressed here. To engage in this task, some brief historical context is required.

The tomato has been an internationalized food commodity for over four hundred years. In 1544 the plant was introduced to Spain, and by 1570 the fruit had spread from Italy to England (Khoo, 2000). In the intervening centuries tomatoes were diffused all over the world. It is impossible to imagine Italian pasta without tomato-based sauces, Mexican food without salsa, or a New York hot dog without ketchup. However, the internationalization of tomatoes was neither without controversy or resistance. In England (which is far from an ideal growing climate) the tomato was introduced as a decorative garden plant. For many years, in England and elsewhere, tomatoes were alternately regarded as an aphrodisiac, a poison, or a medicine (Smith, 1994, p.14).

The United States provides the best-documented case of the slow and socially contested diffusion of the tomato. Although tomatoes apparently were introduced into Florida in the 1600s (Khoo, 2000), they made minimal impression on the dietary habits of Americans until two centuries later. The first historically recorded cultivation of tomatoes outside Florida occurred in a Pennsylvanian jail garden in 1802 (Smith, 1994, p.19), although the fruit had apparently been grown and used in small quantities for a couple of decades beforehand. In any case, most Americans at this time seemingly regarded the tomato as poisonous, a myth that was only dispelled after a brazen public display of their consumption on the courthouse steps of Salem, New Jersey, in 1820 (Smith, 1994, pp.1–5).[2]

From its earliest culinary days, the tomato was consumed both as a fresh and a processed product. By 1780 the English diet had been introduced to tomato sauce, which 'was put on anything sent to the table' (Smith, 1994, p.20). Some cynics would say nothing has changed since! Through the nineteenth century, an extensive artisanal and household-based production system developed for the manufacture of tomato sauces, pickles and other products. Ketchup, which differed from tomato sauce because it was intended for preservation rather than being served shortly after preparation, seems to have been introduced to American cuisine by French cooks in the late 1700s or early 1800s (Smith, 1994, pp.88–89). Tomato canning also developed in the nineteenth century, though initially, consumers regarded it with suspicion since rudimentary food processing technologies led to high rates of dangerous spoilage. Nevertheless, kitchen apparatus were developed in this era to facilitate household canning and bottling, further familiarizing the consumer with processing tomato products and paving the way for the rise of larger scale industrial processing companies. In 1847, Harrison Woodhull Crosby, Assistant Steward and Chief Gardener of Lafayette College, Easton, Pennsylvania, first developed a process for tomato canning. He converted the college refectory into a laboratory, where he 'soldered his lids onto small tin pails, stuffed some "love apples", or tomatoes, through holes in the lids, soldered tin plates over these holes and immersed the sealed cans in boiling water' (Gould, 1992, p.7). In 1869 H.J. Heinz Co. was incorporated, and such processes became routinized and large-scale. With the passage of the *Pure Food and Drug Act* in 1906, the US Government possessed a mandate to close unsafe operators, bolstering consumer confidence in processed foods and giving increased legitimacy to larger and relatively more reliable firms such as Heinz.

In the Anglo-American countries, the processing tomato emerged during the first half of the twentieth century as a key food within an emerging 'fordist' food consumption regime. Tomatoes were used extensively in canned foods such as soups, baby foods and, in Britain particularly, baked beans. Ketchup, tomato soup and other products became industrialized staples of urban, Anglo-American diets within a 'fordist' system of mass production and consumption (a fact encapsulated neatly by Andy Warhol's 'Campbell Soup' series). The tomato came to be seen as a highly versatile product which, although in botanical terms a fruit, was constructed socially as a vegetable.[3] The revolutionizing effects of World War Two on food technologies, which resulted from unprecedented demands for non-perishable foodstuffs, reinforced these themes. The 1950s and 1960s was a golden era for industrialized consumer food companies with extensive interests in processing tomatoes, such as Heinz and Campbell Soup. By this time, these companies had established pre-eminent positions within North American core markets and had expanded into the like-dietary environments of the United Kingdom, Australia, New Zealand, South Africa and, increasingly, Latin America.

In continental Europe the evolution of the industry followed a very different course over this period. Although Heinz was a major force in modernizing the processing tomato industry of northern Italy from the early 1950s, in Europe the processing tomato industry remained much more in the hands of smaller, regional and family-owned companies. In many parts of southern Europe tomatoes were traditionally bottled on the Assumption Day festival of 15 August (in Italy, *la festa dell'Assunzione*), with the opening of the first bottle having great celebratory significance. In the absence of retail structures with national and continental reach, consumer products were positioned largely within sub-national markets, and to this day, hundreds of tomato canneries, *passata*[4] and tomato paste manufacturers dot the landscapes of Italy, Greece, Spain, France and Portugal. The introduction of European Union subsidy payments for the tomato industry in the late 1970s reinforced these tendencies by giving support to small and otherwise vulnerable firms (see Chapter Five).

Outside North America and Europe, canning and paste production has grown significantly throughout the eastern Mediterranean (Turkey, Israel), North Africa (Morocco, Tunisia, Algeria, Libya) and the Middle East (Jordan, Saudi Arabia, Iran), mainly to service local and neighboring markets. Nevertheless, because processing tomatoes are central ingredients within regional cuisines, and, because populations are steadily growing, these markets remain import-dependent. To service this demand, strong trade linkages have developed for the import of processing tomato products from Turkey and Italy.[5] EU production subsidies and export restitution payments, have facilitated Italian exports. In Latin America, Australia and Africa, production has grown largely in response to rising domestic demand, while during the Cold War the Eastern Bloc maintained self-sufficiency (there were significant processing tomato operations in Hungary, Bulgaria and Georgia). Until recently, there was minimal demand for processing tomato products in Asia, but dietary changes and the wish to develop new agri-food exports have led to a number of major projects in Thailand and, especially, China.

Over the period 1998-2002, global processing tomato production levels fluctuated between 23 and 30 million tonnes of raw product (Table 1.1). In a typical year, California accounts for approximately one-third of global production, while the five EU producing nations (Italy, Spain, Greece, Portugal and France) account for another one-third. The massive North American market is now dominated by a handful of large branded food companies who, in addition to their domestic operations, operate multi-domestic corporate strategies in a number of like-dietary economies. In the equally massive European market, extensive subsidization and distinctive culinary cultures have come to support a highly fragmented, relatively high-cost, production regime. As a result of historical linkages, proximity and the price-depressing effects of subsidies, surplus European production finds markets in northern Europe (the UK, Germany, the Benelux nations, Scandinavia) and the eastern Mediterranean, North Africa and the Middle East. Processing tomato production in Latin America, Australia and Africa tends

Table 1.1 World production of processing tomatoes, 1998-2002 (thousand tonnes)

	1998	1999	2000	2001	2002[1]
Spain	1,182	1,510	1,318	1,463	1,400
France	328	363	314	298	230
Greece	1,248	1,250	1,062	939	850
Italy	4,352	4,932	4,841	4,800	4,000
Portugal	988	999	854	917	850
Algeria	320	400	300	270	260
Israel	270	306	222	144	160
Tunisia	470	730	732	430	540
Turkey	1,790	1,750	1,300	950	1,500
California	8,067	11,103	9,329	7,827	9,600
Other US states	455	488	518	499	557
Canada	510	488	413	482	544
Mexico	290	380	216	136	100
Hungary	220	130	128	100	100
Bulgaria	156	150	90	30	30
Morocco	150	155	100	150	170
China	780	800	1,800	1,000	1,800
Brazil	1,017	1,290	1,200	1,000	1,200
Chile	867	950	925	725	550
Argentina	224	330	260	255	210
Australia	334	309	367	380	375
South Africa	170	203	180	204	215
India	45	120	140	120	120
Thailand	121	188	124	140	160
Others	197	274	328	276	276
Total world	24,551	29,598	27,061	23,535	25,797

Note 1: figures for 2002 are estimates, at October 2002.

Source: World Processing Tomato Council (2002).

to be dominated by transnational companies and a few large national firms, who seek to satisfy local, and occasionally export, markets.

Developments in the 1990s, and especially towards the end of the decade, have challenged this economic geography. Although most processing tomato production remains wedded to continent-based, production-trade complexes, the persistence and stability of these arrangements is not as assured as it once seemed. The current impetus for change is coming from many directions, is being manifested at different points within processing tomato value-chains, and is generating new scales of engagement by industry actors. The impulse to change is being embodied in a host of institutional transformations, including shifts in national government policies, technological changes, the manipulation of the tomato's biological properties (either by hybridization or genetic modification), new consumption practices, and corporate restructuring. Debates about the strength and relative importance of these various factors notwithstanding, there seems little doubt that the period since the mid-1990s has witnessed a profound qualitative and quantitative shift in the global economic geography of this industry.

Two broad, inter-related developments capture the essence of these transformations. First, the industry's production-consumption geographies are being reworked, as previously undreamt-of trade and production options have become possible. Unlike some agri-food industries (for example, wheat or sugar) which historically have been characterized by immense global trade flows (Morgan, 1980; Mintz, 1985), the processing tomato industry has had a low incidence of international trade. Regional or national markets have been serviced by regional or national production. However, developments in the late 1990s suggest the possibility of fundamental changes within these relationships. Later chapters of this book document many of these processes but, for the moment, a single example will illustrate the point. Over recent years in the Naples Bay area of southern Italy, significant quantities of tomato paste has been imported from China for re-export to third markets. To some degree, these imports have displaced previous trade flows involving the purchase of paste from northern Italy and neighboring Mediterranean countries. This led to the situation whereby, in 2001, a southern Italian canning company fulfilled a contract for private label canned baked beans for a British supermarket, using Mexican-sourced beans and Chinese tomato paste, bringing into effect a highly flexible and footloose set of production arrangements: the British supermarket's decision to source its canned baked beans from the southern Italian canner was made on the basis of short-term price-based contract, and the southern Italian canner in turn sourced its inputs on the basis of short-term price-based contracts.

Second, the place of processing tomatoes within global food consumption practices is being transformed. Within the West, there is a lengthening of agri-food chains and increasing monetary value is being added to foods beyond the farm gate, as prepared foods and out-of-home dining account for a larger share of expenditure on food. Because of their versatility and associations with particular regional cuisines (Italian, Creole, Mexican, North African), tomatoes are central ingredients in foods such as pizzas, pasta sauces and salsas, which are at the forefront of these consumption shifts. Heinz's North American President has identified processing tomato products as pivotal components of a 'hip-food' diet for teenagers and young adults because of their role in take-out, snack and 'grazing' foods (a marketing term for this is: 'cell foods for the cell phone generation') (Jiminez, 2000). Perhaps not surprisingly, these foods are at the leading edge of the Westernization of diets globally, especially in urban Asia. Growth in the consumption of such foods generates steadily increasing international demand for processing tomato production, especially paste. It is estimated that US pizza consumption in the year 2000 utilized at least 10 per cent of the entire Californian tomato crop (Fargeix, 2000). Franchised chains such as Pizza Hut dominate the market for pizzas, and these players demand standard grade products at bulk prices. Consequently, a significant component of this industry is being restructured to satisfy the need to supply paste as a price-dependent, industrial input. Yet at the same time that products such as paste are being *devalued* as they are converted to standard grade industrial inputs, other processing tomato outputs are being *revalued*. This is especially true within middle-class Western diets, where perceptions of 'quality' and 'health' are a key force driving consumer choices. With such trends, tomatoes are being reinserted into diets as a 'functional food' (Heasman and Mellentin, 2001) and (along with olive oil) are represented as components of a 'healthy Mediterranean diet'. Closely associated with these trends is the growing demand for organic tomato products also grew rapidly in the second half of the 1990s.

Such changes are destabilizing established structures in the industry and are creating uncertainty within regional production complexes. For industry actors and observers of global change in agri-food sectors, there is a pressing need to understand how the various elements of change are connected with one another, in order to better appreciate the industry's future directions.

Generating answers to these questions is not a matter simply of measuring trade and production flows. Although quantitative statistical data provide important evidence of change, they represent only the surface manifestations of industry restructuring. Researchers have long recognized that the *understood possibility* of sharpened international competition is a powerful force in the restructuring of established production systems (Webber et al., 1991). For example, in the case at hand, restructuring is not simply being driven by the importation of large amounts of relatively cheap Chinese tomato paste into certain markets, but by the threats of such an eventuality. The point is that globalizing forces carry considerable discursive power, whatever their material effects. To find answers to the question of how international restructuring is affecting this industry, we must tell the stories of the ways in which companies, growers, traders and retailers, among others, are responding and adapting to the current situations they face.

Globalization and agri-food restructuring

Participants in the processing tomato sector are interpreting and responding to current industry challenges in diverse ways. On the one hand, some industry participants see an emergent sectoral geography based around what can be labelled the 'globalized agri-food model', that is, the development of highly flexible and footloose production systems, in which supply arrangements and regional futures are dictated increasingly by short-term shifts in competitiveness. In this model, supply chains are detached form their territorial roots in a volatile bidding war staged at a global scale, and resulting in a 'race-to-the-bottom' as competitive pressures drive down prices. Large buyers such as retail chains, international franchise outlets, and the food processing industry are identified as primary instigators of these tendencies. On the other hand, some industry participants are more sanguine about the wider relevance of the globalized agri-food model. According to their reasoning, such developments will remain on the margins of the commodity system, because price is not the sole arbiter of competitiveness. Such arguments emphasize the importance of quality and reliability as chief determinants of the geographies of processing tomato commodity chains, and are based on the belief that these attributes are best managed within territorially and commercially embedded supply chain relationships based on notions of trust, knowledge and regional distinctiveness.

In these arguments we find broader narratives on the character of contemporary global agri-food restructuring. In the past decade, many social scientists have embraced the concept of globalization as a totalizing explanation for restructuring processes across the world. It is obviously tempting to identify the emergence of radically new and global-scale processes, such as the sourcing of Chinese tomato paste by Italian contract canning companies, as emblematic of 'a new era of globalization'. Such theorizations presuppose what has been labelled 'an end of geography' (O'Brien, 1992) where international production systems are constructed with less regard to the spatial position of actors.

Debate on the McDonald's Corporation provides an entry point for these arguments to be applied to research on global agri-food restructuring. In 1993, the sociologist George Ritzer used the McDonald's Corporation, and the fast-food industry generally, as a metaphor to illustrate the direction of change in global society and, in particular, in world food and agriculture. According to Ritzer (1993), 'McDonaldization' relates to a system of food provisioning centrally based around themes of efficiency, calculability, predictability and control. It is supposedly 'dehumanizing' in that consumers have no scope for deviation from corporate menus and recipes, and this standardization, it is argued, benefits fast-food corporations more than it benefits consumers. Moreover, according to Ritzer the system is sustained only because various environmental, social and public health costs inherent in the system are not fully paid for, leading to an outcome, described by Atkins and Bowler (2001, p.98), as 'the rationality of the irrational'.

It is difficult to deny that Ritzer's arguments capture some important elements of contemporary global agri-food restructuring. The McDonald's Corporation experienced a compound annual rate of return of 17 per cent over the 1990s, and by 2001 there were 28,000 McDonald's restaurants globally in 120 countries (McDonald's Corporation, 2001). Evidence presented at the celebrated 'McLibel' trial in the United Kingdom alleged the unsustainability of the company's global operations, and these arguments have been reinforced by the recent publication of

Fast Food Nation (Schlosser, 2001) a best-selling account of the alleged social, environmental and public health costs of McDonald's and other fast-food chains.

Yet to what extend does, 'McDonaldization' provide an explanatory model for wider trends in global agri-food restructuring? Despite its immense size and geographical scope, the McDonald's Corporation provisions less than one per cent of the world's population daily (McDonald's Corporation, 2001). 'McDonaldization' implies a unlinear direction of change towards the McDonald's exemplar, which suggests that over time, the remaining 99 per cent of the world's population will be provisioned through systems that echo the McDonald's model. This is a totalizing discourse that gives insufficient weight to the agency of consumers to shape their own food consumption practices, or to the dynamics of capitalist competition that present alternatives to the standardized model supposedly exemplified by the McDonald's Corporation.

Academic scholarship on globalization during the late 1990s has critiqued these perspectives. Rather than conceptualising 'globalization' in terms of a single model (for example, the McDonald's method of provisioning), this research emphasizes that globalization is best understood as a *process* that is forever incomplete, contested and dynamic. Business theorists have great difficulty in pinpointing a single transnational corporation that conforms to the model of a 'global corporation' (Pritchard and Fagan, 1999) and, despite the longings of some neo-liberal writers, it is naïve to assume the powerlessness of nation-states as regulators of economic activity. As Clark (1993, p.107) observes: 'It is as if [globalization] theorists have skipped over the real-time processes of restructuring to get to the end of the process and thus the proclamation of a new reality'. Moreover, extensive research from a range of perspectives also identifies what are referred to as 'the limits to globalization', defined in part by the continuing relevance of commercial and cultural relationships embedded in social spaces of various kinds (Kelly, 1997). Such perspectives reject the so-called 'steamroller metaphor', whereby globalization is viewed simply as a force that crushes local culture and distinctiveness. Although globalization processes are undeniably challenging economic activities in profound ways, we live in a world in which 'geography matters' still, in the sense that many economic and cultural processes tend to be organized, defined and embedded within politically and culturally defined regional and national spaces. In the newer globalization scholarship of the late 1990s, emphasis is given to the interactions of 'global' processes with those at national, regional and local scales. As differently-scaled processes interact, new levels of diversity and difference are created (Hay and Marsh, 2000). According to Tomlinson (1999, p.16), globalization should be viewed not as simply 'coming in' from somewhere else, but as being produced and reproduced in and through spaces:

> if globalization is understood in terms of simultaneous, complexly related processes in the realms of economy, politics, culture, technology and so forth, we can see it involves all sorts of contradictions, resistances and countervailing forces.

These perspectives intersect with other recent research which uses the concepts of 'flows' and 'networks' to explore contemporary socio-economic change. Castells (1996) argues that what is generally termed 'globalization' is more accurately identified as the 'rise of a network society' whereby new connections and linkages are transposing traditional ones. The task for researchers, hence, is to identify and document the ways places, products and ideas influence one another. McGrew (1992) encapsulates these arguments with his claim that socio-economic and cultural processes are being 'syncretized' because of four contradictory tendencies that are occurring simultaneously: (i) universalism in concert with particularism; (ii) homogeneity accompanied by differentiation; (iii) integration running parallel with fragmentation, and; (iv) centralization offset by decentralization.

Research on the McDonald's Corporation itself has provided significant support to these arguments. Although McDonald's might appear as an icon of global uniformity, studies have shown that the operations of the McDonald's system can differ significantly in across national and cultural environments. McDonald's menus may cater for significant variation among national consumer groups and the cultural meaning of consuming McDonald's products is constructed in different ways across the globe. Consumption patterns are also marked by *reflexivity*, meaning that perceived global influences may generate reflexive responses, which at times can lead to new constructions of 'the local'. Indeed, one of the key marketing strategies of McDonald's at an international level is to 'localize' products through the use of cultural signifiers (hence, in Australia, the McDonald's menu offers a 'McOz' burger; in Malaysia it offers a 'McRendang' burger, etc). Detailed anthropological research into consumption practices within McDonald's stores in East Asia reveals inter-cultural differences as well as similarities in the way tensions between 'the global' and 'the local' are negotiated. According to this scholarship, McDonald's restaurants are not simply inserted as a 'global Other' into East Asian foodscapes, but their patronage involves various elements of contestation and cooption by local consumers:

> The process of localization is a two-way street: It implies changes in the local culture as well as modifications in the company's standard operating procedures. Key elements of McDonald's industrialized system - queuing, self-provisioning, self-seating - have been accepted by consumers throughout East Asia. Other aspects of the industrial model have been rejected, notably those relating to time and space. In many parts of East Asia, consumers have turned their local McDonald's into leisure centers and after-school clubs. The meaning of 'fast' has been subverted in these settings: It refers to the *delivery* of food, not its consumption (Watson, 1997, pp.36–37).

These arguments do not deny the immense economic power of the McDonald's system, nor suggest that McDonald's passively accepts the consumption whims of the marketplace. McDonald's, along with other fast-food chains, engages in a wide range of strategies (such as advertising and promotions) with the aim to actively shape food consumption practices. Yet, this scholarship calls to attention the persistence of *difference*, and the creation of new *hybridities*, in processes that may otherwise simplistically be assumed as constructing global homogeneity.

Perceiving globalization in this way shapes the approach that is adopted in analyzing global agri-food restructuring. It underlines the need for utilizing a framework that gives full recognition to the socio-spatial fractures and fissures in the world's economic structure:

> As global capital transcends national boundaries in search of profit opportunities, its behavior is influenced by the landscape it confronts. Whilst the shift to globalization may be a prevailing feature of recent agri-foods restructuring, its concrete manifestations are necessarily complex and diverse (Pritchard, 1996, p.50).

These arguments represent a major theoretical plank of this book. Globalization is presented as a process that is forever incomplete and contested, and which contains apparent complexities and contradictions, including reflexivity and hybridity. Yet while such arguments inform an understanding of the concept of globalization, they alone do not explain the particular tendencies occurring at this point in history, especially with respect to agri-food industries. To this end, the literature on globalization needs to be augmented by that which focuses on the world-historical contexts of the global economy.

World-historical researchers identify the period since the early 1980s as an era of transition within global agri-food systems (Friedmann and McMichael, 1989; Friedmann, 1993; McMichael and Myhre, 1991; McMichael, 1994; McMichael, 1996). This is because this period reveals that

> in the agro-food sector there exists the largest gap between national regulation and transnational economic organization (Friedmann, 1993, p.30).

In other words, there exist a series of unresolved struggles between the nation state and the global market as a regulator of world agri-politics. As suggested by Ufkes (1993, p.219):

> Contemporary struggles over agricultural trade liberalization can be seen as an elaboration of the tension between national and international politics, or to put it another way, between efforts to preserve agrarian forms and accordant structures of national regulation and efforts to promote global accumulation.

In the terminology of food regimes theory, these unresolved tensions correspond to the shift away from the so-called *second food regime*, the relatively stable era of international food relations of the middle decades of the twentieth century. This regime was associated with the extensive national regulation of agriculture, via restrictions to trade through quotas and tariffs, and the payment of farm subsidies which, in turn, encouraged production surpluses which helped to depress world agricultural prices and to establish import dependencies of less developed nations on (subsidized) agricultural exports from developed nation producers. Because agriculture was not included within the multilateral institutional framework of the General Agreement on Tariffs and Trade (GATT), few restrictions were placed on the trade policies of major agricultural producers. However, by the mid-1980s, these producers faced escalating costs of agricultural subsidy payments, and came to be locked into reciprocal trade wars as each sought to capture new export markets in order to offload surplus produce. In this environment in 1986, the 'Cairns Group' of nations,[6] who supported freer agricultural trade, pushed for the incorporation of food and agricultural sectors within the GATT Uruguay Round of trade negotiations. The eventual completion of the Uruguay Round in 1993, saw agriculture and food fully incorporated within the nascent structures of the World Trade Organization (WTO), but with little change to pre-1986 subsidy levels (Oxfam, 2002).

Accordingly, the structural conditions of global food and agriculture remain indeterminate. On the one hand, regulatory and institutional changes to the world food system during the 1990s greatly facilitated the emergence of globalized agriculture. These processes involved the incorporation of world agricultural trade within the WTO, and the harmonization of trading regimes via initiatives within the private sector and international agencies, including Codex Alimentarius and the International Standards Organization. Yet at the same time, there remains extensive subsidization of agriculture within the developed countries, albeit within programs restructured in order to comply with WTO rules.

In recent years, researchers have debated the merits of various approaches to describe and account for these struggles over the political conditions of global food and agriculture. In the early 1990s, some New Zealand researchers identified the emerging WTO structures as constituting a so-called 'third food regime' (Le Heron, 1993; Le Heron and Roche, 1995) based around new, privately regulated, agri-commodity systems. It is not surprising that such perspectives emanated from New Zealand, the world's leading proponent of deregulated, free market agriculture. However, the continuation of extensive agricultural protectionism in northern economies has brought into question the relevance of extrapolating from the 'New Zealand experience'. Moreover, other researchers have problematized what this case actually demonstrates, suggesting that the 'New Zealand experience' reflects unconnected 'crisis experiments', rather than a paradigmatic shift in agri-food organization (Campbell and Coombs, 1999).

Recent research by McMichael and colleagues (McMichael, 1999; McMichael, 2000; Araghi and McMichael, 2000) provides valuable insights to the contemporary political struggle over food and agriculture. Two observations are central to McMichael's critique. The first is that diversity and instability are hallmarks of contemporary global agri-food restructuring. The second is that these developments are taking place within a global political context of uneven trade liberalization involving, pre-eminently, the maintenance of agricultural protection in the north in the context of liberalization in the south. The emergence of globally 'hyper-mobile' finance capital is pivotal to these two processes. As finance capital has gained hyper-mobility during the past two decades, a global 'casino economy' has been created in which 97.5 per cent of world financial flows are speculative (McMichael, 1999, p.17). Such financial speculation has been held to account for the economic crises in Mexico (1994), East Asia (1997), Russia and Brazil (1998) and Argentina (2002). On the one hand, processes of speculative capital flight have provided the subsequent justification for International Monetary Fund (IMF) 'bailout' packages that have restructured southern economies in line with ideologies of neo-liberalism, thereby paving the way for agricultural trade liberalization. On the other hand, globally mobile finance capital has itself played an enabling role as an agent of restructuring. For agri-food activities to attract investment capital, they must hold the potential to generate rates of return that are competitive with alternative investments, in a global context. These financial transformations, which effectively amount to the destabilization of national economic systems in favor of the logic of global financial systems, intensify pressures to extract profit and value from economic activities. Referring to studies of the restructuring of contract farming relations in the Californian strawberry industry (Wells, 1996), the feminization of agricultural labor forces, and the global expansion of contract farming, McMichael (1999, pp.17–20) suggests that each of these seemingly separate processes has common roots, namely, in the destabilization of wage relations within national economies, as a result of the ways mobile finance capital is intensifying competitive relations at a global level. Hence, diversity and instability in global agri-food restructuring needs to be understood as a *structural outcome* from contemporary global political-economic change.

The appeal of this framework is that it provides an approach that links literatures on globalization and global agri-food restructuring to the historical circumstances of the world economy. As the literature on globalization has emphasized, the so-called ‘McDonaldization’ model presents a simplistic portrayal of global agri-food restructuring, because of processes of reflexivity and hybridity. Yet as world-historical literatures have indicated, assessment of these issues needs to take place within the historical frames of the contemporary world economy.

This view of global change is given voice in this book. The widespread contemporary restructuring this is evident in the processing tomato industry (and in other industries as well) is symptomatic of tensions between the impetus for, and limitations of, the globalization of economy and culture. At the practical level, the task for researchers is to document and assess how these varied forces are interacting, and with what effects. This purpose requires an *epistemological framework* (a way of relating new information to existing knowledge) and a *methodology* (a way of collecting and reporting information). These issues also require a sense of *praxis*, meaning the way in which research findings are organized so that they intervene within social contexts, and thus provide insights which are relevant to the analysis of the current situation. The remainder of this chapter establishes these credentials for the current study.

Where are we coming from? The motivations for this book and our position as researchers

This book treads an unfamiliar path amongst the voluminous research on agri-food globalization published in recent years. In choosing to examine a single commodity system at a global scale, this book has an ambitious mandate. Of course, there already exists a series of landmark studies charting the global evolution of particular agri-commodities over time, including sugar (Mintz, 1985), coffee (Pendergrast, 2000) and potatoes (Salaman, 1949; Zuckerman, 1998). But the over-riding concern of those studies is to assess the historical conditions under which global commodity complexes have emerged:

> how a society learns to consume food differently: eating more food (or less), eating different food, differently prepared, in different context, with the social (and perhaps the nutritive) purpose of the consumption itself revised or modified' (Mintz, 1995, pp.4–5).

Our concern is somewhat different. Although stressing the importance of world-historical context, our primary interest is with the contemporary global *geography* and *political economy* of the commodity under review. Aside from a brief historical overview of the emergence of the processing tomato, our analysis is rooted in the post-1945 period and, particularly, from about 1990 onwards.

Although limited in time to this period, the global-scale focus of our inquiry nevertheless presents serious challenges. Most analysis of agri-food restructuring utilizes case studies which operate within regional or national parameters. At the risk of being criticized for selective citation, we would include as examples: Miriam Wells' (1996) research on the labor processes that underlie the Californian strawberry industry; research by Alessandro Bonanno and Doug Constance (1996) on yellowfin tuna; Kathleen Stanley's (1994) research on the beef industry of the US Mid-West; Neils Fold's research on copra (2000) and cocoa (2002); Megan McKenna, Michael Roche and Richard Le Heron's (1999) work on the New Zealand pipfruit industry; Jane Dixon's (2002) work on the Australian poultry sector; research on the Washington State apple complex (Sonnenfeld et al., 1998); and Bill Friedland's (1994) research on Latin American fresh fruit and vegetable exports, and the Californian lettuce and tomato complexes (Friedland, Barton and Thomas, 1981). In contrast, this book encompasses the analysis of change in places as diverse as California, Italy, Xinjiang Province of China, the Murray-Darling Basin of southeastern Australia, Turkey, Thailand, and Ontario Province, Canada. Examining this global politico-economic geography requires the documentation of a vast array of processes. In addition, our research necessarily focuses on events within the processing tomato system with reference to a broad set of economic, cultural, political and institutional landscapes.

The scope, complexity and inter-relatedness of these changes – including shifts in agronomy, corporate strategy, trade economics, geopolitics, logistics management, biotechnology and genetics, mechanical engineering and marketing, among others – demands a broad, *interdisciplinary* approach. Contemporary agri-food research, however, tends to remain fractured along specialist disciplinary lines. In the social sciences, at least three broad areas of inquiry generally vie for relevance. *Agricultural economics* mostly interpret agri-food restructuring through the abstract analysis of market performance and efficiency. This approach gives emphasis to quantitative analysis and predictive modelling, within a positivist epistemological frame. *Business and organizational theories* utilize empirical and conceptual tools to model and analyse corporate structures and behaviour, often with the view to assisting profit-generation for particular corporate entities. Finally, *critical agrarian political economy*, a relatively new inter-disciplinary field, attempts to integrate sociology, geography, political economy and anthropology in order to build an understanding of structural changes in the global agri-food economy through informed case study analysis.

Each of these three areas provides its contributions and insights. However, cross-fertilization of ideas between these fields has been weak. Research journals and conferences remain wedded to particular disciplinary orientations, with rigorous inter-professional communication being the exception rather than the rule. This book attempts to build bridges across these divides. Although generally working within a framework of critical agrarian political economy, our separate disciplinary backgrounds (Economic Geography [Pritchard] and Politics and Technology Policy Studies [Burch]) also allow us to analyze agri-food globalization with the benefit of cross-disciplinary viewpoints. Through this book we seek to demonstrate the need for a wide frame of reference, if the subtleties and complexities of agri-food globalization are to be captured.

The global frame of reference we employ in this book has required an extensive and flexible research approach. In order to examine the *global connections* of regional production complexes, this book has amassed a wide range of data and information about various regional production areas, and utilizes this comparative and networked perspective to build insights into the processes by which globalization is constructed. Methodologically, the research for this book has utilized the collection and interpretation of secondary sources (statistical data, articles in industry and academic journals, etc) as well as information gathered through interviews with key informants. Thus, this book is itself a product and artifact of an increasingly globalized world. The preparation of this book entailed field research in eleven countries: Australia, the United States, Canada, France, Italy, Spain, Turkey, Belgium (for meetings with European Commission bureaucrats), Thailand, China and Japan.

The use of a multi-site research method is a relatively new development in critical agrarian political economy, although the approach has been applied in other research fields, such as comparative international politics. The first significant methodological review of this approach for agri-food research was Friedberg's (2001) analysis of her work on the green bean trade complexes linking Africa and Europe. As Friedberg notes, it is inevitable that internationally mobile researchers with limited time and budgets face immense challenges in 'getting across the detail' of a situated production complex with extensive global linkages. To this end, a core element of this method utilizes an active 'snowball' approach whereby contacts made in one part of the world are used to facilitate connections in another. For example, in our research a meeting with a tomato seed company in Parma, Italy, helped establish a link with another seed company in California. Frequently, we found that key informants travelled extensively, creating arenas for discussion and flows of information about other places.

The information we collected through research trips was validated through ongoing email communication, which with mutual agreement was often copied to third parties and which led, in turn, to new information networks being established. Because of our ongoing professional commitments (including university teaching and other research activities), field research took place in short, sharp bursts. This timing was itself important, because it encouraged an emphasis on making face-to-face contacts (thereby establishing an understanding of, and legitimacy for, the research) that would form the basis for ongoing communication through email. Hence, the research method for this book depended on the existence of electronic communication and databases. In this sense, it differed somewhat from the traditional fieldwork practice, by which data would be collected through extended periods in the field.

These approaches challenged the standard researcher-subject dichotomy. Our access to, and participation within, these networks of individuals (through travel) and information (through communication technologies) diminished the conventional distinction between the identities of ourselves (as researchers), and our key informants. On occasions, the key informants we sought to interview used their access to us (as researchers) to inform their own perceptions of the industry. In many interviews, we were asked as many questions as we ourselves asked. In the eyes of many industry participants, we were positioned as both seekers and holders of knowledge. While we were certainly not the first researchers to be situated in this kind of way, these positions influenced the context of our research.

Although our key informants understood our status as independent 'outsiders', our associations with a large number of industry participants, in combination with our knowledge of industry trends gained from these associations, at times produced a perception that we held an 'insider' status. Crucial in this regard was participation at the *IV World Congress of the Processing Tomato*, at Sacramento, California in June 2000. Attendance at this conference provided a means of establishing contacts and networks, and helped provide legitimacy to the project. These issues of positioning and access, as Friedberg (2001, p.363) notes, raise epistemological, ethical and strategic issues. To navigate through this complex set of issues, we engaged in extensive cross-checking of information, and made early drafts of this manuscript available for examination and review.

At a different level, issues of how we were positioned and constructed as researchers raises an important point with respect to the preparation of this book. As noted earlier, this book has been prepared with two clear audiences in mind. On the one hand, it is intended to complement and extend existing academic research on agri-food globalization. However in addition, this book also seeks to engage with individuals and organizations in the processing tomato industry, who would not normally read the academic literature.

Although these two readerships are not always mutually exclusive, the task of reconciling them raises questions of written expression and content. Many academic readers of this book would be interested primarily in our analysis of how this industry study informs wider arguments and models of agri-food globalization and restructuring. Industry audiences however, may be more interested in the specifics of the case study. Furthermore, between the two audiences there are different expectations in terms of vocabulary and jargon. Analysis of agri-food globalization within the framework of critical agrarian political economy employs a particular discourse based around a collection of ideas, concepts and approaches, such as *food regimes*, *actor network theory*, *subsumption*, *appropriationism* and *substitutionism*, to name just a few. While the academic audience for this book would be familiar with these concepts, the industry audience might not and, consequently, we have minimized our use of academic jargon. In short, a central objective of this book is to 'talk about' the global restructuring of processing tomato commodity chains in a way that builds upon the framework of critical agrarian political economy, while making it accessible to audiences not familiar with the terminology and structure of argument.

Finally, the origins of this book need explication. We were initially drawn to this industry in the context of two crises which afflicted the Australian processing tomato industry during the 1990s. The first of these involved a substantial increase in the import of canned tomato products in the early 1990s (especially from Thailand, China and the EU), which were being marketed by Australian supermarkets under 'own brand' and 'generic' labels, and which provided a major competitive threat to the local industry. Prompted by this crisis, we analyzed the future of the Australian industry (Burch and Pritchard, 1996a), in a paper circulated throughout the Australian processing tomato community and re-published internationally in the industry journal *Tomato News* (Burch and Pritchard, 1996b). The second crisis occurred in 1997-98 in the context of Australian Government regulations prohibiting collective bargaining arrangements between processing firms and tomato growers. According to the Australian competition authority, collective bargaining was deemed anti-competitive in that it implied collusive behavior on the part of borth processors and growers. Implementation of these competition regulations (under the guise of 'deregulation') led to a major destabilization of the local industry. In discussions with industry participations following these events, it became apparent that the challenges presented by neo-liberal re-regulation were just one component of a wider competitive scenario facing the industry. The Australian industry was being re-positioned internationally because of rapid restructuring of the industry at a global scale. The practical imperative to document and assess these changes provided the fundamental inspiration for this book.

The structure of the book

While the practical interests of the Australian processing tomato industry provided one important reason for this book, it also owes its origins to our desire to critically evaluate and contribute to current debates on agri-food globalization. Despite a vast literature on the restructuring of agriculture and food, framed around the concept of globalization, there remains little consensus on the pace, dimensions, impulses and implications of contemporary changes. Clearly a range of institutions and processes are involved in current global restructuring of food and agricultural sectors. For researchers, there is a need to document and explain these interactions in ways that identify the general principles they illustrate.

To document the global restructuring of the processing tomato industry and interpret what these developments imply for debates on agri-food globalization, this book is divided into six following chapters. In brief, the organization of these chapters responds to three challenges we faced in the course of our research:

(i) to present and analyze the parameters of the processing tomato system (Chapters Two and Three);

(ii) to provide evidence for the general argument of this book, namely, that contemporary global agri-food restructuring is a contested historical process, featuring a complex interplay of developments at varying scales and in different geographical territories (Chapters Four, Five and Six), and

(iii) to assess and interpret the significance of these arguments in the light of broader debates about the processing tomato sector and agri-food globalization (Chapter Seven).

In Chapters Two and Three, the 'processing tomato system' is defined as the activities by which a relatively standardized agri-commodity (processing tomatoes) undergoes a series of transformations that successively add value and profit. Therefore, the processing tomato system reveals distinct differences from agri-commodity systems such as coffee or iceberg lettuce, where a raw material undergoes relatively simple processing to create a single, clearly identifiable, final product. In the processing tomato sector, a number of first-tier processing activities create commodities that can represent end-products in their own right (such as canned tomatoes or tomato paste), or can contribute towards the manufacture of a rich diversity of consumer food products, such as baby foods, canned soups, frozen pre-prepared meals, ketchup and pizzas (Table 1.2). In this sense, the processing tomato sector has some similarities with the broiler chicken commodity system in which, as Dixon (2002) and Kim and Curry (1993) have observed, a mass-produced 'fordist' agri-commodity is transformed into a highly diversified set of food outputs.

Table 1.2 Simplified model of processing tomato chains

Growers' sphere	First-tier processing	Second-tier processing	Final retail product	Consumption
Processing tomatoes	Whole/pieces diced/peeled canned	---	Retail/ wholesale trade	Home kitchen and restaurant consumption
Processing tomatoes	Paste, stored in aseptic drums	Food service industry	Various end-products (e.g., soups)	Home kitchen and restaurant consumption
Processing tomatoes	Paste (pizza sauce), stored in aseptic drums	Food service (franchise chain) industry	Pizzas	Restaurant and take-out consumption
Processing tomatoes	Paste (hot-break) stored in aseptic drums	Food industry (ketchup manufacture)	Retail ketchup	Additive to kitchen or restaurant prepared meals

In reviewing the system's key characteristics, these chapters reflect and reiterate the issues and arguments of this book, namely, which aim to 'unpack' global agri-food restructuring by critically analyzing globalization within a world-historical context. Applying these frameworks to the processing tomato system necessarily involves the presentation of a considerable volume of empirical data and background analysis, although these chapters are not intended to be encyclopaedic: their intent is not to chart the progression and destination of the world's entire annual processing tomato output via a country-by-country survey. Rather, the objective is to point to the key factors shaping the system. To this end, Chapter Two starts by examining the global tomato seeds complex, then moves onto issues relating to tomato cultivation and the introduction of mechanized harvesting, which is identified here as a catalytic development in the restructuring of on-farm activities in the processing tomato industry. Much of this chapter concentrates on experiences in California, where mechanized harvesting was initially developed and took root. In the later parts of this chapter, the focus is widened to take into account the ways in which the 'California model' of high input-high output tomato growing has diffused internationally.

Following on from this, Chapter Three analyzes the contemporary attributes of the sector's processing industry, and reaches into issues and debates surrounding the consumption of processing tomato products. An important element of this chapter is the recognition of the co-relationship between production and consumption. Much research on agri-food commodity systems has traditionally given pre-eminence to the task of documenting production flows, leading to a relative neglect of consumption issues, a tendency that Lockie (2001) has referred to as 'the invisible mouth'. It is clearly not sufficient to treat consumption as a residual category, since the demands and preferences of consumers actively shape production relations. In the case at hand, this is demonstrated by the increased consumer demand globally for pre-prepared and fast foods such as pizzas, as well as new consumption modes relating to organic farming, functional foods and genetic modification. Acknowledging these issues however, does not necessitate a research position that uncritically accepts 'the sovereignty of the consumer'. Consumption patterns are constructed through the interactions of an array of agents including food scientists, nutritionists, advertizing and marketing agencies, the media, large companies and international bodies such as Codex Alimentarius. For instance, in the late 1990s the demand for processing tomato products in Anglo-American dietary communities was bolstered by widespread publicity relating to the positive health effects of lycopene. However, this development was not simply an episode of 'neutral' food science providing consumers with information that in turn helped transform production systems. Consumer awareness of lycopene was built through the establishment of new networks of cooperation and coordination among a diverse set of agents, including large tomato processing companies. Institutions of production therefore, actively constructed new consumption relations. Accordingly, the relationships between consumption and production are viewed as co-dependent, and characterized by tensions and ongoing cycles of action and response.

Chapters Four to Six provide three empirical accounts of the complexities of contemporary global restructuring in the processing tomato sector. The subject matter for these chapters reflects our intention to interrogate agri-food globalization, so each of these three chapters addresses an important question linked to what can be labelled 'the globalization thesis':

(i) Chapter Four critiques the suggestion that intensified global competition leads inevitably to an erosion of the roles of national and sub-national spaces as arenas for industry regulation, especially with respect to grower-processor relations. In this chapter, a comparative analysis of Australia and Canada is used to bring these differences to the fore. Although these production sites have certain similarities (that is, they are both English-speaking, with high input-high output farming styles) comparative analysis reveals major differences in regulatory institutions and practices.

(ii) Chapter Five critiques those arguments which suggest that the global agri-food industry is inevitably heading towards a scenario in which global markets will replace national and regional ones. The particular focus of this chapter is on changes in Europe in the context of restructured agricultural producer subsidies. The late 1990s witnessed a historically significant juncture in European agriculture. In the wake of the WTO Agreement on Agriculture, the EU formulated its *Agenda 2000* program to restructure farm subsidy payments in line with concepts of the 'multifunctionality of agriculture' and the 'post-productivist transition'. These initiatives were aimed at shifting the emphasis in farm support from quantitative measures to more sophisticated strategies entailing the explicit valuation and preservation of the social and natural elements of the rural environment. In the processing tomato industry, these policies led to the complete re-drafting of the tomato support payment regime in December 2000. Coinciding with these developments, there has been extensive restructuring within the European tomato industry. However, we argue that this does not yet seem to be leading towards an erosion of what we call 'the distinctly European tomato', namely an industry based around production and consumption practices firmly embedded within various national and regional arenas.

(iii) Chapter Six analyzes the claim that contemporary global agri-food restructuring is witnessing an 'across-the-board' expansion of export competition from developing countries. Since the late 1980s, considerable attention has been focused on the shift of agri-food production to low-cost production sites in Asia, Latin America and Africa. The extent to which these processes have materialized in the processing tomato industry is explored through detailed case studies of Thailand and China. In both these cases, the impetus to shift production to these sites was contingent upon particular historical moments when economics and politics conspired to encourage the development of this industry. Yet Thailand and China have had widely divergent experiences. In Thailand, tomato production has experienced significant difficulties in recent years, while in the western Chinese province of Xinjiang, production volumes grew rapidly in the late 1990s, leading to significant restructuring of the industry at an international scale.

These arguments on the international restructuring of the processing tomato system are brought together in Chapter Seven. In this chapter, we discuss the issue of the extent to which the processing tomato industry is 'globalized', and re-introduce the theoretical discussion of agri-food globalization, presented earlier in this chapter. These perspectives are vital to the aims of this book because this conceptual framework focuses attention on the ways in which (i) restructuring processes execute a reorganization of profit and value within processing tomato supply chains, and (ii) the various roles of spatial flexibility help to bring about these outcomes. Processes of restructuring are multi-scaled, contain reflexivity and agency, and reflect the varied cultural, historical, political and economic factors which currently operate within the global context. As a result, the restructuring of the processing tomato industry will continue to take many (and unexpected) paths.

The point is, however, that a strong political-economic theme underlies these many paths of restructuring. Although spatially and socio-temporally diverse, restructuring in this sector reflects an international reworking of the ways in which profit and value are produced. Although the manifestations of global restructuring are varied, they are underpinned by economic processes that identify, separate and exploit profit opportunities. In short, the diversity and apparent complexity of restructuring processes in this industry does not reflect a chaotic formulation of territorial 'difference', but a structural outcome from global-scale shifts in the creation of economic value.

Placing these developments within a world-historical context, moreover, emphasizes how diversity, uncertainty and an accelerated pace of change should be interpreted as systemic outcomes from the contemporary global political-economic regime of mobile capital and increasingly liberalized trading relations. Processes of globalization are not converging towards the ascendency of a single globalization 'model'. Rather, they are creating a global environment of experimentation and change. Taken together, what we are seeing, in many varied sites across the world, is a restructuring of activities in line with perceptions of newly emerging, *globalized* opportunities for profit, that themselves are in a constant state of evolution.

Notes

1 The countries represented at the *IV World Congress* were: USA, Canada, Mexico, Dominican Republic, Argentina, Brazil, Chile, Venezuela, Peru, Italy, Portugal, Spain, Greece, France, Turkey, Israel, Morocco, Tunisia, Cameroon, Senegal, Ghana, United Arab Emirates, Saudi Arabia, Thailand, China, Japan, Australia, New Zealand, Poland, Hungary, Sweden, UK, Netherlands, Germany.

2 In 1987, in a quest to appropriate the tomato's history as its own, civic leaders in Salem decided to re-enact this scene as an annual event. Ironically, this came just eight years

after Campbell Soup closed its Salem County factory, and effectively brought to an end the County's tangible association with the commodity.

3 The vexed question of whether the tomato is a fruit or vegetable goes back to 1893 when, in an act of Lamarkism, the US Congress resolved a trade dispute by declaring the tomato 'a vegetable' (Khoo, 2000).

4 *Passata*, an Italian word, refers to a thick sauce of crushed peeled tomatoes. It is generally used as the base for an accompaniment to pasta.

5 Italy is the pre-eminent EU exporter to the non-EU Mediterranean and Middle East region. However some other, interesting, trade flows exist. According to industry sources, Libyan President Moammar Gaddafi has interests in a Greek processing tomato facility that owns import quota rights to Libya.

6 The Cairns Group is a group of fourteen agricultural trading nations: Australia, New Zealand, Malaysia, Thailand, Indonesia, The Philippines, Brazil, Argentina, Uruguay, Chile, Canada, Hungary, Colombia and Fiji.

Plate 2 High input, high output tomato cultivation, Australia, 1999

Photo: Bill Pritchard.

Chapter 2

Manufacturing red gold: growing the world's tomatoes

The heart of the Australian processing tomato industry is in the Murray and Goulburn Valleys in the northern part of the State of Victoria. This is flat alluvial country, worn down over the centuries. In winter, grey fogs often shroud the landscape until midday. In summer the land is baked dry, with temperatures regularly rising to 40 degrees Celsius.

The land is not pretty. This is a working countryside and its towns reflect these priorities. Shepparton, the main regional centre, has a main street dominated by companies catering for the local business and farming population. There are accountancy offices, rural supplies stores, regional town retailers, Chinese restaurants, an 'Irish' pub and, in the distance, the flumes of food processing factories. In this blue-collar rural town household incomes are relatively low but there is little unemployment. The unemployment rate in Shepparton is the lowest in the State (Heinrichs, 2000).

Superficially, the social and economic landscape of this region seems as durable as its topography. Although the main streets of Shepparton (and those of other regional towns) have been braced by business turnover, their mix and style continues to reflect the commercial culture and imperatives of an agri-industrial regional economy. For the passing traveller, the rural landscape appears stable in a world of change.

Yet the apparent stability within this landscape masks deep and profound changes in the nature of the local farming sector over the past twenty years. The processing tomato industry, one of the region's key agricultural activities, typifies these changes. In the early 1980s there were over 300 processing tomato growers in Australia, the vast majority of whom were located in the Murray and Goulburn Valleys. By 2002 there were just 33. These 33 growers moreover, were producing a considerably larger volume of output than that generated by 300 growers just twenty years previous. These changes (and comparable ones in other agricultural sectors) have reverberated widely through the regional economy, with implications for everything from the distribution of land ownership and purchasing patterns for farm input supplies, to the political character of the region. Whereas these rural regions once solidly supported Australia's rural conservative party, the Nationals, their restructured agricultural economies now encourage a greater political diversity (Pritchard and McManus, 2000). Between 1996 and 1999, the National Party's primary vote for the State seat of Shepparton fell from 66 per cent to 39.9 per cent (Victorian Electoral Commission, 2001).

The changes in processing tomato growing, that have seen an 85 per cent drop in the number of Australian growers within twenty years, have parallels across the globe. World demand for processed tomato products is higher than ever, but is being met by an ever smaller number of growers. Documenting and explaining these trends is the key concern of this chapter.

The industrialization of agriculture

The commercial cultivation of processing tomatoes is a highly sophisticated practice. Over time, growers have become increasingly dependent on circuits of knowledge and capital from external (that is, off-farm) sources. In recent years much has been written about the 'industrialization of agriculture' and its implications for the future of farming (Lawrence, 1987; Buttel, Larson and Gillespie, 1990; Goodman and Reclift, 1991). At the heart of this literature is a concern to explain a central paradox. At one level, the industrialization of agriculture has meant that farming practices are circumscribed increasingly by technologies, knowledge and power held outside the farm. With the advent of more technically precise output specifications and the complex inter-relationships between cultivation variables and agri-industrial supply chains, farmers appear to have less scope for independent decision-making. In the case of processing tomatoes, for example, decisions on seed choice, agri-chemical usage, product transport and harvest timing are often placed out of the realm of individual farmer choice, and moved into the realm of supply chain coordination. Although individual farmers may remain wedded to notions that they are independent business operators and custodians of the land, their capacities for autonomous decision-making are increasingly constrained. Yet at the same time – and this is the crux of the paradox of industrialized agriculture – farming remains dominated by family-owned enterprises. Although the practice of farming has become industrialized, its ownership has not. Large agribusinesses shy away from taking direct ownership and control of farms; the evolution of industrialized agriculture has *not* created an industrialized labor system whereby individual farmers have become wage-laborers to corporate interests. In contrast, the industrialization of agriculture has frequently resulted in farmers operating their businesses within the context of contract farming systems, in which they hold subservient positions within networked supply chains (that is, their practices and decisions are increasingly determined by off-farm actors). But, at the same time, farmers are also self-employed owner-managers of (increasingly) large enterprises who bear the risks and gain profit from their activities.

Social scientists use the concept of *subsumption* to account for this paradox. Firstly developed by Karl Kautsky in *The Agrarian Question* (1899), subsumption explains the processes by which off-farm capital (including actors such as processing companies, agri-chemical suppliers or financial institutions) shape productive activities on farms. Kautsky's concern was to explain the restructuring of European agriculture in the late nineteenth century and, in particular, the question of whether agriculture would follow the example of manufacturing.

During the eighteenth and nineteenth centuries much manufacturing was transformed from a system of craft-based artisanal production, to a system of capitalist factory production. This transition effected a transformation in the role of labor. Guild structures (where tradesmen were self-managing) were replaced progressively by factory-based wage-labor (see Mathews, 1989, pp.10–25). The 'agrarian question' posed by Kautsky was whether agriculture would be similarly restructured, resulting in corporate farming systems replacing family-farming arrangements. Kautsky's key insight was to reject claims that agriculture inevitably would follow in this path:

> Kautsky concluded that industry was the motor of agricultural development... but that the peculiarities of agriculture (its biological character and rhythms – see Wells, 1996), coupled with the capacity for family farms to survive through self-exploitation (i.e., working longer and harder to in effect depress 'wage levels') might hinder some tendencies, namely, the development of classical agrarian capitalism. Indeed agroindustry – which Kautsky saw in the increasing application of science, technology and money to the food processing, farm input and farm finance systems – might prefer a noncapitalist [*i.e., non corporate-owned farming*] sector (Watts, 1996, p.232).

Kautsky's work has received considerable attention since the 1980s. The partial translation of Kautsky's work into English (Banaji, 1980) generated considerable interest within rural sociology. In the 1990s, Kautsky's insights were popularized further (Little and Watts, 1994; Watts, 1996; Goodman and Watts, 1997; Rosset et. al, 1999). For such latter-day followers, Kautsky's analysis 'has a striking saliency for rural and agrarian development theory as the twenty-first century draws near' (Watts, 1996, p.230). This is because the key issues facing European agriculture at the end of the nineteenth century – agricultural crisis in the face of rapidly increasing productivity, low prices and the import of grains and meats from the 'settler' states of the New World – bear a striking similarity to processes associated with the globalization of agriculture in the last few decades of the twentieth century. Watts and his fellow researchers suggest that the requirement of contemporary farming to generate farm produce in evermore competitive environments encourages agri-industrial capital to exert increasing control over, *but not direct ownership of*, on-farm activities. Thus, agri-industry 'saturates and controls the [farm] production process through appropriation (machines, seeds, biotechnologies, credit)' (Little and Watts, 1994, p.250). Farm enterprises survive through 'self-exploitation' (i.e., working harder, longer, or making more use of unpaid family labor) and a willingness to accept low economic returns on labor and capital (Atkins and Bowler, 2001, p.57). In the terminology of this literature, agri-industrial capital does not fully incorporate farm activities within wage-labor structures (so-called *formal subsumption*), but instead makes farm activities subservient to its needs via indirect means (so-called *real subsumption*) (Mabbett and Carter, 1999). These concepts have particular relevance to the discussion of contract farming in the processing tomato industry, pursued in Chapter Four.

If the concept of real subsumption explains why the industrialization of agriculture does not lead inexorably to wage-labor corporate farming, it also highlights the diversity, or 'multiple trajectories' (Watts 1996, p.236) of agricultural restructuring. Family-farm enterprises respond to the demands of a sharply competitive marketplace in a variety of ways depending on their size, structural characteristics and level of commodity dependence. These responses may include the use of increased debt to finance capital expenditures or farm expansion, farm-business diversification, the attempted development of niche or specialist products, or strategies to redefine grower-processor relationships via mechanisms such as cooperatives or sole grower contracts. Additionally, farmers across the world are increasingly dependent on off-farm sources of income such as the wages of spouses (typically wives), farm tourism ventures and machinery hire which, taken together, constitutes *pluriactivity* (Atkins and Bowler, 2001, p.58). The point is that as agriculture becomes more industrialized there emerges a greater incentive (if not desperation) on the part of growers to actively restructure their farm enterprises and sources of income, producing a landscape of change characterized by diversity and experimentation.

The significance of these themes for the processing tomato industry is analyzed below. Dramatic changes in the technological basis of tomato cultivation and the emergence of mechanized harvesting have redefined levels of technical efficiency in the industry. In most contexts this has led to an increase in average tomato tonnages per farm, and absolute reductions in the number of tomato growers. However, there is considerable variation at a global scale concerning the manner in which these broad processes have been played out. The point of the preceding discussion is to argue that diversity in farm-level restructuring is a systemic outcome from contemporary agri-food globalization. Notwithstanding broad trends towards larger tonnages per farm, global convergence towards one 'farming style' is unlikely, because individual farm enterprises are inserted within their production landscapes in differing ways and, as a result, there will be continued divergence in the economic and social relations between on-farm and off-farm interests.

To consider these issues in greater depth, the starting point for analysis must involve a discussion of the role of nature in tomato production systems. In large measure, the biophysical properties of the tomato plant dictate this industry's agri-industrial character. Common tomato varieties grow in a range of climates and local environments, however optimal production conditions are obtained within specific climates and soils. Tomato plants require three to four months from seeding to harvest and, during this time, plant growth benefits from extensive sunlight, dry weather and uniform, warm conditions. Humidity encourages foliage diseases, discouraging production in wetter or tropical climates. However, because of the high level of water required to cultivate tomatoes, irrigation plays a critical role in this industry. Tomatoes are versatile with respect to soils, but higher yields tend to be generated with slightly acidic loams, rather than lighter soils. Nutrients within the soil play an important role in tomato yields.

These characteristics tend towards the global processing tomato industry being concentrated around 40° latitude North and South, in generally drier regions either on continental west coasts or inland locations. Of all the major growing regions, only Goias (Brazil) departs from this basic geographical principle. These biophysical factors are best expressed in the Californian San Joaquin Valley, resulting in approximately one-third of the world's processing tomatoes being grown within a 300 kilometre radius of Sacramento. These factors also play a significant role in explaining some of the difficulties in maintaining activity in some potentially low-cost production areas, such as Thailand and Senegal.

Clearly, these biophysical factors are pivotal to the global restructuring of the processing tomato sector. In large part, they determine where, and how, restructuring occurs. Much recent scholarship in agri-food studies has emphasized the role of biophysical factors in making the character of agricultural restructuring 'different' to that of manufacturing industry, and with this we concur. But there is little reason to labor this point. As Friedland et al. (1981, p.6) have argued: 'the peculiarities of weather and soil in the making of food and fibre crops should not ... obscure the universal character of commodity production in capitalist agriculture'. Hence, we now move forward, and analyze the ways that recent agri-industrial processes have impacted on biophysical characteristics and reshaped global tomato cultivation.

High input, high output tomato cultivation

Extensive scientific research over recent decades has reshaped the practices of processing tomato cultivation. To optimize the key outputs of yield and soluble solid content, research has investigated numerous variables including soil types and fertilizer treatment, cultivars, irrigation practice, and disease and pest management. Each of these areas (and their associated sub-fields) is the subject of an enormous and expanding literature, much of which is recent and subject to professional debate. This book does not pretend to provide a comprehensive analysis of developments in each of these fields, and the ways in which they impact on yield and soluble solids. However, some background to these developments is a necessary precursor to the prime goal of this chapter, namely, the assessment of key recent changes to the growing of processing tomatoes.

Yield, or tonnage per unit of area, is the most elementary measurement of tomato output. It is an especially important measurement for growers, because the fixed capital costs of land impose a heavy burden on processing tomato cultivation. Hence, greater outputs per unit of area result in generally higher growing margins and profit. International comparative data on yields has been published only in the past few years and their accuracy is open to question (Table 2.1). Clearly, there is a wide disparity in yields among growing regions, from the extremely high yields of Israel and California, to the extremely low yields of Algeria, Hungary and Morocco. Although generalizations are dangerous, regions with low levels of mechanization tend to yield between 40 to 50 tonnes per hectare, while those that are mechanized tend to yield above 60 tonnes per hectare. Because of mechanization, improved grower practices and the effects of new seed varieties, yields appear to have increased over time for most grower regions. Longitudinal evidence for these trends is made difficult by lack of data, but output statistics from California since 1965 attest to the sustained nature of yield improvement (Figure 2.1).

Table 2.1 Reported processing tomato yields, 1999

Country	Tonnes per hectare
France	69.0
Greece	45.0
Italy	67.3
Portugal	60.0
Spain	49.4
Algeria	16.0
Israel	100.0
Tunisia	43.0
Turkey	42.5
Morocco	26.0
Hungary	23.6
California	81.4
Canada	73.8
Brazil	63.0
Australia	69.0

Source: Tomato News (2001a).

However, measurement of yields is only a partial indicator of processing tomato output. An increasing proportion of processing tomatoes is used for conversion to paste, which then becomes the raw material input for the production of soups, ketchup, sauces and other food products. For the optimization of paste output per tonne of raw tomatoes, the all-important measurement of output is the level of soluble solids, which refers to the concentrated content of tomato fruit. This is measured in terms of percentage of fruit volume, or degrees brix, which shows the percent of soluble solids calculated as sugar (Gould, 1992, p.223). This is important for paste manufacturers because of the role of evaporation in this process: paste manufacturers are interested in the quantity of tomato solids per tonne of raw product, rather than the quantity of raw product *per se*. Tomato paste tends to be produced at approximately 30 brix, meaning that when dried, tomato solids comprise 30 per cent of the paste's weight. In the 2000 season, major production areas generated net soluble solids rates of between 4.8 to 5.3 per cent (Table 2.2). In general, solid rates have not grown in tandem with yields; in fact, Australian data points to a decline in solids over the past decade (Table 2.3).

There is extensive international evidence that for most varieties, soluble solids levels vary inversely to yields (Bryce, 1999). At the level of individual irrigation farm practice, cultivation strategies are often perceived in terms of 'yield versus solids'. Most growers perceive their cultivation strategies as resting on whether they 'grow for yield', or 'grow for solids'. Because most processing company payment regimes are skewed heavily towards payment for tonnage delivered at the factory gate (discussed further in Chapter Four), growers traditionally have opted to 'grow for yield'.

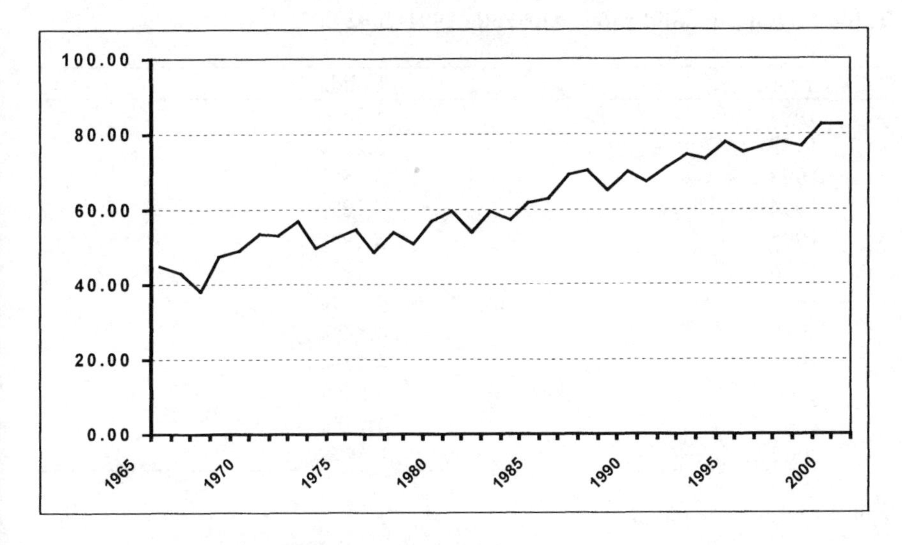

Figure 2.1 Californian processing tomato yields, 1965-2000 (tonnes per hectare)

Source: Californian Tomato Growers Association (2001).

Table 2.2 Net soluble solids rates, major production areas, 2000

Country	Solids	Country	Solids
China	5.0%	Greece	5.0%
Spain	5.1%	Portugal	4.8%
Turkey	5.0%	South Africa	5.0%
Italy Parma	5.0%	California	5.3%
Italy Foggia	4.8%	Australia	4.8%
Italy Naples	4.8%	Chile	5.0%

Source: Unpublished data from a major processing tomato firm, reproduced with permission.

Table 2.3 Soluble solid rates, Australia 1991-2002

Year	Solids
1991	5.25%
1992	5.22%
1993	5.38%
1994	5.10%
1995	5.19%
1996	4.99%
1997	4.76%
1998	4.87%
1999	4.57%
2000	4.69%
2001	4.77%
2002	4.64%

Sources: Horn (various years).

Among the numerous variables influencing the relationship between yields and solids, irrigation plays a critical role. Decisions on how to irrigate and, in particular, when to cut-off irrigation (that is, how soon before harvest) influence yield/solids ratios. In general, there is an inverse relationship between these outputs when irrigation is varied: later and/or more extensive irrigation may increase yields, but this may be to the detriment of solids levels. The type of irrigation plays a critical role here, though it must be considered in the context of cultivar types and crop management practices. Furrow (or flood) irrigation does not have the precision of drip irrigation, where water is delivered to plant roots directly through drip tape. Because of the relative scarcity and cost of their irrigated water, Australia and Israel have been world leaders in the introduction of drip irrigation technologies. Recent Australian research suggests that the first year implementation costs of drip irrigation are roughly 50 per cent higher than for the use of furrow irrigation, but that due to higher yields achievable under drip, these costs have a three-year payback period (Hulme et al., 2000). Evidence from Australia bears out the suggestion that drip irrigation is associated with higher yields then furrow irrigation (Table 2.4). Because the benefit of higher yield exceeds the effect of lower solids, drip irrigation generates a higher rate of solids per hectare (Table 2.5). Moreover, whereas there appears to be an historical correlation between the adoption of drip irrigation, higher yields and lower solids (Stork et al., 2000), a recent survey of 119 cultivated blocks in Australia suggests that when compared with furrow irrigation, the adoption of drip techniques has negligible impacts for solids (Hulme et al., 2000).

Debate over the relationships between yield, solids and irrigation types brings into focus questions on the impact of seed variety. Across the world, the adoption of drip irrigation has occurred alongside the introduction of a large number of new hybrid seed varieties, because many of the more popular of these varieties are characterized by root systems that are more efficient in searching for soil moisture.

Table 2.4 Californian and Australian yields, 1991-2000 (tonnes per hectare)

	California	Australia		
	all fields	Furrow irrigation	Drip irrigation	All fields
1991	71.06	n/p	n/p	44
1992	74.20	n/p	n/p	57
1993	73.30	n/p	n/p	44
1994	77.56	n/p	n/p	58
1995	75.10	n/p	n/p	61
1996	76.44	55	83	61
1997	77.79	54	89	64
1998	76.44	72	99	79
1999	82.49	57	78	69
2000	82.49	54	83	71

Note: n/p = not published.

Sources: Californian data: 1991 from *Tomato News*, December 1992; 1992-93 from *Tomato News* March 1994. Australian data: 1991 to 1995 (Rendell McGuckian et al., 1996, p.41); 1996-2000 Horn (2000).

Table 2.5 Relationships between yield, soluble solids and irrigation practice, Australia

		Tonnes sampled	Yield (hectare)	% soluble solids	Tonnes SS per hectare
1998-99	Furrow	78,157	57	4.64	2.65
	Drip	124,983	78	4.39	3.42
1999-00	Furrow	102,113	54	4.85	2.62
	Drip	145,298	83	4.58	3.80

Source: Horn (2000, p.11).

Thus, the analysis of the relative advantages in terms of yields and solids of drip versus furrow irrigation is mediated heavily by the choice of cultivar.

In many production sites, government departments may play a key role in the field trials of cultivars. This role has become increasingly central to the industry with the expansion of hybrid seed types. Industry acceptance of proprietary claims about particular cultivars rests on the perceived impartiality of independent research. These activities are a critical component of what we refer to (in Chapter Four) as the 'mode of industry regulation' for the processing tomato sector. Field trials of cultivars are extensive. In Australia since the late 1980s, State Departments of Agriculture have undertaken observation field trials of approximately 150 to 200 processing tomato cultivars annually (Ashcroft et al., 1990; Wade et al., 2000). These trials assess seed varieties according to a wide range of criteria. In addition to measures of yield and solids, field trials can be used to help identify such factors as disease resistance, performance for early-season, mid-season and late-season plantings, and percentage of green fruit and fruit size. Accordingly, the social practice of farming becomes implicated ever more deeply within agri-scientific knowledge systems.

As noted earlier, since the 1980s researchers in the field of critical agrarian political economy have given considerable attention to questions relating to the increasing role of scientific knowledge within the farming sector. According to David Goodman and Michael Redclift in their book *Refashioning Nature* (1991), agri-industrial processes and knowledge can be understood via the inter-relationship of two conjoined concepts: appropriationism and substitutionism. *Appropriationism* refers to the processes whereby farming is transformed by the historical accretion of technology. For example, plant-breeding techniques have progressively conditioned natural biophysical processes to suit modern agriculture. The development of processing tomato varieties to suit mechanical harvesting (discussed below) provides a classic example of appropriationism. Thus, appropriationism 'describes the partial but persistent transformation into industrial activities of certain farm-based labor processes' (Atkins and Bowler, 2001, p.60). The concept of *substitutionism* refers to the wholesale replacement of agricultural outputs for synthetic (or non-agricultural) ones. An example of substitutionism is the replacement of natural for synthetic fibres. Thus:

> [agri-industrial companies] have responded by adapting to the specificities of nature in agricultural production. Within the limits defined by technological progress, discrete elements of the production process have been taken over by industry and the products of agriculture were likewise substituted by outputs from a growing complex of food and fibre industries, in a similarly discontinuous but enduring process, to achieve the spectacular growth of the industrial and mass production of food and fibre (Le Heron, 1993, p.37).

These concepts are seen readily in the cultivation of processing tomatoes. In recent years, traditional direct seeding methods have been challenged by transplant techniques in a number of regions across the world. Transplanting involves growing out seedlings in greenhouses for two-to-three weeks prior to their being planted. The cost-effectiveness of transplant techniques however, rests on relationships between seed prices, the economics and agronomics of weed and pest control, and managerial strategies of large family-farming enterprises. In Australia, a key factor in popularizing the transplanting of processing tomato seedlings was the high price of hybrid seeds during the second half of the 1990s. Higher seed prices encouraged more judicious use of seeds, which in turn encouraged growers to use greenhouse seedlings rather than direct seeding methods (which incur significant seed wastage). Thus, the widespread use of transplants new circuits of association between the farm sector and off-farm actors (e.g., producer organizations and transplanting companies) which further redefined the extent to which 'farming practice' actually occurs on farms. For example, in 2002 the Italian tomato transplanting company, Masterplant, cultivated half a billion tomato seedlings across the world, in line with a production technique that produces up to

4,000 plants per square metre (Boscolo, 2002). As a result of these types of developments, typical cost structures for processing tomato cultivation are dominated by expenditures on agri-inputs and capital, with growers' returns varying considerably. Table 2.6, which illustrates two estimates of grower cost structures from Australia during the 1990s, shows both the potential variability in growers' returns and the differences in agri-input cost structures between the direct seeding method (1994-95) and transplanting (1997-98).

Table 2.6 Two estimates of growers' costs, Australia

	Estimate 1: Average costs, 1994-95	Estimate 2: 'Best practice' costs, 1997-98
Transplants	n/a	8.71%
Fertilizer	8.20%	6.39%
Chemicals	8.88%	7.94%
Fuel	3.71%	2.69%
Repairs and maintenance	6.83%	4.78%
Irrigation	2.73%	4.57%
Labour – weeding	4.00%	2.15%
Labour – permanent	5.07%	6.28%
Overheads	4.68%	4.49%
Harvest costs	15.12%	16.80%
Capital costs/depreciation	25.66%	19.61%
Other	8.98%	n/a
Total costs	93.85%	84.42%
Surplus and grower income	6.15%	15.58%
Total	100.00%	100.00%

Note: n/a = not applicable.

Sources: Estimate 1 Rendell McGuckian et al. (1996, Appendix 4.1). Estimate 2 from an unpublished report by a major Australian processing tomato firm, 1999, reproduced with permission.

The global tomato seed complex

The development of hybrid seeds has been pivotal to the emergence of high input-high output tomato cultivation, and deserves analysis in its own right. Tomato seeds can be generated through open pollination or hybridization, which means that seeds are either sourced from a previous tomato crop, or are specially produced under controlled conditions. The advantage of the latter is that it gives greater certainty to growers and processors. The expansion of mechanized harvesting (see below) has been a strong impetus for the increased use of hybrid seed varieties. Because mechanized harvesting requires uniform ripening and robust fruit characteristics, there is a co-dependence between seeds and harvest methods. The importance of developing specific traits was recognized in tomato research conducted in the 1940s, but the demands of industrialized agriculture have underlined the significance of such characteristics. The prospects for genetic manipulation provide a further recognition of the strategic role of seeds in this industry.

Growth in demand for hybrid tomato seeds over the past two decades has created a new international political economy of seed production. The key to understanding the role of seeds within the processing tomato industry (and, in fact, any horticultural sector) relates to the way in which hybridization establishes proprietary rights over seed resources. In horticulture, seeds became the 'catalytic element in the convergence of chemical and mechanical technologies after 1930' (Goodman and Redclift, 1991, p.169). By the late 1980s, with genetic modification on the horizon, the seeds industry assumed enormous significance within agri-food company strategies. This period witnessed intense merger and takeover activity associated with the control of the commercial seed industry, as a prelude to corporate expectations of the incipient biotechnological revolution. Seeds are crucial ingredients for agricultural biotechnology, because genetic manipulations of plant structures are marketed to growers through new seed varieties. From the late 1980s, transnational corporations including ICI and Unilever, moved aggressively into the seeds industry. The forces of competitive restructuring gathered pace through the 1990s, encouraging the integration of different sub-sectors of the agri-supply business, and promoting widespread strategic alliances of seeds, agri-input and biotechnological companies (Buttel, 1999). Research by Pistorius and van Wijk (1999, pp.119–22) reveals that from the mid-1990s, major firms in this area, such as Monsanto, Du Pont, Pulsar, Novartis and Zeneca, were transformed into multi-faceted but integrated conglomerates, typically with interests in six fields: (i) plant breeding; (ii) biotechnology; (iii) genomics; (iv) software; (v) chemicals and pharmaceuticals, and; (vi) food processing.

As will be seen, control of hybrid seed varieties in the processing tomato industry has undergone significant change in the past decade in line with these developments. Moreover, alongside shifts in the corporate ownership of hybrid seeds have come major changes to the global economic geography of hybrid seed development and commercialization. The hybrid tomato seeds industry is part of a larger global seeds complex, but it is also distinctive in that it operates with specialist multinational enterprises, its own industry institutions, and its own issues of regulation. With respect to the latter point, recent restructuring in the hybrid seeds industry is tied inextricably to a set of regulatory changes at national and international scales relating to intellectual property protection and plant breeders' rights.

Given that the focus of this book is on the processing tomato industry, rather than the seeds industry, these issues will be considered only briefly, with most attention being placed on the ways in which changes in the hybrid tomato seeds sector have influenced the wider economic geography and political economy of the processing tomato industry. Importantly, extracting information about the hybrid tomato seeds industry is also made difficult by the fact that although the scientific literature on hybrid tomato seeds runs to many thousands of publications (often taking the form of field trial results), there is an almost complete absence of critical social science analysis of this industry. The one exception to this is the work of Rosset et al. (1999), which analyzes this sector in northeast Thailand. Official data on hybrid seed production and trade are also scarce, and companies in this sector are notoriously secretive. Exclusive rights to commercial and scientific knowledge is the critical element of competitive advantage in the seeds industry and, perhaps not surprisingly, many actors in the industry are loath to discuss their company or industry publicly.

The commercial development of hybrid tomato seeds involves a series of stages based around seed research and development, trials and actual production. In the first stage, hybrid plants are created through breeding two selected but unlike parents. There is a complex science in selecting which parents will create hybrids with particular characteristics, the details of which are outside the ambit of this book. Needless to say, this component of the hybrid seed industry is dominated by specialist scientific labor under strict corporate control and management. Plant breeding takes place in selected fields or corporate greenhouses, with pollen from one parent inserted into the flowers of the other. The second stage of trials with new hybrid varieties may be carried out by the company itself or by knowledgeable farmers. Departments of Agriculture may also play a role in these trials by acting as 'honest brokers' to verify plant characteristics. In the third stage, when a decision is taken to commercialize a hybrid, the seed from parent lines is provided to contract growers who grow out the plants, pollinate them and harvest the fruit for seed extraction. There is no mechanized process for either pollination or seed extraction, meaning this final stage is highly labor-intensive. Once hybrid seeds are extracted, they are made available to contract tomato growers under commercial terms.

With heightened demand and profitability in the global hybrid seeds industry, this production system has become increasingly flexible and complex. In particular, an international division of labor has emerged whereby knowledge-intensive (first and second stage) activities have remained in the major (Western) tomato production regions, but labor-intensive (third stage) activities have been made internationally mobile and, in general, been located in low-cost production sites such as Thailand.

The rise of this international division of labor raises some key issues that reflect developments in other sectors and commodities. The emerging structure of the hybrid seed industry appears to execute the same kind of flexible geographical strategy associated with industries such as athletic footwear. In the case of hybrid seeds commercialization, as with companies such as Nike Inc., knowledge and capital-intensive activities remain in the developed world while labor-intensive activities are characterized by an intense, price-based 'race to the bottom' competition. There has, of course, been considerable debate on the extent to which such 'manufacturing sector' models can be applied to agri-food systems, and while it is inappropriate to generalize on this issue across the wider agri-food sector, the hybrid seeds sector is certainly one agri-food industry that parallels the flexible manufacturing model.

A significant and diverse number of low-cost nations are involved in third stage hybrid seed extraction for the processing tomato industry, including Venezuela, Chile, Mexico, India, Thailand and China. Among these, only Thailand has been subjected to any scrutiny. A 1995-96 survey of the industry (Rosset et al., 1999) was complemented by a series of key informant interviews for this study in mid-2001 (see Chapter Six). Thailand's tomato seed industry is located in the northeast region, centred around Khon Kaen and the Lao border. In addition to producing seeds, this region, also known as Isaan, is home to the Thai processing tomato industry. The Thai processing tomato industry is discussed extensively in Chapter Six, but for the moment it is worth emphasizing the importance of cost-competitiveness as the driving force in this industry.

The northeast is Thailand's poorest region and its low wage agricultural economy was the critical factor in attracting the seeds industry to this area. Rosset et al. (1999, p.85) provide a glimpse into the highly labor-intensive nature of this industry:

> [For hybridization to occur] [T]he actual crossing of the lines demands intensive, back-breaking labor. The two lines are grown in separate plots or rows. The labor involves collecting the pollen from the 'father' line and keeping it dry and viable until the cross is to be made, usually a matter of days. All flowers belonging to the 'mother' line, those plants designated to receive the pollen and produce fruit and ultimately the hybrid seed, are emasculated. Emasculation entails opening the flower about three days prior to its natural flowering date and removing the anthers. This operation ensures that no self-pollination occurs, a process that would produce non-hybrid seed and affect the purity of the final product. Diligence and honesty on the part of the farmer, and any family or hired labor involved in this stage, is of utmost concern to the [seed] company… Once hybrid seeds are harvested and processed, which involves holding the picked fruit for three days in bags, screening and washing the seeds with water, another washing with carbolic acid, and drying them for three days in the open and dusting them with fungicide, a grower delivers the seeds to the company.

The major companies in this sector are each positioned differently with respect to their geographies of seed production and commercialization. In the low-cost seed extraction business, it is often the case that intermediary agribusiness companies manage seed production on behalf of large firms. In the tomato seed sector a few transnational companies play the key role in new seed development and distribution. In the second half of the 1990s, Heinz Seeds emerged to become the most important of these.

Heinz Seeds is an extremely interesting case because it serves to connect debates on the hybrid seeds and (mainstream) processing tomato sectors. Of course, Heinz has been a major player in the processing tomato industry for over a century, and since 1936 has operated tomato-breeding programs. However, it was only in 1992 that the company formally established a discrete tomato seeds business (Heinz Seeds). This timing coincides with the emergence of the key strategic role which seeds have come to have in the contemporary global agri-food sector. Through the establishment of globally integrated seed research and development programs, coordinated via the company's seed technology headquarters at Stockton, California, Heinz has been able to quickly expand its interests in the global tomato seeds business. Heinz Seeds has since become the world's largest tomato seeds company and has a high level of market share in particular sites. In California in 1999, eleven separate Heinz varieties were in the top 20 selling seeds. The top selling Heinz variety (H8892) had 13.7 per cent market share (Tomato News, 2000b, p.43). In Australia, Heinz increased its share of the processing tomato seed market from 25 per cent in 1995, to 72 per cent in 1999 (Horn, various years).

Geographically, Heinz Seeds operates a diverse international production system. While the business is headquartered in Stockton, California, seed breeding programs are conducted in the northern summer in California and Ontario, and in the southern summer in Victoria, Australia. Field trials are then conducted in a wide range of countries including Portugal, Spain, Greece, Hungary, Chile, Venezuela, and New Zealand, as well as in Australia, Canada and the USA. Once field trials confirm proposed varieties for development, Heinz Seeds grows tomato plants for seed extraction in Mexico, Thailand and India (Heinz Seeds, 2001).

The international division of labor in hybrid seed production is also of interest because it involves the construction of extensive intra-firm commercial relations, which raises important issues of pricing. The past decade has witnessed steep increases in hybrid seed prices, as new and supposedly improved varieties are developed. In Australia, according to industry sources, the price of 'typically used' open-pollinated seeds was about $70 (AU) per kilogram in the early 1990s, but hybrid seed prices were about $500 (AU) per 100,000 seeds – equivalent to $1,500 (AU) per kilogram – in 2002.

The high prices of hybrid seeds have been debated widely over recent years within grower communities. Certainly many newer varieties are acknowledged as having superior characteristics and growers welcome ongoing hybrid research. The extent to which improvements justify increased prices, however, is a subject of contention. Of course growers have little influence over the pricing decisions of hybrid seed providers, because growers are price-takers in the market. In the US, Canada and Australia, production contracts typically specify which seed varieties are allowable, meaning that growers are contractually obliged to use particular hybrids, under legally specified terms and conditions. Moreover, general acknowledgement of the superiority of new hybrid varieties (or at least, an acceptance that hybrid research will generate long-term improvements) has in the past tended to silence grower complaints over price.

To the extent that grievances are aired, they relate to the transparency of seed companies' pricing decisions. From the perspectives of seed companies, there is a legitimate commercial difficulty in pricing these knowledge-rich products. Sunk costs in research and development dominate the cost of hybrid seed production. The survey by Rosset et al. (1999, p.88) of Thai seed growers revealed that the farm-gate price for hybrid tomato seeds ranged between 1,600 to 3,000 Baht per kilogram, equivalent to $64 (US) to $120 (US) per kilogram. In all probability, this would equate to just five per cent of the final price for hybrid seed paid by US contract growers. The 2,000 per cent mark-up on the Thai farm-gate price is presumably justified by the fact that for seed companies to attain a commercially appropriate rate of return on the sale of hybrid seeds, pricing decisions need to recoup not only the productive expenses of generating these commodities as tangible goods, but the intangible, intellectual property embedded within them. The problem is, however, that these calculations necessarily take place within the byzantine context of international corporate accountancy, well beyond any public or governmental scrutiny. Faced with escalating seed prices and little insight into how prices are formulated, it is hardly surprising that this area at times raises complaint from tomato growers.

These issues lead to questions of rates of return in the hybrid seed industry. Public information in this area is non-existent. However, two general points seem to be able to be made about the general profit conditions in this industry. First, the role of research and experimentation suggests considerable profit risk among companies. The successful commercialization of a hybrid seed variety may generate sizeable profits, but this has to be set against the possibility of a series of unsuccessful hybrid trials. In principle the costs of unsuccessful varieties are factored into commercial prices of successful ones, however it is unclear how this materializes in the practical world of corporate strategy and seed pricing.

Second, the international division of labor within this industry presents a range of issues regarding intra-firm pricing and global profit sharing. The latter issue is of particular significance and has to be treated in greater length in order to fully understand the dynamics of pricing and the returns to seed companies.

Where a single company operates seed development activities in different countries, it most likely engages in 'related party transactions.' Usually, these transactions involve the parent company and one or more of its overseas subsidiaries, and relates to the sale by the parent of tangible goods such as intermediate inputs in production, or intangible assets such as access to, or rights to use, intellectual property residing in patents, trademarks and/or brand names, etc. In the course of making available these inputs, the price upon which the parent company settles may not necessarily reflect the costs of production (plus a margin for profit), or even the need to ensure that overseas subsidiaries operate efficiently on the basis of market prices. The profitability of the individual sites of production is not so important to the transnational corporation as the profitability of the group as a whole, and one of the critical factors in this relates to the nature of the tax regimes in different countries. All things being equal, companies will rationally seek to maximize profits in low-tax centres and minimize profits in higher taxing environments, subject to taxation laws, which in some developing country contexts, may be poorly enforced. For example, in paying for the purchase of necessary intermediate inputs or the use of patents and brand names, a subsidiary company may be required to pay the parent company a sum which reflects the monopolistic conditions of the transaction rather than the costs of production. In a classic study of transfer pricing in the Andean Pact region, Vaitsos (1973a, 1973b) found that extensive over-charging was applied by foreign parent companies, when supplying their local subsidiaries. This occurred in numerous industries, including pharmaceuticals, chemicals, electronics and rubber products. The most extensive case involved a pharmaceutical company supplying substance of valium to its Colombian subsidiary at a price which was 6,600 per cent greater than the price charged to other purchasers. What this meant was that the parent was able to disguise the extraction of 'super-profits' as a normal commercial transaction, which would enable the company to minimise overall taxation liabilities.

The introduction of more globally integrated and complex structures in the agri-food industry raises the question of the extent to which these issues relating to tax policy and performance are now emerging (Pritchard, 2001). Whereas many agri-food companies possess great international flexibility and an extensive range of related party transactions, tax authorities also have become more sophisticated in their formulation of taxation rulings in these areas. Because of the absence of public documentation of this industry, it is unknown how hybrid tomato seed companies deal with the complex pricing and tax issues associated with their related party international transactions. By definition, the discussion of these issues is necessarily speculative. In principle however, it would be surprising if corporate financial advisors had not closely investigated the tax implications involved in these activities. Typically in this industry, parent seed stock is exported from seed companies in developed nations to related party subsidiaries in developing nations. The hybrid seed then produced by contract growers is re-exported from seed company subsidiaries in developing nations, to their parent companies in the developed world, for on-sale to contract tomato growers located (mainly) in developed nations. Superficially, it appears that in the hybrid seeds industry, extensive (loss-making) research and development occurs in (relatively high taxing) developed nations, while (potentially highly-profitable) seed extraction occurs in (relatively low-taxing) developing nations. As *related party* dealings, the terms of these transactions are outside the purview of interested observers, and so questions remain about how international profits are made and distributed within this industry.

Mechanized harvesting

Clearly, the on-farm activities of processing tomato growers are closely related to scientific knowledge systems, the expertise and use of off-farm agents, and the increasingly tight networking of supply chain actors. The catalyst for these transformations was the development of mechanized harvesting. This process acted as the touchstone for a wider set of changes within the growing and processing sectors, creating closer interconnectiveness between the spheres of growing and processing within the tomato industry, and encouraging changes to farming and processing practices.

Given the significance of these developments, it is surprising to realize that the development of mechanized processing tomato harvesting originated with the scientific practices and aspirations of a single individual, working essentially outside the mainstream of corporate and university research. This is in contrast to the way much research and development occurs, being mobilized through complex networks of research laboratories, extension programs and corporate-based research teams. Individuals, or even individual teams of researchers, often play incremental roles within larger complexes. Contemporary agri-industrial science funding is strongly biased towards integrative research activities (such as field trials) that test and/or extend existing knowledge in marginal ways.

Shortly after America's entry to World War Two, an extension officer attached to the University of California at Davis (UCD), G.C. (Jack) Hanna, developed an interest in breeding a processing tomato capable of being harvested mechanically. At the time, this ambition was seen as 'bizarre and eccentric' by researchers in the Department of Vegetable Crops at UCD. Because some colleagues 'believed that his quixotic venture would bring the department into ridicule' (Friedland and Barton, 1976), Hanna was required to pursue his quest with very little funding or university support. Consequently, although Hanna's interests in mechanical harvesting began in 1942, the practicality of mechanical harvesting was not confirmed until 1966.

Hanna and colleagues who joined him later faced skepticism from mainstream researchers because of the complex inter-relationships involved in the mechanization of processing tomato harvesting. The two key problems facing researchers were: (i) the need to develop a tomato able to withstand the considerable stresses associated with mechanized processes; and (ii) the need to develop a tomato that ripened uniformly so that a mechanical harvester could extract all fruit at the time of harvest (with hand-picking, this was not so important as pickers could return for unripened fruit at a later time). Consequently, the development of mechanized harvesting required both the establishment of tomato varieties with robust fruit and uniform ripening attributes, and the construction of harvesting machinery capable of picking tomatoes in a way that minimized fruit damage. Hanna, then, faced an inter-disciplinary 'applied systems' research problem, which involved his working simultaneously on both plant genetics and agricultural engineering.

Although 'the notion of a mechanically harvested tomato was, at the time, unthinkable in the agricultural science network of the United States' (Friedland and Barton, 1976), Hanna continued to work on this project through the war years, and gradually accumulated a wealth of knowledge on tomato breeding attributes. This research was undertaken without the formal departmental approval at UCD (Friedland and Barton, 1975, p.23). In 1947 he took leave for a period of six weeks (again without the formal support of his department) in an attempt to identify a tomato variety with appropriate shape and ripening characteristics. During this leave he identified an appropriate variety in Geneva, New York, and returned to UCD with seeds to continue his plant-breeding program. Eventually, Hanna's personal commitment to this project was recognized by the chairman of the Agricultural Engineering Department at UCD, who assigned an agricultural engineer, Coby Lorentzen, to work with him. For the next ten years, Hanna (in plant breeding) and Lorentzen (in mechanical engineering) worked together on this project.

The plant-breeding component of the project proceeded steadily over the late 1940s and 1950s. Hanna's aim was to breed a tomato that had a tough skin, a capability of relatively easy abscission (separation) from the vine, and which ripened in large volumes at the same time.

The mechanical engineering component of the project also gathered pace in the 1950s. By the mid-1950s, it had been determined that mechanical tomato harvesting was best pursued through a system that cut the tomato plant just below the surface of the earth, then elevating the vine onto a series of shaker arms that shook the tomatoes through the arms onto chutes from which they slid onto conveyors. This system however, did not entirely dispense with labor. Because the conveyors would collect unripened and overripe tomatoes, as well as clods of earth, workers would be stationed on each side of the harvester taking out these materials. Good tomatoes would remain on the conveyors where they would be transferred to bins being towed behind the harvesting equipment. In 1956, Hanna and Lorentzen entered into discussions with Ernest Blackwelder, a manufacturer of farm machinery, to build this machine (Friedland and Barton, 1975, p.25).

In addition to the initial lack of support provided to Hanna and Lorentzen by colleagues at UCD, California tomato growers were also slow to be convinced of the need for mechanical harvesting. According to Friedland and Barton:

> Grower attitudes towards labor had become thoroughly fixed over the decades. Traditionally used to having cheap, abundant and docile labor, most growers felt that agricultural labor required little skill, but lots of muscle and docility. Most growers (like most people) had distinct ethnic prejudices, and agricultural hand labor has always been highly ethnic. Because growers tend to be tradition-bound, few looked to mechanization as a possible solution (Friedland and Barton, 1975, pp.21–22).

However, such an explanation ignores the fact that the process of innovation is a complex process, which has its roots in economic and social imperatives rather than technical capabilities. In other words, growers were probably making a rational choice in the early 1950s, when expressing some skepticism about the need for an expensive technology as a substitute for cheap labor. This was to change, though, as the conditions governing the entry of cheap labor into the US grew more restrictive, and growers began to reconsider their attitudes towards new technology.

Lester Heringer, a board member of the California Tomato Growers Association (CTGA) was a key player in this respect. Although minutes of CTGA meetings indicate that the Association was considering the issue of mechanized harvesting from early as 1951 (Anonymous, 1965, p.4), it was not until the late 1950s that Heringer came to champion mechanization. In 1956 the CTGA provided funds to UCD to help develop the mechanical harvester, and in 1958 Heringer offered his farm (ranch) for pilot studies of mechanized harvesting. In 1960, at a landmark event at the Heringer ranch, 2,000 growers, processors, bankers and industry experts witnessed the first public demonstration of mechanical harvesting (Anonymous, 1965, p.5). For this reason, Heringer has been described within the grower community as 'the father of mechanized harvesting'.

In 1958, Hanna and Lorentzen, in cooperation with Blackwelder, built a prototype mechanical harvester. Early demonstrations of this prototype were not impressive, with most growers considering the project as 'naïve and impossible' (Friedland and Barton, 1975, p.22). However, changes to the original prototype improved its performance and in 1959 UCD patented the machine. In the following year, a licence was granted to the Blackwelder Manufacturing Company to build machines for commercial use. At about the same time, Hanna's ongoing work in the area of plant breeding saw the development of the VF-145 tomato, with characteristics which were appropriate for mechanized harvesting. Various strains of the VF-145 dominated the processing tomato industry throughout the 1960s and much of the 1970s (Sims, Zobel and May, 1979, p.6), until the emergence in 1976 of the UC-82, the so-called 'square tomato'. Thus, at the onset of the 1960s, the key technological breakthroughs needed to assure the viability of mechanized harvesting, had been made.

Notwithstanding these developments, mechanized harvesting would not take off in California for another four to five years. As noted earlier, the reasons for the delay are connected with wider issues concerning farm labor and, in particular, California's access to low-cost Mexican workers.

From the start, Hanna and Lorentzen's pioneering work had been underpinned by their wider perception of looming economic and political pressures on the Californian processing tomato industry. In a 1974 interview with the University of California Santa Cruz (UCSC) sociologist Bill Friedland, Hanna emphasized that his research was driven by his recognition that the United States 'would eventually exhaust the foreign labor pools from which it had drawn the successive waves of ethnic recruits for agricultural labor' (Friedland and Barton, 1975, p.23).

For the previous century the evolution of Californian agriculture had been interwoven with the political economy of immigrant labor. Until the *Chinese Exclusion Act* of 1882, Chinese immigrant labor had underwritten the expansion of Californian agriculture (Majka and Majka, 1982). Successive waves of Japanese, Filipinos and Mexicans followed the Chinese. The US-Mexican border was poorly policed until the 1950s, allowing relatively easy entry for Mexican labor. Along with these flows came internal US migration, especially from the 'dust bowl' states of Oklahoma, Arkansas, Missouri and Texas. Low agricultural prices and environmental degradation in the 1920s and 1930s encouraged the steady westward migration of displaced farmers and sharecroppers. During the Great Depression, 300,000 dust bowl émigrés moved to California (Majka and Majka, 1982, p.106). To migrants such as the fictional Joad family, immortalized in John Steinbeck's *The Grapes of Wrath*, California was 'the golden state' – an unbounded land of opportunity and prosperity. The reality for many of these migrants was very different. The huge supply of poorly skilled, often illiterate and financially desperate farm laborers, led inevitably to low piece-rate wages and periodic unemployment.

By the 1940s, however, these conditions had changed. The manpower needs of the Second World War led to labor shortages in Californian agriculture, resulting in the US Congress instituting temporary working visa arrangements for Mexican workers. The *bracero program* as it was called (*bracero* means 'hands' in Spanish), was sustained in 1951 through the adoption of *Public Law 78*. Under this legislation, the US Secretary of Labor was empowered to authorize the temporary importation of labor so long as evidence was provided of a labor shortage and that no American jobs were lost (Galarza, 1964).

By the mid-1950s the future of the *bracero* program was a topic of great debate within Californian agriculture. Research by the US Department of Labor in 1959 found that the bracero program did not meet the legislative requirement to produce 'no harm' for domestic US labor. It was found that the *bracero* program underwrote cheap wages and employment uncertainty for US agricultural workers. However, faced with a strong campaign by Californian agriculturists to maintain the program (and in a Presidential election year, as well), the US Secretary of Labor gave a stay of execution to the *bracero* program for five years, until 1964. Although lobbied heavily by growers through the early 1960s, Congress agreed not to extend the *bracero* program past 31 December 1964, although transitional arrangements for the 1965 harvest meant a continuing presence of some *bracero* labor for that season. In the early 1960s *bracero* labor accounted for approximately 80 per cent of the annual 50,000 strong tomato harvest workforce (Holt, 1965, p.9).

It was the demise of the *bracero* program, rather than the practical possibilities of mechanized harvesting, that determined the timetable for the introduction of mechanized harvesting. Although field trials of the Hanna-Lorentzen-Blackwelder prototype in 1960 had demonstrated its practicality, it was not until 1964, with the end of the *bracero* program in sight, that Californian tomato growers began to embrace this technology. In 1964, just 3.8 per cent of the Californian processing tomato crop was mechanically harvested. This rose to 24.7 per cent in 1965, 65.8 per cent in 1966, 81.8 per cent in 1967, and finally, 95.1 per cent in 1968 (Friedland and Barton, 1976). Thus, in just five seasons, the incidence of mechanical harvesting rose from almost negligible proportions to overwhelming dominance.

The advent of mechanical tomato harvesting had profound social impacts. From the mid-1960s these were intensively studied by Bill Friedland, Professor of Community Studies and Sociology at the University of California Santa Cruz (UCSC). Friedland's work during this period (see *inter alia*, Friedland and Barton, 1975; Friedland and Barton, 1976; Friedland, Barton and Thomas, 1981) is a powerful testament to the need to analyze the social impacts of agri-industrial technologies. Friedland's work shows that, although mechanical tomato harvesting 'saved' the Californian tomato industry after the demise of the *bracero* program, and thus helped create tens of thousands of tomato-related jobs in the State over the subsequent decades, it was also the case that these technologies were associated with social and economic marginalization of farm workers and some growers. In the words of the agricultural economists Schmitz and Seckler:

> We point with justifiable pride to the fact that now only a small percentage of the total population produces our food needs. But we tend to forget the painful process of adjustment that accompanied the transition from a rural to an urban society. We have forgotten that for many people the transition was involuntary; that many people have been forced off the farm only into an economic and social limbo in rural towns and urban ghettos (Schmitz and Seckler, 1970, p.569).

Schmitz and Seckler's research, published in the prestigious *American Journal of Agricultural Economics*, employs an economic analysis which attempts to measure the social benefits and costs from tomato mechanization. Net social returns from this technology are computed by estimating increased productive efficiency less the social cost of displaced workers. The researchers estimate that the development of the mechanized harvester incurred costs of $3.25 million (US), in compounded 1967 values. For that investment, harvesting costs were reduced by between $5.41 (US) to $7.47 (US) per ton and displaced 91 man-hours per acre of harvest. Consequently, they estimate gross social returns to public-private investment in developing mechanized harvesting were 'in the vicinity of 1,000 per cent'. However, because no mechanisms existed to transfer those benefits to the 'losers' from technological change (that is, displaced workers and some growers), they equivocate on the question of measuring the size of social costs and benefits to 'society as a whole' (Schmitz and Seckler, 1970, pp.569–75).

The number of workers employed in California's annual processing tomato harvest fell from over 50,000 in 1964, to 18,000 in 1972 (Friedland and Barton, 1976). In this same period, production tonnages increased by 50 per cent. The initial social impacts arising from these processes can be summarized under three headings: a shift in the gender composition of the workforce from male to female; a shift from Mexican to American labor; and a shift from a transitory to a settled workforce. Prior to mechanization, tomato harvesting was predominantly men's work. In addition to *bracero* and other immigrant labor, itinerant males were an important component of tomato harvesting. Friedland and Barton (1976) identify this social category as 'largely nomadic individuals, many semi-alcoholics, who are found in sizeable numbers in many Central Valley towns and cities'. Writing from the perspective of growers, Becket (1966, p.22) notes that '[W]hen sober, he [the itinerant male] is fairly acceptable'.

The shift in the gender of the tomato harvest workforce can be explained at two levels. First, by reducing the physical requirements of the labor process, mechanical harvesting undermined the previous assumption that this was 'male work'. The first harvesters required manual sorting of tomatoes, tasks that were analogous to assembly line food processing work in which females traditionally dominated. At another level, the replacement of male by female employees could be seen as a strategy by Californian agriculturists to construct a more compliant workforce in the face of increased farm unionization. Fears of strikes by workers were a major influence over the industry in the 1960s (see below). By the early 1970s, it was probable that 80 per cent of the processing tomato harvest labor force was female. The transition to a predominantly female labor force occurred via the enrolment of informal networks, rather than the systematic use of labor contracting (Friedland and Barton, 1975, p.61).

In a 1966 study of tomato pickers in Yolo County, California, Becket provides a profile of the agricultural farm labor force. This is an important historical study because it was carried out at just the time when mechanical harvesting was transforming the industry. 1965 was the last year in which the majority of the Californian processing tomato crop was hand picked, and when significant numbers of *braceros* (approximately 10,000) were involved in this activity. Through interviews with 751 workers during August and September 1965, Becket reveals the highly transitory nature of the tomato harvest labor force. With the phasing out of the *bracero* program, Mexican nationals comprized just 22 per cent of the farm labor force, while students aged under 18 comprized a further 26 per cent.[1] Some 49 per cent of the farm labor force were temporary residents, and almost half had worked less than three months during the period from January to August/September 1965. Furthermore there was considerable labor turnover within the industry during harvest. One association hiring workers for tomato and melon growers in the San Joaquin Valley reported to Becket that 44 per cent of its workers were employed for only two days at each establishment (Becket, 1966, p.15).

Becket's study also suggests that Mexican nationals and Mexican-Americans harvested 11.6 boxes per hour, compared with 7.6 boxes per hour for non-Mexican workers. This difference seems to be the result of different age profiles between these two groups. Whereas a considerable proportion of workers of Mexican descent were prime working age, non-Mexican cohorts were dominated students under 18 years of age, or older 'single men'. Thus the demise of the *bracero* program not only reduced labor supply for tomato growers, it took away many of the industry's most productive workers.

During the 1970s, mechanical harvesting underwent further transformations. In the early 1970s, bulk-handling techniques replaced the traditional 50 pound (approximately 20 kilogram) bins. Traditional bins had evolved in line with the physical labor requirements of hand picking. Mechanized harvesting however, allowed the use of larger bins and, eventually, the introduction of bulk-handling using gondola hauling techniques. This in turn, necessitated more precise scheduling arrangements. Because gondolas contained much larger volumes of fruit with the possibility of spoilage through delays, these technologies encouraged processors, transporters and growers to develop tighter information systems (Friedland and Barton, 1975, p.16).

Furthermore in the late 1970s, electronic sorting of tomatoes was introduced, with major implications for the farm labor force. Prior to electronic sorting, each mechanized harvester was accompanied by a team of manual sorters, usually female. When electronic sorters first came into commercial use in 1975, they allowed four workers to sort twice the volume of tomatoes as sixteen manual sorters had done previously. However, it is very significant that the introduction of this technology was accompanied by a high level of militancy and industrial confrontation, which was notably absent when mechanized harvesting was initially introduced. Why was this the case?

The industrial militancy of this period dates back to struggles in the 1960s between Californian agriculturists and unions, notably the United Farm Workers (UFW). Access to *bracero* labor subdued wages growth in the 1950s and early 1960s, and the economic strength of the UFW grew concomitantly with the decline of the *bracero* program. In 1948, Californian agricultural wages were 64.7 per cent of national average manufacturing wages; by 1959 they had fallen to just 46.6 per cent (Majka and Majka, 1982, p.141). Hence, with the ending of the *bracero* program, the time was ripe for a major wages push. Aggravating these pressures were the poor living and working conditions for farm workers. In addition to wage rates, the initial UFW campaigns targeted such issues as the provision of cool water during working hours, access to toilets and an end to alleged profiteering from meal provision.

The processing tomato industry was not a central target for the major UFW industrial campaigns of this period. Nevertheless, the industry was caught up in the general industrial unrest of the times. Leading the UFW was Cesar Chavez, the son of immigrant farm workers who moved to California during the Great Depression (Majka and Majka, 1982, p.106). Variously described as a 'charismatic labor leader', 'militant folk hero' and, by his opponents, 'a scoundrel' (Ferris and Sandoval, 1997, p.3), Chavez organized the largely non-white farm workforce and catapulted their struggles into the national arena via his strategy of developing community alliances. Between 1967-70, the UFW led a national boycott of Californian table grapes in protest at working conditions in this industry. At the height of this boycott, the UFW had sympathy committees in 50 US cities and had placed intense pressure on supermarket chains to discontinue stocking this product. The boycott was further publicized by staged marches to Sacramento and fasts by Chavez, emulating the tactics of Martin Luther King and Mahatma Gandhi. By 1968, the UFW had won the support of Presidential aspirant Senator Robert F. Kennedy, who nominated Chavez as a Californian delegate to the 1968 Democratic Convention (Majka and Majka, 1982, p.189). However, Kennedy's assassination,[2] the election of Republican Richard Nixon as President, and the sustained anti-union stance of Californian Governor Ronald Reagan dented the UFW by the early 1970s. At one level, the Nixon administration sought to overcome the table grapes boycott by dramatically increasing Department of Defense purchases of this commodity. In the last year of the Johnson administration, the Department shipped 252,000 kilograms of table grapes to its troops in Vietnam; in the first year of the Nixon administration this was increased to 985,000 kilograms (Majka and Majka, 1982, p.193). At another level, the Teamsters Union (which had endorsed President Nixon) began organizing farm workers. Bitter disputes between the UFW and the Teamsters in 1970 led to three UFW members being shot, a UFW attorney being beaten unconscious, and the UFW Watsonville office being bombed (Mooney and Majka, 1995, p.166). Hence, just at the time when the UFW had succeeded in making most grape growers capitulate to its demands, the union itself was being undercut by violent assault by the Teamsters, and a consequent loss of membership (Ferris and Sandoval, 1997, pp.156–60). These pressures dramatically weakened the influence of the UFW. A prolonged strike in 1973 cost the UFW $3 million (US), and by the following year it represented less than five per cent of Californian farm workers (Majka and Majka, 1982, p.241).

Despite the UFW's political marginalization, campaigns continued through the 1970s on the issues it had highlighted. In August 1976, some 75 members of the UFW marched in protest to the Agricultural Engineering Department at UC Davis, where electronic sorters for the processing tomato industry had been developed, to highlight the plight of displaced workers. According to a union organizer at the protest: 'We're not stopping mechanization. We understand the necessity for it. But we also understand that they [machines] shouldn't displace workers' (Duscha, 1976). Sensing the mood of the times, President Carter's Secretary of Agriculture, Bob Bergland, told an audience in Fresno in 1979 that he would prevent Federal funds being used in agricultural research that put farm laborers out of work, and that private agribusiness should fund its own interests (Taper, 1980, p.83). This new policy however, was never implemented, and in any case became redundant with the advent of the Reagan Presidency.

At about the same time, legal action was also instituted against UCD for its role in developing tomato varieties used in mechanized harvesting and, by extension, in the loss of farm jobs (Friedland, 1991; Blackburn, 1979, p.3; Taper, 1980). The so-called 'mechanization suit' involved an attempt to make the land-grant universities consider the social consequences of their funded research programs. Under the *Hatch Act* of 1887, land-grant universities were established to foster agricultural development under broadly Jeffersonian principles (Friedland, 1991, p.25). However, according to the plaintiffs[3] in the 'mechanization suit', the research priorities of the land-grant universities favored agribusiness companies whose interests were diametrically opposed to the interests of small farmers and farm workers. As a consequence, the land-grant universities were alleged to be contravening the statutory basis under which they were funded. In 1986, some eight years after the inception of this case, Judge Marsh of the Superior Court of the County of Alameda ruled that the plaintiffs' case had merit. He ruled (*inter alia*) that: 'There must be some University-level process designed to insure consideration of each of the legislatively expressed interests [of the *Hatch Act*] with primary consideration given to the small farmer' (cited in Friedland, 1991, p.31). This decision compelled the University of California's land-grant institutions to deal with these issues, although subsequent appeals and the reluctance of plaintiffs to seek enforcement mitigated the long-term significance of this ruling.

Labor unrest was also widespread in the tomato regions of the US Mid-West during the late 1970s and early 1980s. The Mid-West was slower than California to embrace mechanized harvesting, resulting in substantial job shedding taking place in the 1970s (Rosset and Vandermeer, 1986; Hurt, 1991). In August 1978, the Farm Labor Organizing Committee (FLOC) led a strike of 2,000 (out of a total of 8,000) tomato harvest workers in Ohio. The strike was targeted at fields contracted to Campbell Soup and Libby's. As a result, only 45 per cent of that season's tomato crop was harvested. According to a journalist covering the strike '[T]he growers organized truck caravans to block FLOC's access to the fields; some carried rifles to intimidate the strikers; others sprayed pesticides on them, burned their flags and threatened to kill them' (Barkan, 1978, p.10). As a result, the FLOC's ultimate attempt to establish a three-way agreement between farm labor, growers and canners failed miserably, and had the unintended consequence of accelerating the introduction of mechanized harvesting. As Charles Hess, Dean of Agriculture at UCD observed in 1978: '[T]he thing that drives growers to mechanize is the fear of a strike' (Schrag, 1978, p.51).

The successful introduction of mechanized harvesting and electronic sorting not only provided another step in the evolution of capital-intensive tomato cultivation, but also demonstrated the increasingly integrated and interdependent nature of the emerging agri-food system. The introduction of mechanized harvesting involved a synergy between the mechanical basis of harvesting and the biological basis of tomatoes. Similarly, the introduction of electronic sorting involved a recognition of a growing interdependence of between formerly discrete segments of the production system. According to a 1976 report in *California Farmer*: 'To make these [sorting] units pay, a grower must be well organized, use the equipment night and day, and have plantings in more than one location' (Anonymous, 1976, p.12). However, Friedland and Barton note:

> [I]n historical retrospect, the development of the tomato, the machine, the cultivation practices, and the dispersal of information can be viewed as 'sleepwalking'. The actors involved were rarely conscious of the systemic elements involved in their research (1975, p.28).

This recognition of the *systemic* relationships between technology, economics, politics, labor relations and environment has also profoundly shaped subsequent academic research on agri-industrial change. Prior to the work of Friedland and Barton on tomatoes, and their later work on the Californian lettuce industry (Friedland, Barton and Thomas, 1981), rural sociologists and other social scientists viewed these processes largely within their separate fields. By emphasizing the systemic nature of change however, Friedland and his colleagues instituted a new approach to rural sociological research – *commodity systems analysis* – which, through various permutations, formed the kernel of the *new political economy of agriculture* approach that dominated much of rural sociology in the 1990s (Buttel, Larson and Gillespie, 1990). According to one contemporary observer: 'Friedland has consistently opposed the notion of an institutionally separated rural sociology and has helped to fashion a more theoretically-informed, holistic, radical and critical rural sociology' (Newby, 1982, p.468). Hence, it would not be inaccurate to state that the tomato – or at least, the research by Friedland et al. on this commodity – played a seminal role in the establishment of important new approaches to the analysis of rural restructuring.

Size matters: mechanical harvesting and farm management

Mechanical harvesting was the change-agent that transformed farm management and made possible the increasing farm size, which now characterizes this industry. Small acreages tend to be correlated with the hand-picking of tomatoes and on-farm diversity. However, mechanized harvesting is associated with a shift to larger acreages and a tendency towards specialization. This is allied to both the agronomic and financial implications of mechanized harvesting.

At the agronomic level, mechanized harvesting requires the introduction of new cultivation practices. 'Precision' is the key element of these practices, in planting, row spacing, irrigation and scheduling. Agricultural extension officers in each of California's major tomato counties have played a critical role in promoting these new cultivation practices, thus realizing the potential cost savings from mechanized harvesting. It was not until 1962 that an information bulletin relating to these practices was produced for tomato growers in California (Friedland and

Barton, 1975, p.28). This original edition (with a length of five pages) had evolved to a comprehensive manual of 28 pages by 1968 (Sims, Zobel and King, 1968).

Where growers own their harvest machinery there is a compelling financial case to make optimal use of that capital. Early harvesting machines cost growers $25,000 (US) each (Friedland and Barton, 1975, p.53), meaning that the transition to mechanized harvesting also implied a transition to larger production volumes. Although there were demonstrable financial benefits from mechanized harvesting – a 1965 study of 63 harvesters operated in 15 California counties indicated that the average cost of machine harvesting was $9.84 (US) per acre compared to between $17.04 (US) and $17.19 (US) per acre for hand-picking (Parsons, 1966) – these benefits were realized most effectively on larger allotments. In any case, there is a clear statistical link between the introduction of mechanized harvesting and the emergence of large-scale tomato farming. In California, just prior to the introduction of mechanized harvesting in the early 1960s, there were over 4,000 growers; by 1973 when virtually 100 per cent of the crop was mechanically harvested, there were less than 600. During that same period, production tonnage doubled (Friedland and Barton, 1976).

The intersections of technology and politics which saw the introduction of mechanized tomato harvesting in California now spawned a major industry in the manufacture of harvesting machinery. Blackwelder, which was granted the initial licence to use the UCD patented technology, built 15 machines in 1960, and a further 155 between 1961 and 1965 (Rasmussen, 1968, p.536). A number of other companies quickly followed Blackwelder's lead, including FMC in 1960 (Anonymous, 1965, p.8), Hume, Benner-Nawman, and Ziegenmeyer in 1961, and Massey-Ferguson in 1962 (Hurt, 1991, p.90). At about the same time, a separate team of researchers from Michigan State University under the leadership of an agricultural engineer named B.A. Stout and a horticulturist called S.K. Ries, also developed a crude mechanical tomato harvester (Rasmussen, 1968, p.536). Flaws in initial models were largely overcome by 1966 when, in a single year, over 250 Blackwelder harvesters were manufactured and sold. In that year there were 760 tomato harvesters in California (Rasmussen, 1968, p.540). From these modest beginnings there has emerged a sizeable global industry, and the huge international demand for tomato harvest machinery is currently serviced mainly from agricultural machinery companies located in the United States and Italy, including FMC, Johnson, Guaresi and Gallignani.

In addition to encouraging reduced labor requirements and larger farm sizes, mechanical harvesting also led to geographical shifts in the industry. The transition to tomato varieties which encouraged uniform ripening placed a premium on climatic predictability. Because all tomatoes planted at the same time would ripen together, it was important that the tight harvest timetable was not interrupted by poor weather. Thus, the advent of mechanical harvesting implicitly encouraged geographical shifts in the industry towards regions where the climate was more predictable. In the US, this resulted in the relative growth of production in California *vis-à-vis* other states. Whereas in 1949, California accounted for just 39.8 per cent of total US processing tomato production, by 1974 it accounted for 83.3 per cent (Friedland and Barton, 1976). Thus the mechanical tomato harvester, a Californian invention, helped cement California's pre-eminent production site for the processing tomato industry.

Within California, mechanized harvesting has also encouraged a series of regional shifts in production. The second half of the 1990s witnessed significant shifts in the geography of Californian processing tomato growing. In 1995, southern counties grew just 38 per cent of the Californian processing tomato crop. By 2000 this had increased to 48 per cent (Table 2.7, Figure 2.2). These shifts relate to increased yields being generated in southern counties, resulting from newly introduced varieties and agronomic practices.

Table 2.7 Regional shifts in Californian processing tomato cultivation, 1995-2000

	1995	1996	1997	1998	1999	2000
			Tons (millions)			
North	4.61	4.30	3.72	3.39	4.75	3.11
Central	1.99	2.02	1.95	1.84	2.54	2.21
South	4.01	4.34	3.68	3.66	4.95	4.96
			Percentage of total			
North	43.5%	40.4%	39.8%	38.2%	38.8%	30.3%
Central	18.7%	19.0%	20.8%	20.7%	20.8%	21.5%
South	37.8%	40.7%	39.3%	41.2%	40.5%	48.3%

Note: Volumes are measured in US tons. 'South' includes: Fresno, Kern, Kings, Madera, Monterey, San Benito, Tulare, Santa Barbara and Imperial counties. 'Central' includes: Alameda, Contra Costa, Merced, San Joaquin, Santa Clara and Stanislaus counties. 'North' includes: Colusa, Glenn, Sacramento, Solano, Sutter, Yolo, Butte and Yuba counties.

Source: Processing Tomato Advisory Board (2001).

These regional shifts generate profound social consequences for affected communities. To illustrate these, it is worth considering the experience of one county in recent years. Colusa County lies at the northern extremity of the Californian Central Valley, and hence is the prime candidate for an examination of the social consequences of the southwards shift in Californian tomato production. During the second half of the 1990s, the county produced an average of 800,000 tonnes of processing tomato output annually, with a high of just over a million tonnes in 1999 followed by just 750,000 tonnes in 2000. With the general shift of tomato cultivation southwards, Colusa tomato growers have become increasingly reliant on the nearby Morning Star paste plant at Williams. According to the Agricultural Extension Officer for Colusa, an estimated 80 per cent of processing tomatoes grown in the county are sold to the Morning Star facility (Murray, 2001). The propinquity of this plant and the general support for regional growers by

Morning Star has stabilized the County's agricultural economy over recent years, but problems of general economic malaise are nevertheless apparent. On 1 January 2000, the county's resident population was estimated at 18,750, and there had been no population growth in recent years (Department of Finance California, 2000).

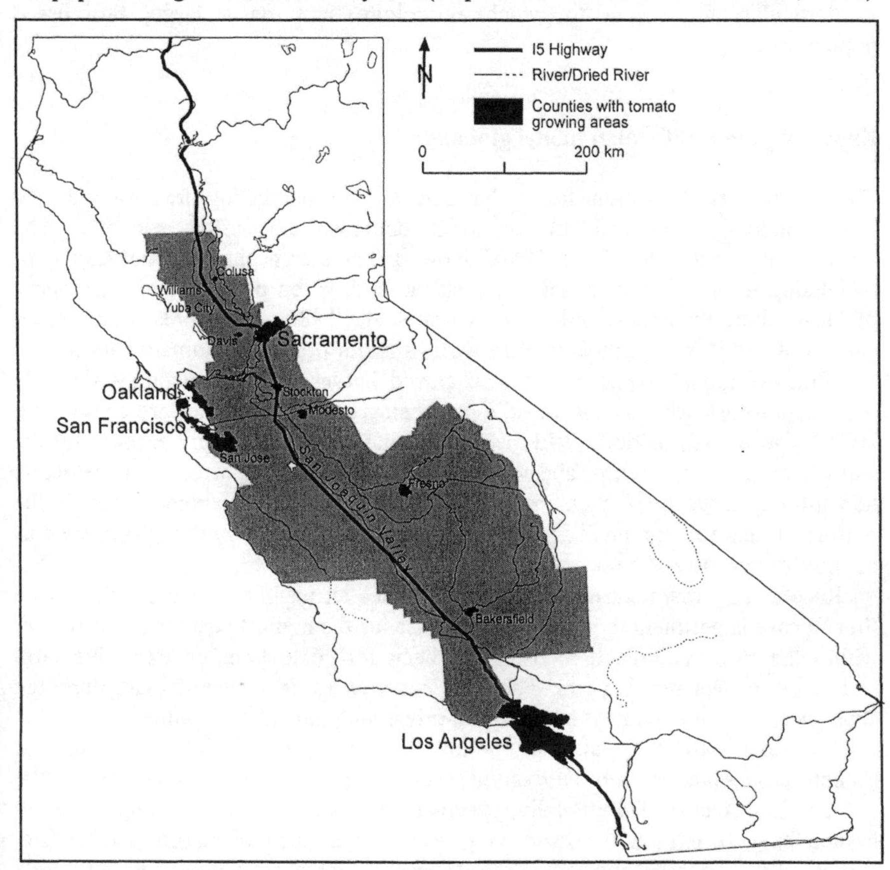

Figure 2.2 Californian tomato growing regions

Poverty levels were also considerably above the state average. The US Census Bureau (2000) defined an estimated 18.1 per cent of the Colusa population as living in poverty, compared with 16.0 per cent state-wide. Employment opportunities in the county were highly seasonal, with non-harvest periods according few opportunities for stable work. Moreover, tomato cultivation was increasingly occurring via leased land arrangements. Although no firm data exist, leased land accounts for an estimated 50 per cent of the County's tomato crop. Typically, landlords are paid according to shared-rent arrangements and receive between 12 per cent to 18 per cent of the gross value of production (Murray, 2001).

Hence, the shift to high output, mechanically harvested processing tomato cultivation has generated an extremely efficient mode of agriculture which greatly assisted California's economic development and growth, but which in the process, has thrown down fundamental challenges to the nature of the State's rural society, in particular, the extent to which agriculture supports a large, farm-based population.

Exporting the Californian model globally

The Californian transition to mechanized tomato harvesting transformed that state's industry and had an important demonstration effect globally. The Californian experience of the 1960s showed that the mechanized harvesting of processing tomatoes was technically possible and, given particular circumstances of farm labor, financially attractive. Through the 1980s and 1990s, mechanized harvesting left its mark in many of the world's major processing tomato nations.

With the global expansion of mechanized harvesting over recent years, it is legitimate to ask whether the world's processing tomato industries are converging on 'the Californian model'? Will mechanized harvesting ultimately account for the world's entire tomato crop, and would such a change bring with it an inevitable transition to large grower enterprises? To put this question another way, is the writing on the wall for small growers who dominate production in regions, such as the Mediterranean and some areas of the Third World?

Recent farm restructuring in France provides an insight to this question. The French case is pertinent to our inquiry because, as the highest cost tomato producer within the EU, the French industry has been forced to undergo extensive farm restructuring over recent years. Thus, the experiences of France arguably presage changes in the wider context of European processing tomato cultivation.

To understand the basis for the introduction of mechanized harvesting to French processing tomato cultivation necessitates a wider perspective on the French rural sector. Traditionally, survival strategies for small family farms in France have rested on the need to generate high rates of return per hectare, combined with internal diversification. Processing tomatoes have come to provide a distinct commercial niche within this agricultural context. Traditionally, France's processing tomato growers have been mixed horticulturists growing mainly for the fresh market. To reduce their reliance upon the inherently volatile market for fresh produce, farmers have sought to secure reliable (but lower) commercial returns on a portion of their farm via production contracts with processing firms. In other words, many French farmers use the contract production of processing tomatoes as a means of generating a relatively secure income stream for their farm, and thus to act as an insurance for their potentially higher return/higher risk production of fresh market horticulture. It has not been uncommon for French horticultural smallholders to generate 80 per cent of their farm incomes through fresh market sales, and 20 per cent from contract processing tomato cultivation (Vallat, 2001).

This agrarian heritage is central to the French processing tomato industry's current restructuring and its adoption of mechanized harvesting. The diversification strategies of individual farmers created an industry with a large number of small suppliers. As recent as 1993, some 65 per cent of French processing tomato tonnage was supplied by farmers each growing less than 100 tonnes of tomatoes (Sonito, 2000). Moreover, industry sources suggest that many of these growers would have supplied volumes of 50 tonnes or less. For example, in 1997, the French cooperative Le Cabanon had 68 growers supplying less than 50 tonnes each, out of a total of 400 suppliers.

The proliferation of small farmers in the French processing tomato industry encouraged the emergence of a particular economic geography for this sector. Because the cultivation of processing tomatoes was, for most farmers, a relatively minor activity, closeness to processing facilities has not been a major factor determining the location of growing areas. Growers of processing tomatoes have not clustered near to factories. Rather, they have tended to be dispersed in accordance with the different spatial logics associated with the supply of fresh market horticulture.

The restructuring of the traditional diversified small farming component of the French processing tomato industry began in the 1980s, following sharpened competition from Italy and the (then) new EU member states of Greece, Spain and Portugal. Faced with the need to become more competitive, mechanized harvesting began to be introduced into France. In the mid-1980s the French inter-professional agency for processing tomatoes, Sonito, developed a prototype mechanical harvester suited to French conditions. Compared with US models, this was smaller and better suited to operating on undulating terrain.

There have been profound impacts in France from the introduction of mechanized harvesting, with industry sources suggesting that it initially reduced the labor input for harvesting a hectare of tomatoes by one-third, and by the year 2000, it required just one-fifth the labor input that it did prior to mechanized harvesting (Vallat, 2001). These developments came at a critical time for the French industry. The collapse of the Italian lire in 1992-93 (when financial speculators pushed the currency out of the European monetary system) made French tomato growers extremely uncompetitive *vis-à-vis* the Italian industry. However, higher farm productivity in France resulting from the implementation of mechanized harvesting enabled French growers to meet this challenge in a more effective way than what would otherwise have been the case. Cooperative strategies allowed the costs of mechanized harvesting to be spread amongst the grower population, while the French *Cooperative d'Utilisation Nationale Agricole* (CUNA) undertook contract harvesting for many growers, thus facilitating additional resource efficiencies.

The competitive challenge from Italy coupled with wide-scale mechanized harvesting – by 2000 some 98 per cent of French processing tomatoes were harvested mechanically – encouraged the emergence of new farming structures. Following the crisis in French broadacre agriculture during the 1990s, some wheat and maize growers diversified into processing tomatoes. In general, these new entrants grew larger volumes per farm than traditional mixed horticulturists. Also, the entry of these new growers encouraged further geographical dispersal of the industry, northwards along the Rhône Valley and into new production areas near the mouth of the Rhône.

The extent of these changes is revealed in statistics collected by the French inter-professional agency, Sonito. Between 1993 and 1999, France experienced a 56 per cent increase in processing tomato production, but also saw the number of growers decline by 45 per cent (Table 2.8). Tonnage grew in each of the major French production zones, although was slowest in Provence Alpes and Cote d'Azur (PACA). The extent and rapidity of these changes is striking; two-thirds of the processing tomato growers in PACA who were producing in 1993 had left the industry by 1999, highlighting the degree to which French (and indeed European) agriculture was in transition.

Table 2.8 Regional patterns in processing tomato production, France 1993-99

Production zone	Tonnes 1993	Tonnes 1999	Change	Number of growers 1993	Number of growers 1999	Change
PACA	64,429	77,959	+21%	627	207	-67%
Rhône Alpes	67,431	121,376	+80%	376	308	-18%
Languedoc-Rousillon	58,646	91,487	+56%	192	142	-26%
Aquitaine	42,818	70,650	+65%	550	286	-48%
Other	4,706	9,854	+109%	28	32	+14%
Total France	238,030	371,326	+56%	1,773	975	-45%

Note: PACA is Provence Alpes and Cote d'Azur.

Source: Sonito (2000).

The recent experiences of the Spanish processing tomato sector reinforce the general points made about France, while also illustrating the existence of national and regional differences within Europe. The Spanish processing tomato industry is located mainly in the province of Extramadura, southwest from Madrid, and adjacent to the Portuguese border. This region, along with Andalucia, has traditionally been the poorest region in Spain. Agriculture dominates this region because of the availability of irrigated water from the Giardiana River system. Irrigation infrastructure was developed during the Franco dictatorship, via the *Plan Badajoz*, part of the regime's strategy to avert rural discontent. The history of small peasant landholdings in this region has enabled Extramadura to attract significant EU structural adjustment funds, although there has remained a significant income gap between Extramadura and Andalucia and the rest of Spain. High economic growth rates in Spain since the 1980s have fed primarily into the burgeoning regional economy of Catalonia, Valencia and the industrialized north, rather than the agricultural interior.

In the year 2000, Extramadura accounted for approximately 70 per cent of Spanish processing tomato production. Of the remaining amount, approximately 20 per cent to 25 per cent was accounted for by production in the Evora River region, in the north of the country. Production in Extramadura is focused on paste, whereas in the Evora River region output is divided between paste, whole peeled canned, and *passata*, with each accounting for approximately one-third of production.

As Spain's major production region, Extramadura has been at the forefront of agricultural transformations aimed at generating greater efficiency in production. Because of the importance of the cost of raw tomatoes to the competitiveness of paste (raw product comprises approximately 65 per cent of paste costs), Extramadura's fate has rested on its ability to deliver competitively priced tomatoes to the factory gate (although in the European context, of course, the effective level of competition is reduced by the effect of EU subsidies and tariffs). Consequently, Extramadura witnessed the introduction of mechanized harvesting and other agricultural innovations in Spain before the Evora River region.

The harvesting of processing tomatoes in Extramadura had become wholly mechanized by the mid-1990s. In contrast, in the year 2000 at least 50 per cent of the processing tomato crop in Evora was still hand-harvested. The introduction of mechanized harvesting in Spain displaced traditional farm laboring populations, such as seasonal ethnic Romany peoples, many of whom lived semi-permanently in Portugal. In this same period, the industry adopted hybrids almost wholly, and 70 per cent to 80 per cent of the processing tomato crop is transplanted rather than being directly sown.

As with other regions of the world, the introduction of mechanized harvesting in Spain was associated with farm-level restructuring. Although there is no accurate statistical data on these trends, industry informants suggest that average farm size grew consistently at a rate concomitant with the introduction of mechanized harvesting. Twenty years ago much of the Spanish processing tomato industry consisted of farmers cropping just 0.5 to 2.0 hectares of processing tomatoes. By the late 1990s, median farm sizes had increased to between 5 to 10 hectares, and a number of larger farm enterprises, with 20 to 50 hectares of processing tomatoes, had emerged. This last group were so-called 'professional' farmers who managed their crop investments through the use of sub-contracted farm labor and land. The growth of mechanized harvesting has been accompanied by extensive use of leased land for tomato production.

These transformations are important not only from an agronomic point of view, but because of the way they feed into rural politics. In southern Europe during the twentieth century, there was an indivisible distinction between the structures of agricultural production, and rural political institutions of power and control. In Italy, a considerable proportion of agricultural production was organized through large federations of cooperatives based around political groupings. This included the (*rossa*) ANCA-Lega group (representing the political left) and the (*bianca*) Catholic group (Moyano-Estrada et al., 2001, p.243). In Spain, the emergence of agricultural cooperatives in the 1920s and 1930s saw the rise of the Catholic CNCA (National Catholic Agricultural Confederation), the socialist FNTT-UGT

(National Federation of Agricultural Laborers) and the anarchist CNT (National Labor Confederation), although the latter two organizations were abolished with the ascent of Franco. Spanish democratization and entry into the EU saw a re-emergence of politically-based agricultural cooperatives in the 1970s and 1980s (Moyano-Estrada et al., 2001, pp.247–49). Evidently, the expansion of mechanized harvesting, larger farm sizes and fewer farmers brings with it a new agricultural logic that makes the traditional agri-political structures of southern Europe seem dated. As such, these agricultural technologies are enmeshed with a wider set of rural political and institutional shifts.

Summary

The world consumes increasing amounts of tomato-based products, but these are grown by fewer and fewer farmers. Starting with California in the 1950s and 1960s, a new regime of tomato cultivation emerged. This production regime was underpinned by the agronomic and financial implications of mechanized harvesting. New hybrid tomato varieties, the more extensive application of transplant methods, widespread use of leased land, and the general capital intensity of agricultural practice were part-and-parcel of this regime. Growers were positioned within complex networks of off-farm advice and expertise. The circulation of knowledge and information ensured that the actions of growers were contextualized within tightly orchestrated relations, largely governed by the needs of the agri-food corporations which increasingly have determined the parameters of production. Decisions on what seeds to use, when and how to control weeds and pests, and when and how to harvest, no longer were the sole responsibility of growers. In California and Australia especially, but increasingly in Europe also, growers are in a paradoxical position. They control large farm enterprises, in many cases worth several millions of dollars. Yet the management strategies they are obliged to use are constrained by the demands of other actors within processing tomato supply chains.

In Australia, sustained grower restructuring over a 15-year period has resulted in just 33 growers growing almost 400,000 tonnes of processing tomatoes. This output is sold to three paste factories and two canning factories. The extensive use of drip irrigation technology generated record yields and helped sustain an industry that was under considerable price competition. The growers that remain in this industry are knowledgeable, asset-rich and globally focused. They are also highly financially dependent on the future viability of the industry (and the industry is highly dependent on the fortunes of ever-fewer growers), creating new elements of vulnerability.

There is a sense of inevitability about the restructuring trajectories of processing tomato cultivation. The social, economic and technological implications of the farm practices first instituted in California in the 1960s are leaving their mark globally. They are the foundation for a system that produces larger volumes of tomatoes at increasingly competitive prices, but with fewer growers. Yet at the same time, whereas the Californian model of mechanized harvesting has taken root across the world's processing tomato regions, it has *not* created carbon copies of Californian agriculture. Because the introduction of mechanized harvesting occurs in vastly different agricultural, ecological, political and cultural contexts, its impact and influence varies. Thus, whereas mechanized harvesting leaves similar footprints whenever it is introduced, there is continuing diversity in the size, shape, depth and impact of those footprints. In turn, this diversity constructs and perpetuates new arenas of difference and competitive advantage within the global processing tomato sector.

Notes

1 Becket (1966, pp.24–25) makes the following observation about student labor: 'His productivity does not match that of an adult … On the other hand, his natural energy and exuberance are seldom completely exhausted, and many a field supervisor has been driven to distraction by the speed with which a tired lad who can hardly drag a lug box out of the row can become involved in a tomato fight.'

2 Cesar Chavez was one of the last people to speak to Robert Kennedy before his assassination, leaving Los Angeles' Ambassador Hotel just minutes beforehand. UFW organizer, Dolores Huerta, was walking alongside Kennedy's wife, Ethel, a few steps behind the Senator, when the shooting occurred (Majka and Majka, 1982, p.190).

3 The plaintiffs were the California Agrarian Action Project and a number of individual farm workers. The case was run by California Rural Legal Assistance.

Plate 3 A 'state-of-the-art' processing tomato factory, Xinjiang, 2001

Photo: Bill Pritchard.

Chapter 3

Pizzas, pasta sauces and ketchup: producing and consuming tomato products

Once harvested, tomatoes are transported to factories for first-tier processing. The vast majority of the world's processing tomatoes are either canned or converted to paste. Other 'first-tier' transformations, mainly juicing or drying, tend to be undertaken as niche activities, within specialist firms or regions. Accordingly, the focus of this chapter rests squarely with the production and consumption of canned tomatoes, tomato paste and derivative products.

Tomato canning involves a relatively simple set of technologies. The basic techniques for tomato canning were formulated in the nineteenth century and remain much the same today, despite the obvious differences of scale, efficiency and product safety brought about by a century of progress in food technology. Because tomatoes do not undergo significant transformation in the canning process, there is relatively little change in weight between raw and finished products. According to European data, the production of one tonne (net) of whole peeled Roma tomatoes requires approximately 1.16 tonnes of raw product, while one tonne (net) of canned peeled tomato pieces requires 1.23 tonnes of raw product (AMITOM, 2000, pp.33–39). The relative simplicity of this technological process is illustrated in Figure 3.1.

Once received by processing plants, tomatoes are transferred for washing and sorting by manual, mechanical or hydraulic means. Sorting usually takes place via electronic systems, after which the tomatoes are peeled through scalding, using steam or hot water. Under pressurized steam scalding, tomatoes can be heated to 131 degrees Celsius, and are then vacuum cooled. Peel can be re-used as useful by-product from this stage. The peeled tomatoes are then sorted once more and, if being used for cut or crushed product, are diced. The final stage of the tomato canning process is filling, which occurs either through in-container processing, or through aseptic processing. In the former case, tomatoes or tomato particulates (pieces) are conveyed to a can (or glass) specific filling machine. These containers are sealed hermetically then transferred to a sterilizer for heating and cooling. In the case of aseptic processing (used only for crushed or diced product), sterilization occurs prior to filling. Product is heated and held at a temperature to obtain sterility, then aseptically cooled and filled into aseptic bags.

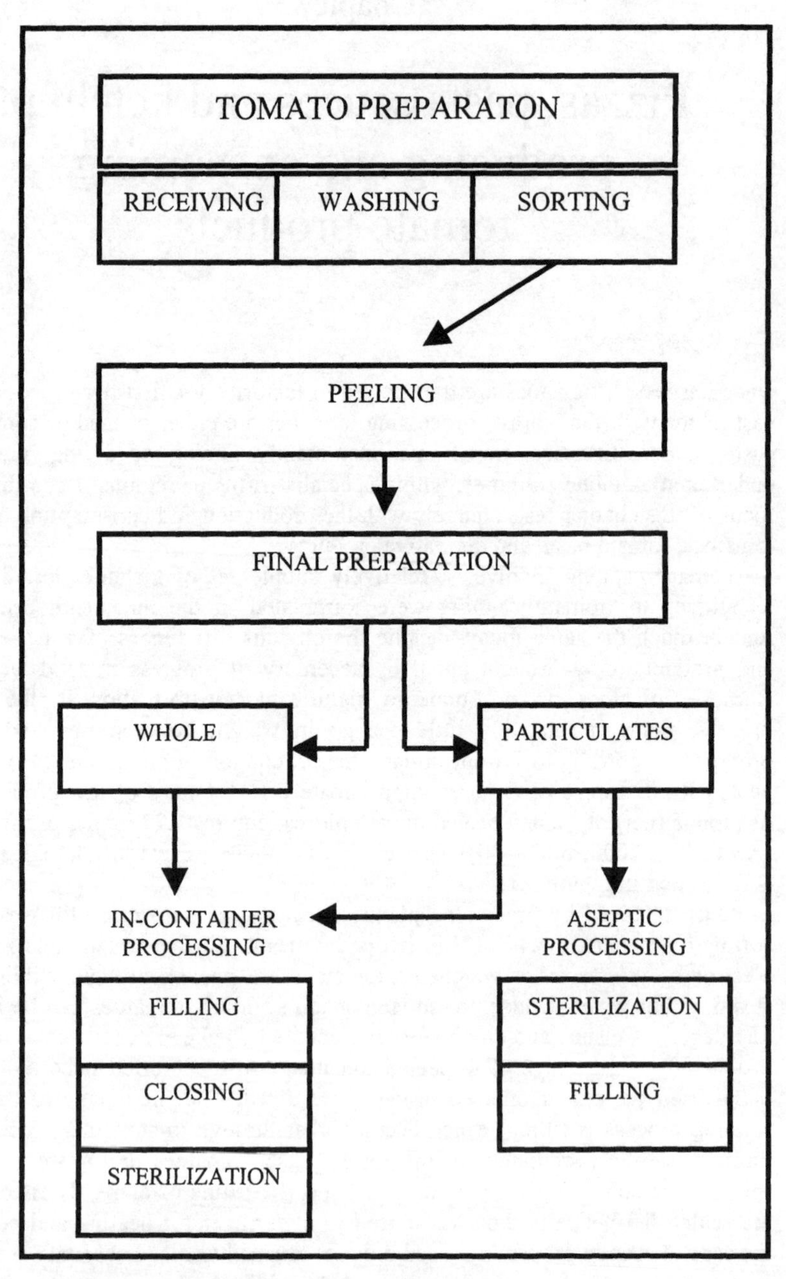

Figure 3.1 The tomato canning production system

The system for manufacturing tomato paste is also, in theory, relatively simple (Figure 3.2). At its simplest, tomato paste is the product of tomatoes separated from their seeds and skins and concentrated to a level of about 30 per cent solids to 70 per cent water (30 degrees brix). Like tomatoes to be used for peeled purposes, the first stage of production involves washing and sorting. Tomatoes are then chopped, before being subjected to rapid heating in a process known as 'hot' or 'cold' break preparation. This procedure is necessary to control pectin, a natural product in tomatoes that acts to thicken and bind the final product. For tomato ketchup production, especially ketchup made without starch, as required in the US and some other countries because of food regulations, a 'hot break' process is used. Under the hot break method, tomatoes are rapidly heated to at least 77 degrees Celsius (and more commonly, over 90 degrees Celsius), which completely breaks down enzymes and thus retains higher pectin levels. Hot break, by involving higher and more immediate heating which destroys enzymes, creates a more viscous product less prone to separation (Gould, 1992, p.202). Under the 'cold break' method, tomatoes are heated to 65 degrees Celsius and held at this temperature to destroy pectin and hence create a thinner liquid that is also brighter in color and fresher in flavor.

The next stage, refining, involves the use of extractors and screens to remove unwanted peel and seeds, leaving only the pulp. Variations to screen sizes can produce paste of different textures. Following refining, the pulp is evaporated. Efficient evaporation involves the removal of water while at the same time maintaining the color and organoleptic properties of the tomato, and significantly reducing its biomass. The paste is then heated for sterilization and cooled, prior to filling. Aseptic packaged paste can be sterilized and cooled using either steam-injected flash cooling or tube-in-tube systems. Flash cooling systems, in general, facilitate superior and more consistent tomato paste production, although tube-in-tube systems are safer to operate from a sterility point of view. Flash cooler systems also do not have the capacity constraints of tube-in-tube systems, a significant advantage in a context in which factory sizes are increasing. For in-container tomato paste, the product is sealed in glass or canned containers before being cooked to ensure sterilization.

Depending on the soluble solids of the raw product and the concentration of paste to be produced, it generally requires approximately six tonnes of raw tomatoes to manufacture one tonne of paste. Although the basic technique for tomato paste production is relatively simple, in practice it exhibits considerable complexity and refinement. Unlike canned tomatoes, which are sold mostly to retail or food service channels, most tomato paste is produced in bulk aseptic drums where it awaits sale to industrial food companies for further processing or re-packaging. To satisfy this demand, tomato paste is produced according to a rigorous and exacting set of quality requirements. The market for paste is fragmented along lines of production technique (hot break or cold break) and degrees brix (the percentage of soluble solids, expressed as sugar, within the product). Paste is manufactured to varying brix levels dependent upon end-users requirements, or according to industry norms. For historical reasons, different regions of the world have tended to produce tomato paste at different degrees brix.

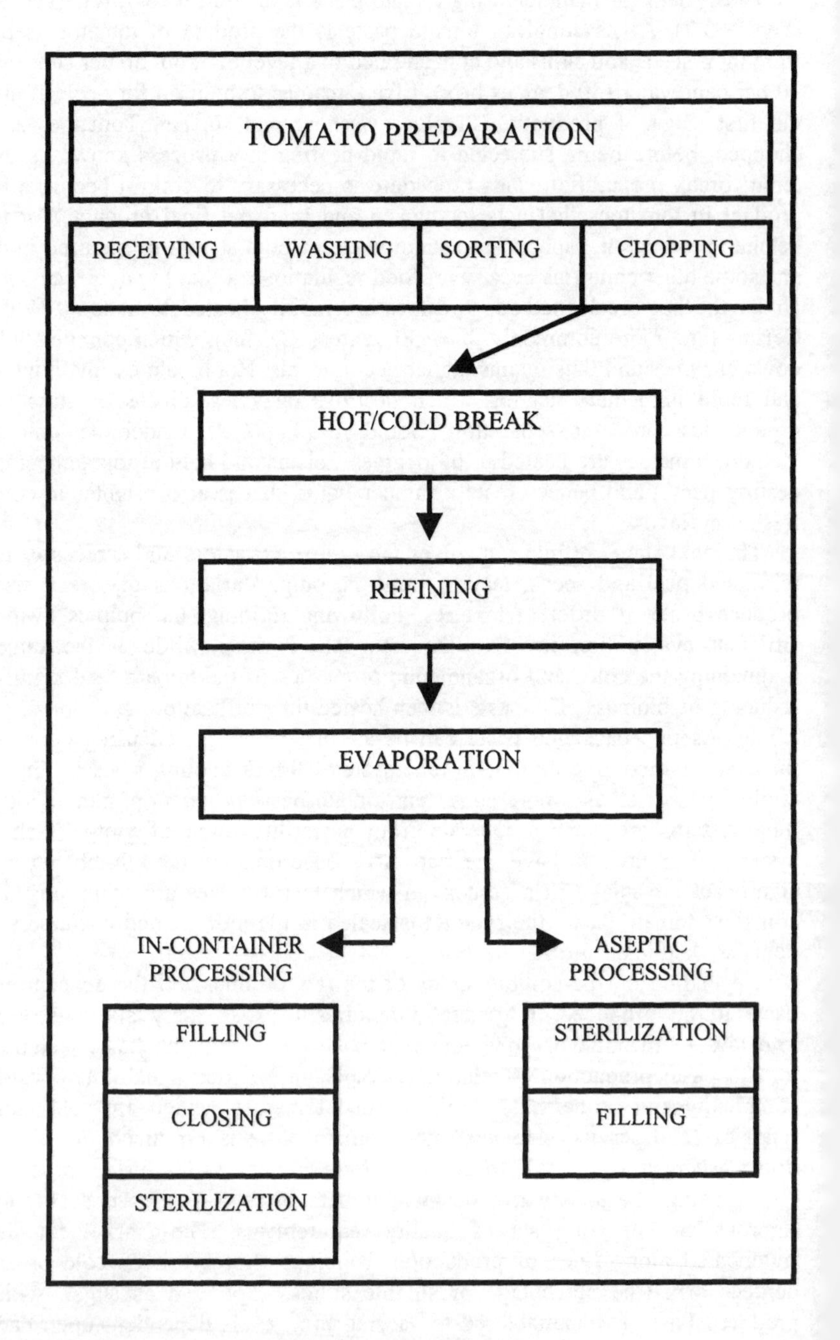

Figure 3.2 The tomato paste production system

The United States traditionally has produced paste at 30-32 degrees brix, whereas much paste in southern Europe is produced at either 28-30 degrees brix or 36-38 degrees brix. This is largely because of processing limitations, with the US system of hot break paste made under flash cooler sterilization only able to produce paste to a maximum brix of 32 degrees.

As with the experiences of many other agri-food sectors, consideration of quality has become an important determinant of supply. In processing tomato factories across the globe, priority has been given to ensuring the smooth and professional functioning of in-house laboratory facilities, which measure a range of product attributes, including mold counts (measuring rot levels) and Bostwick (the viscosity of product). These activities have been complemented by independent assessment of product quality by buyers, and the creation of supplier accreditation schemes. Further formalizing the role of quality within tomato paste production systems has been the role of the International Standards Organization (ISO) accreditation system, along with some culturally specific regulatory schemes. These include Halal and Kosher accreditations, and requirements for organic and non-genetically modified organism production.

The Morning Star Packing Co., of the United States, defines quality as 'that which consistently meets the customer's specifications':

> However, most customers' specifications have acceptable parameters rather than absolute targets. For example, viscosity may be specified as 5 to 7 cm. Bostwick, instead of 6 cm. This is due primarily to historical technological limitations, as well as the relative acceptance of product variability by the consumer. However, if such technology and ability were available to the producer, a specific or narrower range of targets would be advantageous in most users' manufacturing environments ... Aside from individual user's unique specifications for particular finished products, 'high quality' tomato paste is considered to have high color, nutrient retention and serum viscosity (assuming 'hot-break' paste), plus low mold and defect levels (Morning Star Packing Co., 2002a).

Ultimately, the quality attributes of a finished product such as paste are dependent on the raw product. For this reason large processing firms have taken an increasing interest in on-farm activities including seed variety selection, harvest methods and transport.

Underpinning these various developments has been the role of aseptic packaging. The importance of this technology cannot be under-estimated in the construction of the contemporary processing tomato sector. An aseptic storage system implies the sterilization of product on a continuous basis, and its storage within sterilized containers. Typically these containers are 1,000 litre (300 gallon) boxes, or 200 litre (55 gallon) drums. Through aseptic technology, which was introduced to the industry in the early 1970s, tomato paste or particulates can be stored out-of-season and enable finished products such as sauces, soups and ketchup to be made on a year-round basis. Thus, the advent of this technology facilitated the transformation of tomato products into bulk industrial products in which first-tier processing could be separated from subsequent processing activities.

Aseptic packaging and the enhanced role of quality and product specification within the tomato paste sector have led to two commodity chain outcomes. First, the importance accorded to quality has mediated some supply channels. It provides incentives for buyers to limit their sourcing to those suppliers in whom they have confidence over matters of quality. This is especially relevant for large branded food firms, for whom the intangible assets of company reputation and consumer confidence are paramount. Hence, the role of preferred supplier agreements and strategic alliances between first-tier processors and branded food companies has been an important development in the industry in recent years. Second, in apparent contradiction to the first point, with quality and product standards tested and verified, aseptic-packaged tomato paste can become a relatively standardized, tradeable commodity. Hence at the same time that some supply channels have narrowed through strategic alliances, tomato paste has also become an industrial commodity traded on the basis of spot prices. Much of this exchange occurs through the actions of specialist traders, discussed in Chapter Six. However, in recent years the electronic trading of tomato paste has also evolved. In the year 2000, there occurred the first Internet auction of standard-grade tomato paste, with this market being initially developed by AgEx, a small Californian e-commerce business. In the wake of the dot-com shakeout of 2001, this trade was terminated, although as a partial replacement for the trade, The Morning Star Packing Company commenced occasional Internet-based selling of tomato paste in February 2001. Additionally, some Internet trading occurs in the 'tomatoland.com' portal, based in Paris. However, the future of these activities is unclear. As Jeff Boese, president of the California League of Food Processors has noted, tomato paste 'is a consistent, rather standardized product. I would certainly expect some portion of tomato paste sales to move to the 'Net, but I don't think it will replace traditional channels of trade' (Sacramento Business Journal, 2000).

The industrial organization of the processing sector

In most parts of the world the production of tomato products traditionally has been characterized by a proliferation of relatively small processing factories located close to growing areas. These structures evolved because of a combination of technical and commercial factors. Proximity to fields has been important historically because of the rapid deterioration of tomatoes once harvested, and their high weight-to-price ratio. In keeping with these technical considerations, industry ownership structures in many regions have tended to be dominated by family firms and cooperatives often rooted in particular communities via regionally branded outputs and craft traditions. Moreover, transnational food companies have traditionally been shaped by multi-domestic branch plant structures, where individual factories take responsibility for servicing a particular regional or national market with branded products, which makes minimal use of intra-firm trade.

During the past decade these organizational structures have come under challenge. Encouraged by the transformation of tomato paste into a bulk, standardized industrial commodity, corporate strategies have been directed towards the use of larger production scales to generate cost efficiencies. Increasingly, the industrial organization of the processing tomato sector has evolved to reflect a bifurcation between firms specializing in the cost efficient production of tomato products as agri-industrial inputs, and those specializing in branded food products.

The bifurcation of specialist scale producers and specialist branded food companies has been expressed in different ways and with different strengths across the world's major processing tomato growing regions. They have been relatively muted in Europe, because of the impacts of subsidy arrangements (see Chapter Five). In contrast, they have been given strong expression in California where, since the middle 1980s, first-tier processing tomato production has been thoroughly restructured.

Tables 3.1 and 3.2 summarize the structure of the Californian paste sector, as of 2001. Approximately 75 per cent of Californian processing tomato output is converted to paste. There is a clear division within the industry between paste marketers (firms manufacturing paste to be sold to third parties for re-processing) and re-manufacturers (firms manufacturing paste for in-house conversion to branded or value-added products). The shift to a bifurcated industrial structure can be seen by the fact that approximately 60 per cent of Californian tomato paste capacity is located within facilities dedicated to marketing bulk product for other end-users. Overall, dedicated paste marketing firms are larger and newer than those producing re-manufactured product. Reflecting their relative specializations, paste marketers are more likely to be grower-owned, whereas transnational branded food companies dominate re-manufacturing facilities.

The Morning Star model

The Morning Star Packing Company has been central to the evolution of the industrial structure described in Tables 3.1 and 3.2, and has had enormous influence globally. It has been seen to represent a new *techno-economic paradigm* of tomato paste production, involving a sharp break from past structures and practices. Given its size and influence, there is merit in considering the 'Morning Star model' in some depth.

The concept of a techno-economic paradigm refers to a general system of economic production, widely emulated by firms within an industry. According to Carlotta Perez, a leading proponent of this concept:

> the relative cost structure of all inputs to production follows more or less predictable trends for relatively long periods. This predictability becomes the basis for the construction of an 'ideal type' of productive organization, which defines the contours of the most efficient and 'least cost' combinations for a given period. It thus serves as a general 'rule of thumb' guide for investment decisions. That general guiding model is the 'techno-economic paradigm' (Perez, 1985, p.443).

In the late 1980s, Morning Star's CEO Chris Rufer challenged the prevailing techno-economic paradigm in the tomato paste industry. Rufer's key innovation was to recognize that substantial cost efficiencies could be generated from reforms to logistics and scheduling, in the context of larger economies of production. Traditionally, much of the focus for achieving cost reductions in the processing tomato sector involved driving down the price of raw product. This tended to involve long-drawn out price disputes with growers, or attacks on the earnings of seasonal workers. Yet as Morning Star recognized, non-farm factors contribute more than half the cost of tomato paste production (Table 3.3). At its core, the Morning Star model rested on the notion that a new type of relation with growers and haulers, combined with efficiency improvements within the processor sector, could significantly reduce production costs in the sector. The validity of this model is illustrated in Figure 3.3 (generated from data published by Morning Star), which shows that real tomato paste prices have fallen steadily for three decades, and continued right through the late-1990s.

Table 3.1 California tomato paste processing capacity, 2001: paste marketers

Processor	Facility Location	Year Built	Raw tomatoes capacity: Tons/ Hour	Approximate capacity in terms of raw product ('000 tonnes)[3]
Morning Star	Williams[2]	1995	602	1,300
Morning Star	Los Banos[2]	1990	450	970
Morning Star	Yuba City[2]	1975	157	340
Ingomar	Los Banos[1, 2]	1983	296	640
Ingomar	Los Banos[1, 2]	2000	214	460
Los Gatos	Huron[1, 2]	1991	445	960
SK Foods	Lemoore[1]	1990	253	550
Rio Bravo	Bakersfield[1, 2]	2000	229	500
Toma-Tek	Firebaugh	1989	226	490
Colusa Canning (Ralcorp)	Williams	1982	151	330
Pacific Coast Producers	Lodi[1]	<1970	122	260
Atwater	Atwater[2]	<1970	93	200
Stanislaus	Modesto	1942	74	160
Hanover	Colusa[1]	1999	52	110
Total	14		3,364	7,270

Notes: 1. Grower owned and controlled processor; 2. Capacities confirmed by processor. Other capacities are estimates of The Morning Star Company; 3. Calculations by the authors based on Morning Star data.

Source: Morning Star Packing Co. (2002a).

Table 3.2 California tomato paste processing capacity, 2001: re-manufacturers

Processor	Facility Location	Year Built	Raw tomatoes capacity: Tons/ Hour	Approximate capacity in terms of raw product ('000 tonnes)[3]
Ragu (Unilever)	Merced[2]	1973	300	650
Ragu (Unilever)	Stockton[2]	<1970	85	180
Campbell Soup	Dixon	1975	198	430
Campbell Soup	Stockton	1967	164	350
Hunt Foods (ConAgra)	Oakdale	<1970	205	440
Hunt Foods (ConAgra)	Helm	1990	182	390
Del Monte	Woodland	1943	278	600
Del Monte	Hanford	1976	87	190
Heinz	Stockton[2]	<1970	198	430
Signature Fruit	Los Banos[1, 2]	1975	206	440
Signature Fruit	Thornton[1, 2]	<1970	112	240
Signature Fruit	Modesto[1, 2]	1969	102	220
Total	12		2,116	4,560

Notes: 1. Re-manufacturer which also maintains industrial sales activity; 2. Capacities confirmed by processor. Other capacities are estimates of The Morning Star Company; 3. Calculations by the authors based on Morning Star data.

Source: Morning Star Packing Co. (2002a).

Moreover, according to Morning Star calculations, the pace of price reductions has accelerated over time. In the decade 1980-90, inflation-adjusted Californian average tomato paste prices fell by 38.7 per cent, yet in the decade 1990-2000, the same prices fell by 45.1 per cent (Morning Star Packing Co. 2002b). In the four years after the opening of Morning Star's facility at Los Banosin 1990, inflation-adjusted Californian average tomato paste prices were just 63.1 per cent of what they were in the previous four years (Morning Star Packing Co. 2002b). Although these statistics do not indicate any strict causal relationship, they strongly suggest long-term a association between production innovations on the one hand and price reductions on the other. These price reductions have improved the export competitiveness of the US processing tomato industry. Although approximately 90 per cent of US output is consumed domestically, export volumes increased 78 per cent between 1991-92 and 2000-01 (Morning Star Packing Co. 2002a).

Table 3.3 Morning Star's estimation of industry average tomato paste production cost structure, 2000

Fixed expenses		Variable expenses	
Facility capital expenses	8%	Tomatoes with fees	45%
Insurance and taxes	1%	Trucking	8%
Operating overhead	10%	Seasonal labour	3%
		Boiler energy	4%
		Electricity	1%
		Supplies and miscellaneous	1%
		Containers	10%
		Operating interest	4%
		Selling costs	4%
Total	19%	Total	80%

Note: percentages do not add to one hundred due to rounding.

Source: Morning Star Packing Co. (2002a).

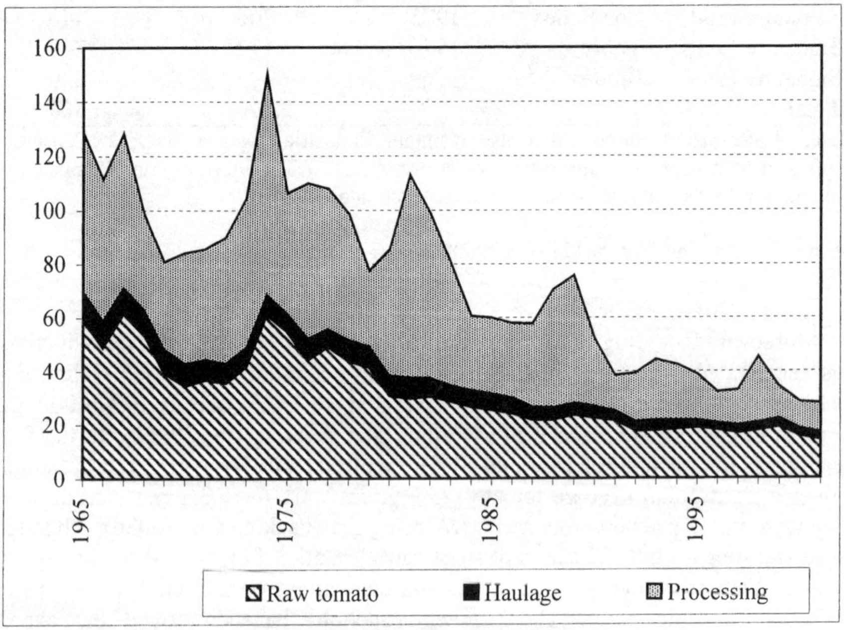

Figure 3.3 Inflation-adjusted cost of producing tomato paste, California, 1965-2001 (cents per lb, 2000 prices)

Source: The Morning Star Packing Co. (2000b).

Figure 3.3 also demonstrates that although the real cost of raw tomatoes has fallen steadily over the past thirty years, the magnitude of cost reductions on the processor side has been far greater. According to the Morning Star Packing Co., inflation-adjusted raw tomato prices fell by 47 per cent between 1970 and 2000, compared to falls of 62 per cent and 60 per cent respectively for haulage and processing. In 1982, processor margins (that is, the price of tomato paste less the cost of raw material and haulage) amounted to 61 per cent of the final tomato paste price. By 2001, this had fallen to 39 per cent. Hence, processor innovations and restructuring, rather than absolute reductions in the price of raw product, have been the main drivers of improved price competitiveness of Californian tomato paste over the past decade.

Morning Star's capacity to identify the potential to generate efficiencies in scheduling and processing rested with its origins in transport. Chris Rufer founded the company in 1970 as a one-truck hauler of tomatoes to San Joaquin Valley canneries. In 1982 Rufer invested 'downstream' (with three grower investors) in a tomato processing plant at Los Banos. This investment served the purpose of securing work for Rufer's transport operations (all the facility's trucking needs were provided by Morning Star), and provided Rufer with an opportunity to put in place major innovations in logistics and plant management. Chris Rufer's interest in this plant was subsequently sold to Ingomar and, in 1990, The Morning Star Packing Co. was established and the first specialist tomato paste facility was built, also at Los Banos. The scale of operations of the two Los Banos factories swamped all predecessors in the Californian processing tomato industry and, with their explicit dedication to industrial paste packaged in either 1,000 litre (300 gallon) boxes or 200 litre (55 gallon) drums, these plants set a new course for the industry. The construction of the Los Banos factory was followed in 1993 by the leasing of a facility at Yuba City (through the Harter Packing Company) for the production of diced aseptic product and, in 1995, by the construction of another facility at Williams. With capacity of approximately 1.3 million tonnes of raw tomatoes annually, the Williams facility remains the world's largest tomato factory.

Three factors are central to Morning Star's expansion in California over the 1990s. First, the company generated unprecedented economies of scale which when combined with state-of-the-art technology and innovations in factory design, produced significant efficiencies. Second, the company used its expertise in transport to create cost savings in logistics and haulage. Morning Star has devoted considerable resources to grower relations and has provided various incentives to larger and more efficient growers. At the same time, the company established a joint venture harvest company (Cal-Sun) with leading growers, intended to streamline grower-factory logistics management. Third, the company's focus on bulk industrial tomato paste was complemented by an attention to detail with respect to quality and efficiencies in marketing. To assist its products to reach US East Coast buyers, Chris Rufer established a re-processing facility (Paradise Tomato Kitchens) at Louisville, Kentucky. This single-mindedness of purpose and strategy has been well suited to an industry characterized by increasing concentration and pressures for cost reduction over the past decade. According to Morning Star:

> To remain competitive for the long term we must be a low cost producer. That is why we rigorously maintain our position as the major force developing and implementing new and improved technology in our industry … Morning Star's facilities are located in close proximity to the world's most productive and efficient tomato growing areas. Today we continue as a private company operating, not only the largest tomato trucking company in California, but also a tomato harvesting company, as well as three tomato processing facilities (Morning Star Packing Co., 2002a).

However, for all the technical innovations of Morning Star, the company's real significance has depended upon the ways in which its strategies have reflected the wider industrial trends in the sector. The fate of any technology rests with its applications in social, cultural, political and economic contexts. As observed by Piore and Sabel (1984, p.5):

> Industrial technology does not grow out of a self-contained logic of scientific or technical necessity: which technologies develop and which languish depends crucially on the structure of the markets for the technologies' products … Machines are as much a mirror as the motor of social development.

In the case of Morning Star's innovations, the key point is that the company's specialization in scale-intensive production of industrial grade paste coincided with the growth of a market for tomato paste. With the corporate strategies of many transnational food companies fixed to concepts of brand management, major opportunities arose for innovative companies to supply bulk tomato paste for re-processing. In California, the shift to bulk tomato paste production has been profound. In 1972 just 20 per cent of the Californian processing tomato crop was packed as bulk paste. By 2000, this had risen to 75 per cent (Morning Star Packing Co., 2002a). In 2000, Morning Star accounted for 40 per cent of Californian paste production, but employed just three full-time salespersons. Fewer than 20 customers bought almost two-thirds of the company's output (University of California, 2000).

By the late 1990s, the 'Morning Star' model had become recognized as the 'best practice' techno-economic paradigm within the global processing tomato industry. With annual industrial tomato paste sales of approximately $350 million (US) and the world's largest processing tomato factory, Morning Star emerged as a major force in this industry. Its innovations in the early and middle 1990s cemented its place in the Californian processing tomato sector, and by the end of the decade Morning Star was looking beyond the tomato paste industry for profit opportunities. In 1998, it established Cal-Fruit, a peach canning business. Then in 2001, Morning Star expanded into fresh-pack tomatoes, through its acquisition of the Sun Garden Gangi's tomato operations, which included a factory at Riverbank, California, and a distribution centre at Modesto. The major outputs at the factory are whole peeled tomatoes, and pizza and paste sauces.

California in transition

The emergence of Morning Star has recast competitive structures in the Californian processing tomato industry, but the size and complexity of the industry in that state makes for difficulties in summarizing the contours of these shifts. In absolute terms the number of Californian processing tomato factories fell during the 1990s. In the second half of the 1990s, tomato factories closed in Vacaville (American Home Foods), Davis (Hunt-Wesson), Thornton (Tri Valley Growers) and south Sacramento (Campbell Soup). While these closures were usually associated with job losses and consequently made the headlines, they were only one element in an intricate mosaic in this industry which also saw new plants established, the transfer of capacity to and from plants resulting from geographical shifts in tomato growing, and the execution of new and innovative corporate strategies. By the year 2000, a relatively small number of firms was responsible for processing the bulk of California's tomato crop. These included Morning Star; the grower-owned industrial paste specialists Ingomar, SK Foods, Los Gatos, Pacific Coast Producers and Rio Bravo, and the transnational branded food companies Del Monte, Campbell Soup, Ragu (Unilever), Hunts (ConAgra) and Heinz. Examining the experiences of these various firms highlights important elements of the restructuring dynamics of the Californian processing tomato industry over the past decade.

The recent histories of grower-owned industrial paste facilities reveal a set of contradictory processes. In broad terms, Ingomar, SK Foods and Los Gatos in particular, have emulated the Morning Star model of scale economies and bulk sales. Pacific Coast Producers, with annual sales of $260 million (US), restructured its operations in 2001 in ways that demonstrate the ongoing relevance of the Morning Star model of scale economies and transport savings. Prior to 2001, Pacific Coast Producers operated a single, relatively small tomato facility at Lodi, near Stockton. In 2001 it purchased a former-Del Monte factory at Woodland, northwest of Sacramento, and invested $18 million (US) in refurbishing and equipping the plant. This action increased Pacific Coast Producers' tomato capacity by 50 per cent and, with 95 per cent of the cooperative's tomato suppliers situated within 30 kilometres of Woodland, was expected to generate $3.5 million (US) in annual freight savings (Jardine, 2001).

Intense debate surrounds the wisdom of decisions by growers to plough capital into downstream processing facilities, in an industry with considerable risk and slim profit margins. On some occasions, unsettled market conditions drive growers to such investments as a means of securing output. In March 2001, for example, three growers in Colusa County formed a partnership and took out a lease on an abandoned processing plant formerly operated by Hanover Foods. Interviewed in the local media, one of the principals argued that:

> Without this (deal), we wouldn't be growing tomatoes ... Hopefully, we have enough sense to realize we're not cutting a fat hog. We know it's a very competitive environment out there, but we aren't trying to compete with some of the large canneries. The business plans to hire a sales force and focus on niche markets, offering, for

> example, some paste from tomatoes grown without using herbicides and pesticides. The company will also use tomato varieties that will provide a higher quality paste than the industry standard (Schnitt, 2001, p.6).

In addition to securing a market for farm output, grower-ownership of facilities at times may generate cost savings because of the ways in which vertical coordination may enhance logistics. Elsewhere in the food industry, considerable attention has been given to these issues with the apparent successes of so-called 'new generation cooperatives' in the Mid-West and Prairie States. However, there is conflicting evidence concerning the performance of grower-owned processing plants. Stewart Woolf, one of the principals of the grower-owned Los Gatos tomato business, has commented:

> A bulk tomato plant is a strategic asset, and it makes more sense to be in the hands of growers than marketers (Linden, 1999, p.17).

But Woolf has also questioned grower involvement in activities further downstream:

> Those who say you have it made when you're vertically integrated, well, that's true when vertical integration is making money. But it may end up generating losses. Greater control doesn't always mean greater profits. If you move downstream, those downstream industries may be under still more pressure (Fresno Bee, 2001, p.7).

Testimony to the dangers of grower investment downstream is seen in the collapse of Tri Valley Growers, in July 2000.

The Tri Valley saga of 1998-2001 provides important insights into the recent restructuring of California's processing tomato sector. Tri Valley was a cooperative enterprise owned by approximately 500 growers who supplied a wide range of fruit and vegetables for procesing. In the middle 1990s, Tri Valley reported gross sales exceeding $800 million (US) and was estimated to make a $1.9 billion (US) annual impact on the Californian economy (Estrada, 2001). This made it California's second largest vegetable and fruit processor, after Del Monte which had gross sales of approximately $1.3 billion (US). Tri Valley employed up to 9,500 seasonal workers and had a year-round work force of 2,000 with eight processing plants in California and one in New Jersey. Processing tomatoes were an important component of Tri Valley's operations, accounting for four factories in California (at Thornton, Modesto, Stockton and Los Banos) and at its New Jersey facility. Tri Valley purchased about seven per cent of California's processing tomato crop, from approximately 70 growers.

After a long (and largely successful) 67-year history, Tri Valley announced a significant slump in profits in 1998. By the middle of 2000, the cooperative had accumulated losses of $200 million (US). According to management and industry analysts, problems in the cooperative's processing tomato business were a major contributor to these results. To some extent, Tri Valley was a victim of cyclical forces. Rapidly increased production volumes over 1998-99 led to surpluses and weaker prices in 2000. There is little doubt however, that the cooperative's losses represented more than merely the effects of these fluctuations. Over the 1990s, Tri Valley failed to match other firms' investment in new and upgraded facilities. During the decade a dozen tomato canneries, boasting the latest technology, were established and lowered the cost structure for the industry. In a high-volume, low-profit margin business such as tomato canning, such a situation was critical for those companies which did not keep pace. At the time of Tri Valley's collapse, John Welty, executive vice president of the California Tomato Growers Association, commented: 'In good market years, less efficient plants can still be profitable. But when it becomes very competitive, and the margins begin to shrink, then those processing plants that are less efficient are going to face financial consequences' (Schnitt, 2000, p.1). The reasons for Tri Valley's lack of capital expenditure in its plants are contested, as is the broader meaning that can be drawn from the cooperative's collapse. As a cooperative, Tri Valley was dependent upon having capital raisings approved by its member suppliers. Yet as a member of the California Legislature and former Tri Valley board member observed: 'Struggling with higher operating costs themselves, the farmers were reluctant to use more of the net income they received from the co-op to invest in capital improvements' (Schnitt, 2000, p.1).

Regardless of these issues, for Tri Valley's owner suppliers the effects of the collapse were immediate. When Tri Valley slid into insolvency, about $145 million (US) in growers' equity disappeared. In 2000, the former Tri Valley operations purchased only half their contracted volumes, and growers were forced to accept prices which were less than 50 per cent of the contract price in order to sell output (Frozen Food Digest, 2001). In 2001, $20 million (US) was appropriated by the US Congress to assist growers affected by the cooperative's collapse. Cannery workers also felt the brunt of the cooperative's closure, but received no such compensation. In 2001, only two of the former cooperative's canneries were operating. In Thornton, where a Tri Valley tomato factory closed, the plant was the largest employer in a town of 1,000. An estimated 85 per cent of the town's population was Latino.

The story of the disposition of Tri Valley's assets following its collapse also sheds light on the dynamics of restructuring in the Californian processing tomato industry. Tri Valley filed for Chapter 11 bankruptcy protection on 10 July 2000, with debts of over $400 million (US). Following protracted legal proceedings and expressions of interest from interested buyers, the cooperative's assets were disposed of in April 2001. The most attractive and readily saleable assets were the former-cooperative's brands. Del Monte paid $35 million (US) for the 'S&W' brand, comprising $25 million (US) for inventory and $10 million (US) for intangibles (Del Monte, 2001, p.4). Red Gold, an Indiana food processing firm, purchased the 'Tutturosso' and 'Sacramento' brands (McKenna-Frazier, 2001). Most of the other assets, including the former-cooperative's factories, were acquired for $87 million (US) by John Hancock Life Insurance, a Boston company that was one of Tri Valley's largest creditors. To operate its new assets, John Hancock established a food business, named Signature Foods. Soon after acquisition, Signature Foods decided to focus exclusively on the former cooperative's fruit businesses, and put the former-processing tomato factories up for sale. One factory, at Los Banos, was purchased by Morning Star. Upon purchase, Morning Star CEO Chris Rufer colorfully described the state of the plant and his ambitions for it: 'To use a term common to residential real estate, that's a tear-down. It will be torn down' (Pollock, 2001, p.15).

In April 2001, one commentator on the Tri Valley saga observed:

> John Hancock, one of Tri Valley's biggest creditors, grabbed the remnants of the co-op reasoning that the wheels may have fallen off Tri Valley, but its brands still have market momentum (Grunder, 2001, p.6).

In recent years, transnational branded food companies in the tomato business have exhibited keen awareness of the profit opportunities available from astute brand management. A central component of this awareness has been the knowledge that brands are increasingly separable from the factories that produce them. The activities of Del Monte Foods' tomato operations since the late 1990s demonstrate the role of brand management in corporate strategies. For the past twenty years, Del Monte's corporate behaviour has been buffeted by a succession of ownership changes. In 1979 R.J. Reynolds acquired the business, as one of a number of strategies to diversify out of tobacco. It was then acquired by the Kohlberg Kravis Roberts (KKR) group in 1988, following the leveraged buyout of Reynolds. As an investment banker, KKR had little interest in the day-to-day operations of Del Monte, and in 1989 it sold the Del Monte fresh fruit business. In 1997, after a series of temporary shifts in corporate parentage, a majority ownership of the Del Monte processed vegetable business was acquired for $800 million (US) by the Texas Pacific Group, a Fort Worth-based investment group. In the middle of 2002, Del Monte again changed hands, being acquired by H.J. Heinz Co.

Del Monte possesses an extensive stable of well-known brand names, the flagship of which is the 'Del Monte' label itself. In 1997 it expanded this range by purchasing the Contadina operations of Nestle for $200 million (US). Along with the Contadina brand came two Californian factories, at Hanford and Woodland. In 1999, with these brand assets bedded down, Del Monte announced a restructuring of production. The company's tomato operations, split between the older Del Monte Modesto facility and the newly acquired Woodlands plant, were forthwith to be concentrated at Hanford, which would be re-equipped via new investment spending.[1] Furthering its expansion of brand assets, in 2001 Del Monte acquired the former-Tri Valley label 'S&W' (see above), although it did not bid for any of the factories previously used by Tri Valley to manufacture this branded product. The point about Del Monte's activities in the late 1990s and early 2000s is that corporate strategy was focused on the capture of brands ('Contadina' and 'S&W') and, once these were acquired, the secondary focus became the need to re-organize the company's production facilities.

H.J. Heinz Co: global brand management

Del Monte is not the only branded processing tomato firm pursuing this strategic emphasis. Indeed, its eventual parent, H.J. Heinz Co., has adopted very similar strategies.

By the early 1990s, Heinz was in command of a corporate food empire with enormous geographic range and sectoral scope, but few operational synergies. In 1996, just 37 per cent of Heinz's turnover was accounted for products carrying the Heinz brand (H.J. Heinz Co., 1997). While corporate growth and rates of return were reasonable by food industry standards, decentralization among the company's widespread operations meant corporate strategy was unfocussed. In 1997, Heinz announced an extensive corporate restructure ('Project Millennia'), which aimed to narrow the company's interests to four core businesses, each with market leadership prospects (ketchup/canned foods, baby foods, pet foods and tuna). For the first two of these in particular, Heinz's competitive strength was identified in terms of brand management. At about this time, Heinz's then-CEO, Tony O'Reilly, emphasized in the business press that Heinz 'only needed the brands, and could do without the factories'. Under Project Millennia, the company announced its intention to sell $250 million (US) of assets, close 25 factories, and slash 2,500 jobs. Not all these goals were fully achieved, so in 1999 the company announced a revised corporate restructure under the title 'Project Excel'.

Implementation of Project Millennia and Project Excel has substantially reshaped Heinz's operations in the processing tomato industry. Although Heinz continues to own significant processing tomato capacity, its size and role has undergone a major change. Prior to the late 1990s, vertical integration operated as a powerful guide for company strategy. Heinz's branded tomato products in the US, such as ketchup, were manufactured in Heinz factories making extensive use of tomato paste sourced in-house. Under Project Millennia and Project Excel, in-house sourcing now figures less prominently in company operations. Heinz continues to operate first-tier processing tomato facilities only to the extent that these generate an acceptable rate of return on capital. For Heinz, it now matters little if the company's tomato paste needs are sourced in-house or not, so long as quality standards are met. Heinz's continued involvement in first-tier processing is understood as a means to an end: to avoid dependence on paste suppliers and to ensure the delivery of product to support the company's protection and development of its branded products.

These strategies assume different shapes, depending on the part of the world in which the company is operating. In Europe, Heinz has embarked on a series of operational changes that give life to its post-1997 strategies. Heinz's European sales occur predominantly in northern Europe (UK, Germany, Netherlands, Scandinavia), using products manufactured in southern Europe (Chapter Five discusses the pan-European processing tomato complex extensively). Heinz has had a long history of involvement in southern Europe. The company began sourcing European tomatoes for the UK market (mainly from Hungary for ketchup and canned baked beans) in the first half of the twentieth century. Disruptions to these sourcing channels during the Second World War and the raising of the Iron

Curtain in the early 1950s, led Heinz to devote considerable efforts to developing large-scale processing tomato production in northern Italy. At this time, Heinz was a key innovator in the peasant economy of tomato growing around Parma and Piacenza. Through the provision of agronomic assistance, new seed varieties and the development of stainless steel processing technology, production levels increased rapidly through the 1950s. In the following decade however, Heinz abandoned Italy for Portugal where, at the time, production was cheaper. Heinz established significant tomato paste facilities in Portugal for export to northern Europe, and soon afterwards also established facilities in Spain, both for paste manufacture and to service the large domestic market. Yet as Heinz's northern European demand for tomato paste and canned tomatoes has increased, it has become progressively less dependent on its own factories in the Iberian peninsula. Heinz's branded food (re-processing) facilities in the UK and the Netherlands source paste from a number of external suppliers, with whom Heinz has generally had long-standing relations. Furthermore, in 1999 Heinz acquired 20 per cent of La Doria, one of southern Italy's largest tomato canning and re-processing firms, thus establishing a strategic alliance involving supply of raw product and co-packing of some branded goods (discussed further in Chapter Five). Through these means, Heinz's European operations have shifted from a model based on vertical integration, towards more flexible structures based around supplier alliances. As a result, Heinz can focus more sharply on brand management rather than production.

These agendas are also apparent in the Asia-Pacific, where Heinz has sought new supplier linkages for first-tier tomato products (particularly paste) rather than investing in these facilities itself. Heinz has established initial relationships within the potentially massive Chinese market through informal linkages with the Tunhe Corporation, one of the two key processing tomato operators in China (see Chapter Six). For the 2001 season, this involved the introduction of Heinz hybrid seed varieties to facilitate improved quality of the Tunhe product. Given the cost advantages of Chinese production, it is reasonable to assume that the long-term aim of Heinz's approaches to Tunhe is the ability to generate large-scale export volumes of Heinz quality tomato paste from China, as well as helping to build the local market. In September 1999, Heinz opened a baby food factory in Qingdao, as part of its market-building agenda in China. Also in the East Asian region, Heinz investigated the establishment of a strategic alliance with Kagome, a Japanese branded food company with an annual turnover of $1 billion (US), and a major re-processor of imported tomato paste in Japan. At the time of the alliance being investigated, Kagome had existing strategic supply relationships with ACIL of Portugal, Cirio of Italy and Industrie de Malloa of Chile, and through a subsidiary, was a major buyer of Californian paste (Tomato News, 2000a, p.18). In recent years however, Kagome has looked increasingly towards China as a supplier of paste. In 1999 Japan imported 25,000 tonnes of Chinese paste, representing approximately one-quarter of China's total paste exports (Tomato News, 2000d, p.41) and about 30 per cent of Japan's total paste imports (Tomato News, 2001b, p.49).

The prospect of using Chinese sourced tomato paste for Heinz branded food products may also be linked to the recent restructuring of Heinz facilities in

Australia and New Zealand. In 1999, Heinz closed its re-processing factory in the Melbourne suburb of Dandenong (which traditionally had serviced the Australian market for Heinz branded products) and shifted production to a refurbished facility at Gisborne, New Zealand. Since 1990, Dandenong had sourced tomato paste from Heinz's specialist first-tier tomato factory at Girgarre, 200 kilometres to the north, in the middle of Australia's premier tomato-growing district. By shifting its re-processing facility to New Zealand, Heinz has restructured its traditional supply dependence on Girgarre, possibly paving the way for significant tomato paste imports from China. As an illustration of how these developments intersect with regulatory arrangements, during this same period, the Australian and New Zealand Food Authority proposed the alteration of food regulations with respect to ketchup manufacturing, allowing starch to be used as an additive to assist viscosity. This change would increase the extent to which lower quality paste could be used for ketchup, opening greater possibilities for the use of relatively cheaper, imported paste. This corporate 'chess game' of factory relocations and closures has also occurred in other production sites, such as Canada and the US, and reflects Heinz's separation of first-tier processing from branded product manufacture, and an increasing geographical flexibility in the sourcing of first-tier product. In June 2002, for example, Heinz further extended this strategy by acquiring Del Monte Foods Co., a leading owner of branded food products, especially in the US.

As far as Heinz is concerned, brands are the company's core assets, and protecting and building the reputation of its brands is central to its contemporary corporate strategy. According to an independent survey of 17,000 European consumers, carried out in 2001, 29 per cent of respondents disclosed Heinz as their 'most trusted' food brand, a result which outpolled the closest rival by 20 percentage points (Hiscock, 2001, p.33). Reflecting this, the Heinz brand has been extended to new food products, especially in the fast-growing snack foods and convenience food sectors. In the UK for example, this has involved the establishment of a 'Heinz tomato crisp' (co-branded with UK snack food company Walkers) and, in possibly one of the most bizarre examples ever of brand extension, a decision in December 2001 to develop Heinz-branded pre-packed sandwiches (to be distributed by the Safeway retail chain) (Marketing Week, 2001).

Heinz's protection of its core brands is nevertheless best illustrated in the case of ketchup. In 2001, ketchup represented 17.5 per cent of Heinz's $9.41 billion (US) of global sales. In the pre-eminent US market, Heinz was dominant with 51.7 per cent market share (Stitt, 2001, p.S25). Yet during the second half of the 1990s, Heinz's ketchup business was under threat. By 1999 total US consumer spending on salsa exceeded that of ketchup and was continuing to grow, whereas ketchup sales were flat. To address this situation, Heinz embarked on an advertising campaign and a series of marketing innovations. Between 1999 and 2001 Heinz raised its US advertising expenditure on ketchup from $6.2 million (US) to $11.5 million (US). A component of this advertizing emphasized the health benefits of consuming tomato products, and focussed on the promotion of lycopene as a possible agent in reducing the risk of cancer (discussed later in this chapter). Then, recognizing that children consumed 55 per cent of US ketchup, the company redesigned packaging with the plastic 'EZ Squirt' bottle and (to the horror of some adult *aficionados*) offered ketchup in green and purple colouring (Reyes, 2001).

In identifying its destiny in terms of expertise in brand management rather than production, Heinz is one of a number of major food companies, including Sara Lee, Unilever and Kraft, which have dramatically shifted their corporate strategies in the late 1990s. Through asset sales, plant closures, mergers and acquisitions, and alliances with input suppliers, these corporations adopted a form and strategy that was significantly different from earlier models in this sector. These strategies moreover, were the counterpoint to the Morning Star model, described earlier in this chapter. Food industry strategy has become increasingly bifurcated between firms specializing in scale production, and those specializing in brand management.

Where's the consumer?

So far, this chapter has analyzed the structures and key debates in the production of tomato products. In broad terms, the intention has been to adopt an approach consistent with Friedland's original articulation of commodity systems analysis. Through this approach, we have covered considerable empirical and contextual territory, and provided an overarching perspective on the composition of this sector. Nevertheless, the discussion so far leaves major gaps, notably concerning the role of consumption.

As discussed in Chapter One, the analysis of production structures cannot effectively take place without a consideration of consumption. As Dixon's (1999; 2002) research on the chicken commodity system reveals, there is a co-dependency between production and consumption. Production arrangements can be understood *only in relation* to systems of consumption because, without the consumer, they would not exist. Dixon articulates this concept via the notion of reflexive accumulation, developed initially by Lash and Urry (1994). Reflexive accumulation links the concepts of production and consumption by seeking to identify the ways consumer attitudes and understandings 'feed back' into productive systems. According to Dixon:

> Reflexive accumulation acknowledges that consumers are increasingly subject to information about foods and are suspicious of food producers. Consumers reflect on foods' functions and foods' relationships to themselves. In short, they take foods' availability for granted, but not its meanings or worth (Dixon, 1999, p.323).

The following two sections identify the ways in which consumers attach meanings and understandings to 'processing tomato products', which in turn shape the commodity system. The first section reviews the role of processing tomatoes in the production of take-out, convenience and highly processed foods. The development and evolution of this market has provided the basis for the long-term growth in demand for processing tomato products over the past few decades. Following this, tomatoes are reviewed in relation to debates on consumption, the body, health and environment. These relatively new (or re-discovered) discourses are redefining the image of the processing tomato for Western, generally affluent, consumers.

The 'pizza economy': convenience and highly processed foods

Processing tomatoes are used as inputs by an enormous range of downstream food companies to produce many thousands of branded food products. In a general sense, this market can be divided into two key segments: products in which processing tomatoes are the key input ('processed tomato products', such as canned tomatoes, ketchup, tomato juice and tomato sauces) and products that use processing tomatoes as one of many food inputs (for example, soups, pasta sauces, baked beans and frozen pizzas). Additionally, the service-trade (notably fast food chains) is a major user of processing tomato products.

In the US, the first of these two sectors was estimated to be worth $5.7 billion (US) in sales in the year to July 2001 (Del Monte, 2001, p.63). These figures suggest that on a per capita basis, American consumers spent approximately $22 (US) per year on these products. Per capita fresh and processed tomato consumption in the US increased by 30 per cent from the late 1970s to the late 1990s, with processed product contributing 81 per cent of the total. Comprehensive utilization statistics are not available, but one published estimate of retail sales suggests that tomatoes are used for sauces (35 per cent), paste (18 per cent), canned products (17 per cent), ketchup (15 per cent) and juice (15 per cent) (Plummer, 2000, p.37). At the *IV World Congress on the Processing Tomato* at Sacramento in June 2000, market research data was cited indicating that per capita spending on processed tomato products was not highly sensitive to factors such as income or ethnicity (Khoo, 2000). The versatility of tomato products has meant that they fill many market positions. Hence, whereas markets for traditional products such as canned whole peeled tomatoes and ketchup have been relatively sluggish, there has been solid growth in demand for products such as pasta sauces and salsas.

The US processed tomato products market is dominated by transnational food companies. For the year to June 2001, Del Monte was US market leader in the solid tomato category (canned, diced, pureed products) with a 19.9 per cent share of the total; Hunt's (ConAgra) was US market leader in tomato sauces with 35.4 per cent, and Heinz was US market leader in ketchup with 51.4 per cent (Del Monte, 2001, p.8). Campbell Soup is the US market leader in canned soups and an estimated one-third of all Campbell factories use tomato products (Khoo, 2000). Hence, each of the key companies in this sector has different strengths. The fragmentation of the US market is not dissimilar to the situation in Western Europe, where company positions in different national markets vary widely (see Figure 5.10).

Because of the diversity of components in this sector, it is difficult to generalize about profit rates, margins and growth prospects. One identifiable trend, however, is a stratification of consumer demand. On the one hand, volume production and co-packing arrangements have intensified the price-competitiveness of tomato products such as paste and canned whole peeled tomatoes. In the US, supermarket private labels account for 32.3 per cent of the solid tomato category (Del Monte, 2001, p.6). The private label sector is characterized by international product sourcing based on slim margins (Burch and Goss, 1999). Since the middle 1980s, private labels have placed intense pressures on branded food competitors, with a number of so-called 'blue-chip' transnational food companies facing considerable market threat (see Pritchard, 2000a, for a case study of Kellogg's). There is a multitude of consumer attitudes towards private labels, and space does not permit a full analysis here of the reasons why consumers may choose private labels over branded products (or vice-versa). Price differences between private label and branded products may be an issue for low-income consumers, but in general, consumers are also aware that private label goods are often manufactured in the same factories as branded competitors. In the wake of 'Marlboro Friday', in April 1993, when the stock prices of branded consumer goods companies fell sharply, there has been a succession of price cuts by branded food companies seeking to restore their sales *vis-à-vis* private labels (Pritchard, 2000a). The apparent growth of the 'working poor' in western nations (Thurow, 1996) may also have contributed to enhanced price-consciousness among consumers during the 1990s. In the US in particular, the market for private label and branded products is also shaped by the extensive use of vouchers, discounts and credit card incentives, meaning that advertised prices may bear little relationship to what consumers actually pay.

Yet at the same time as traditional branded tomato products face intensified competition from private labels, they face a different set of threats at the other end of the product spectrum. Over the past decade, especially in Anglo-American and northern European markets, there has been extensive new product development. In developing such new products, food companies have made considerable efforts to target their market audiences more precisely. As with other elements of the global retail food industry, niche product development and marketing has been prominent in the processing tomato products sector. With niche products have come new branding discourses, often based on concepts of geography, heritage and craft-based production.

The proliferation of new processing tomato brands and product lines such as cook-in sauces during the past decade, points to key elements in the contemporary food industry. Singularities between brands, their owners and the companies that manufacture them, have given way to flexible relationships built around the separations of brands, brand management, and manufacturing. Except for a few iconic products (typified by Heinz ketchup), contemporary branding discourses tend not to embody authenticity, in the sense they inform consumers about ownership, place of production, and product values. In the contemporary food industry, market research and product development professionals develop brands using focus groups, test markets and advertizing campaigns. Of course, profitability in the food industry has always depended on the astute development of brands (in 1900 Henry Heinz famously built his company's profile by constructing New York's first electric advertising sign – a ten metre neon pickle). What appears to have changed in the past decade, however, is (i) more rigorous targeting of product development (food companies use a wide array of research devices, including the interrogation of bar scanning data, to target their products), and (ii) the recognition that via co-packing arrangements, the manufacturing of branded products is potentially contestable and flexible.

Taking the pasta sauce category as an example, the past decade has witnessed a proliferation of relatively specialized, targeted and high value-added brands. These have ranged from the use of celebrity endorsement (e.g., 'Paul Newman's Own') to the wide range of Italian-based ethnic references typically using variations on the theme of 'Mamma's sauce'. To use the advertizers' terminology, the market seems to have 'evolved' from a situation whereby consumers would buy tomato paste for conversion at home into pasta sauces, to the purchase of pre-prepared pasta sauces. This transition has allowed value to be added to this market category. Caught between the intense price competition of standard products on the one hand, and the growth of niche products on the other, transnational food companies have restructured their brand strategies. Del Monte, for example, has described its strategies in this segment in the following terms:

> Del Monte markets its spaghetti and sloppy joe sauces, as well as its ketchup products, under the Del Monte brand name using a 'niche' marketing strategy targeted toward value-conscious consumers seeking a branded, high quality product. Del Monte's tomato paste products are marketed under the Contadina brand name, which is an established national brand for Italian-style tomato products. Contadina also targets the branded food service tomato market, including small restaurants that use Contadina brand products such as finished spaghetti and pasta sauces (Del Monte, 2001, pp.7–8).

The fact that brands increasingly may have separate identities from their transnational food company parentage is particularly relevant for product strategies in the pasta sauce segment. During the 1990s, transnational food companies sought to develop and expand new pasta sauces marketed to appeal explicitly to discourses of peasant, artisanal and home-based food preparation in rural Italy. These brands distanced products from their production actualities (in transnational-owned food factories). Examples are 'Dolmio' (owned by the Mars Corporation), 'Five Brothers' (owned by Unilever) and 'Classico' (owned by Borden Foods until being acquired by Heinz in June 2001). The promotion of 'Classico' typifies the identity-creation:

> At Classico, we take pride in the old world craftsmanship that goes into our own passionately prepared sauces. We only use quality, time-honored ingredients that bring made-from-scratch taste to every jar of Classico Pasta Sauce. Each of our varieties is inspired by the authentic cooking styles and unique ingredients found in regions of Italy (Classico, 2002).

'Five Brothers' promotional discourse, for example, re-writes the technical process of factory-based in-container packaging, in terms reminiscent of backyard gardening: 'Our *kitchens* are built near tomato fields so we can pick, prepare, and pack our tomato sauces the same day' (Lipton Inc., 2002, our emphasis). Moreover, whereas the 'Five Brothers' brand embodies themes of rusticity and tradition, it was mobilized at an international scale as a corporate intellectual property asset. Within the space of a few years, the brand was registered as a trademark in key (Anglo-American) markets. It was registered in the US in 1996 (by Unilever subsidiary Van Den Bergh Foods), in Australia in 1998, and in the UK in 1999. In the Australian case, the establishment of the specialized glass bottling line for this product (the square-shaped and embossed 'Five Brothers' bottle is a key signifier in its attempt to inspire themes of artisanal foods) was Unilever's major tomato-related capital investment in Australia over the years 1997-98. From the perspective of Unilever's head office, 'Five Brothers' appears to have been a key product innovation aiming to compete on a global stage with 'Classico' and 'Dolmio'.

In an economic sense, these developments have relevance because complex commercial and legal arrangements can underpin the various separations of brands from corporate parentage. Transnational food companies are increasingly conscious of the capacity of their ownership of intellectual property, such as brands and patents, to generate new and profitable rates of return. In the case of Nestle for example, trademarks are registered in the company's home country (Switzerland) and used under licence by subsidiaries, creating potential royalty streams that repatriate monies to head office (Pritchard, 1999a; 2000b; 2001). The example of 'Newman's Own' pasta sauces, a product endorsed by actor Paul Newman in which profits are distributed to charities, makes the point that complex legal and commercial relations can underlie what may appear to the consumer as a simple, branded product. When consumers buy 'Newman's Own' sauces, they are in fact buying intellectual property: a recipe endorsed by Paul Newman attached to a commitment that after-tax profits are given to charities. The value of this product, thus, lies in the 'Newman's Own' trademark, which is owned by Paul Newman and registered in nations where these products are sold. The actual product is manufactured under co-packing agreements. (In Australia, it is produced by Meadow Lea Foods Ltd, a division of the Australian transnational food company, Goodman Fielder Ltd.) Hence, the separation of branding from production potentially gives rise to various forms of royalties for the use of trademarks and production fees for co-packing.

Notwithstanding brand and product innovations of various kinds, retail sales of processed tomato products are caught within wider consumption shifts that have occurred over recent decades in western nations. For a major player such as Del Monte, sales have gone backwards in recent years. Del Monte's net sales from its tomato business ('tomatoes and specialty products') fell from $423.8 million (US) in 1999, to $377.4 million (US) in 2000 and $369.8 million (US) in 2001 (Del Monte, 2001, p.63). For Heinz, ketchup sales were flat for a number of years until a massive advertizing blitz and new packaging boosted sales (see above). This is the case because changing social practices have diminished the centrality of the family kitchen and dinner table as sites of food consumption (Bell and Valentine, 1997, pp.57–88). An increased plurality of household types and consumer time-budgets has led to new patterns of meal preparation and consumption. The widespread private ownership of the microwave oven has dramatically changed the temporal and physical character of home meal preparation. These developments offer greater contingency to the role of 'place' in meal preparation so that, increasingly, meals can be prepared in several places over several stages. Distinctions between home prepared and take-away and home delivered meals have become less clear. In Australia, half of all meals consumed take place outside the home (Fidler, 1998, p.265), including one-quarter of all evening meals (Ban, 1998, p.6). These trends are seen most visibly in the growing numbers of cafes, restaurants and fast food outlets across western nations.

At the same time, there is increasing capacity and diversity in terms of retail outlets, which result from the technical basis of new foods. Supermarkets are increasingly responding to the growth of fast food chains by replicating frozen versions of the latter's products for consumption at home. Convenience stores have also come to challenge supermarkets in the supply of such products, and like fast food outlets, are increasingly to be found co-located at service stations and other strategic sites of consumption.

These trends have had considerable significance for the processing tomato industry. Because of the tomato's versatility, processing tomato products have formed a key component of recent dietary shifts in western nations. Increased out-of-home consumption of pasta, Mexican and pizza meals has boosted demand for processing tomatoes and, according to the President of Heinz North America, these trends will be the basis for future marketing and demand within the commodity sector (Jiminez, 2000). The shifts follow demographically-based changes in consumption (with pizzas and Mexican foods being associated with a younger consumer base) and highlight the role of processing tomato products within 'snack', 'grazing' and 'on the move' eating.

The rise of pizza consumption over the past two decades has been the most important component of these trends, and best illustrates emerging aspects of western consumption practices. Pizza sales in the US grew by about five per cent per year during the first half of the 1990s and, for the year 2000, are estimated at between $27 billion (US) to $32 billion (US). There are an estimated 61,000 pizza restaurants in the US, of which 25,000 are owned independently (Frozen Food Digest, 1996, p.24; Stuller, 1997). In the five largest European markets, namely the UK, Germany, France, Italy and Spain, total retail sales of frozen, chilled and fast food pizzas amounted to approximately $2.5 billion (US) in 1999 (Tomato News 2000e, p.27).

Accounting for the massive appeal of pizzas requires consideration of a number of cultural and socio-economic factors. Pizzas embody an unusual and somewhat indeterminate cultural niche within the food industry, having an identifiable geographical and ethnic heritage yet also being regarded (along with hamburgers) as perhaps the quintessential American fast food product. Pizzas are a recent food, being popularized in Naples in the late nineteenth century and introduced to America in 1905, when Gennaro Lombardi opened America's first pizzeria on Spring Street in New York City's Little Italy (Stuller, 1997, p.140). Pizzas expanded from their Italian-American origins in the 1960s, but it was only in the 1980s that they emerged in their current cultural form. With the guarantee of a 30-minute delivery, Domino's Pizzas took a lead role in constructing the pizza as a home-delivered product. At the same time, the pizza industry became increasingly dependent on casualized labor and associated with speed of delivery and consumption. As one commentator has suggested: 'Pizza is tasty. Pizza is filling. Pizza is cheap. Pizza is fast. Pizza is, therefore, American' (Stuller, 1997, p.139). Furthermore, as the following quotes suggest, home-delivered and consumed pizza represents a cultural icon that, in different contexts, satisfies agendas of family-making, convenience, and modern urbanity:

> Rob Doughty, a vice president at Pizza Hut, the world's largest chain, says his company has asked cultural anthropologists to figure out its epicurean allure. 'They suggested that there's a certain power in pizza, because it's a very simple food that people often eat without using utensils, and because it's usually shared … In a way, it's a modern-day version of the ritual breaking of bread. Our researchers found that families often set aside one night of the week for pizza, and use that time to slow down and reconnect as a family.' (Stuller, 1997, p.142);

> The workplace has changed forever, job security is a thing of the past, and the nuclear family has been seriously battered. As Americans struggle to provide for their households, they have been increasingly attracted to pizza because of its ease of preparation, whether it be from having it delivered to one's home or heating it up (Frozen Food Digest, 1996, p.24);

> Style magazine *The Face* recently ran a piece on Pizza Hut home delivery; their journalists and photographer went out on pizza runs, quizzing customers about why they were staying in, eating pizza, on a Friday and Saturday night in London. At 10.20pm on Friday, 24 March, they delivered two small vegetarian pizzas to Simon and James, at James' new bachelor flat in Islington … [James] 'Well … what pizza says is that you're busy but still want to entertain, to have people around. I like the idea of having stuff delivered to your door. It's very modern' (Bell and Valentine, 1997, p.16).

With approximately 90 per cent of US pizzas purchased as hot, ready-to-eat foods (Frozen Food Digest, 1996, p.24), the structure of the pizza restaurant and preparation industry becomes of critical concern. Over the past two decades the global growth of pizza consumption has been closely associated with the expansion of franchized business with centralized buying powers. Major pizza chains specify their tomato paste and sauce requirements through volume contracts with key first-tier processors. In the year 2001, Pizza Hut (a division of the Dallas-headquartered Tricon Corporation) owned or franchised over 8,000 restaurants in the United States, and a further 4,000 in 83 other countries. With revenues of approximately $8 billion (US), the enterprise has become an influential and important player in the world food economy. In the same period its chief global rival, Domino's Pizzas, controlled 6,600 restaurants in 64 separate countries, with annual sales of $3.5 billion (US). These two operations are ranked in the ten largest US chains (Table 3.4). Other significant pizza-based chains in the United States are Little Caesars, Papa John's, Chuck E. Cheese's and Sbarro.

This global expansion has taken pizzas to unlikely locations. In 1980, the first Pizza Hut restaurant opened in Thailand. Fifteen years later there were 85 Pizza Hut restaurants in the country, generating $41.6 million (US) in turnover (Burch and Goss, 1999, p.95). Furthermore, although in many respects pizzas have become standardized, new markets for 'designer' pizzas have been created. Led by the Californian Wolfgang Puck, the pizza base has become a canvas on which celebrity chefs and restaurateurs combine eclectic ingredients to construct post-modern cacophonies: 'tandoori chicken pizza'; 'black ink squid pizza', and 'Thai red curry pizza', among others. In the context of considerable polarization of incomes within western economies, the pizza fills a role as both a cheap and convenient food for poor families, and an up-market icon of sophistication and innovation for higher income groups. The cultural significance of the pizza thus continues to evolve, with further implications for the processing tomato industry.

Table 3.4 Worldwide sales of the largest U.S. restaurant chains, 2000

Rank	Chain	Headquarters	US$m sales	Percentage change, 1999-2000
1.	McDonald's	Oak Brook, Ill.	39,576.0	3.1
2.	Burger King	Miami, Fla.	11,400.0	4.6
3.	KFC	Louisville, Ky.	8,649.0	1.7
4.	Pizza Hut	Dallas, Texas	7,960.0	-1.7
5.	Wendy's	Dublin, Ohio	6,400.0	6.4
6.	Taco Bell	Irvine, Calif.	5,300.0	1.9
7.	Subways	Milford, Conn.	4,072.0	13.1
8.	Domino's Pizza	Ann Arbor, Mich.	3,540.0	5.4
9.	Dairy Queen	Minneapolis, Minn.	2,780.0	2.1
10.	Applebee's	Overland Park, Ks.	2,669.0	13.7
Other pizza chains in the top 50				
20.	Little Caesars	Detroit, Mich.	1,852.0	-5.0
23.	Papa John's	Louisville, Ky.	1,752.0	25.1
49.	Chuck E. Cheese	Irving, Texas	580.4	13.5
50.	Sbarro	Melville, N.Y.	569.0	7.4

Source: Hume (2001, p.36).

New consumption discourses: tomatoes, the body and the environment

The discussion in the previous section of private labels, convenience foods, pizzas and niche pasta sauces emphasizes the co-dependency of production and consumption. The first-tier manufacture of unbranded, industrial tomato products underscores the creation of diversity and convenience downstream. As first-tier products such as tomato paste become cheaper, more reliable and standardized, greater opportunities are presented for the cost-effective production of diverse, convenient, tomato-based foods.

The imperative to generate a lower-cost standardized product is the dominant factor in contemporary production-consumption relations in the processing tomato sector. However, as the 1990s progressed there emerged new consumption discourses based around the role of the tomato in encouraging good health, sustainable environments, and ethical consumption practices. The rise of these discourses, and their associated refashioning of meanings about tomatoes, point to important new connections between food consumption, identity, social structures and politics. A considerable volume of scholarship has recently engaged with these questions, although the conclusions arising from this work differ. At one level, shifts in food consumption discourse seem to suggest an emergent site of resistance to industrialized food production (Lyons, 2001). The rise of these food consumption niches – organics, 'healthy eating', the slow-food movement, anti-McDonald's campaigns, the anti-GM movement – provides an exemplar of the ways in which consumption milieux are strung together by various shared ideas and beliefs; in this case, an opposition to what is perceived as energy-intensive and unsustainable capitalist agriculture. At other levels however, these developments can be theorized as representing new consumption patterns among affluent western middle classes. For example, US research indicates that consumers purchasing functional foods on a regular basis tend to be 'information rich' and relatively more affluent (Spence, 2000). However, recent Australian research on the organics market questions the existence of a strong association between class and consumption (Lockie, 2002).

The point we wish to make is that whatever the underlying origins of these shifts, they have been incorporated within, and edged along by, the mainstream corporate agri-processing sector. To theorize recent consumption shifts thus requires an appreciation of the relationships between consumption and the actions of producers. This is demonstrated in the following case studies on (i) the health benefits of lycopene; (ii) the debates on organics and the environmental properties of foods, and (iii) the controversy surrounding genetically modified (GM) foods.

Leveraging lycopene: tomatoes and the body

According to a recent analysis:

> The health-promoting effects of foods is the food industry's 'big idea' at the start of the 21st century. Most of the world's major food companies are looking to boost their nutritional and scientific expertise in pursuit of the health benefits of food and introduce products and marketing campaigns that tell people how to beat disease and illness through food consumption. The future of food will increasingly be about how it affects our health and well-being and the sorts of products and ingredients that will deliver such health benefits. This big idea has been termed 'functional food' (Heasman and Mellentin, 2001, p.xv).

The promotion of lycopene and its association with the cancer preventative qualities of tomatoes represents a quintessential example of commodity re-positioning in line with trends towards functional foods. Lycopene is an antioxidant that provides tomatoes with their red coloring. In the 1990s, a number of scientific papers were published examining the link between lycopene and reduced risk of prostate cancer in men. An influential epidemiological study (Giovannucci et al., 1995) was supported by later research suggesting 'remarkable inverse relationships between lycopene intake or serum values and risk have been observed in particular for cancers of the prostate, pancreas and to a certain extent of the stomach' (Gerster, 1997, p.109).

Armed with these scientific results, the processing tomato industry leveraged the nutritional values of tomatoes into sophisticated marketing strategies. The first step in this process was the 1996 decision by H.J. Heinz Co., the Campbell Soup Company and Hunt-Wesson to fund the Tomato Research Council, a body dedicated to promoting and communicating research in this area. Heasman and Mellentin (2001, p.251) suggest that these three companies reportedly contributed over $1 million (US) dollars each to fund the Council, and hired a public relations company to ensure communication goals were met. The first major outcome from this investment was a seminar held in New York City in March 1997, which generated considerable media publicity on lycopene. At the end of 1998 the Tomato Research Council was disbanded, with Heinz taking the lead in subsequent promotional strategies. In 1999, Heinz established the Lycopene Education Project and Team Lycopene, a partnership between H.J. Heinz Co., the Heinz Institute of Nutritional Sciences (which operates through financial support from H.J. Heinz Co.), the Cancer Research Foundation of America (a national non-profit health organization) and the Laboratory for Public Health Informatics and Communications Research, of the University of Maryland. In January 1999, Heinz ran a $400,000 (US) advertizing campaign featuring its flagship ketchup bottle with the slogan: 'Lycopene may help reduce the risk of prostate and cervical cancer'. However as Heasman and Mellentin (2001, p.252) point out, following a US Federal Trade Commission investigation, this campaign was dropped on the grounds that the statement suggested a non-authorized health claim. Consequently, during 1999 the Lycopene Education Project distributed scientific papers on lycopene to 4,000 diet and food journalists in the US, which according to Heinz's marketing research, contributed to a doubling of lycopene awareness among consumers (Yeung, 2000). With increased consumer awareness of the health benefits of lycopene, Heinz was able subsequently to use the words 'contains lycopene' on its ketchup bottles without the need for an authorized health claim.

The story of lycopene's promotion raises several issues with respect to debates on the consumption of functional foods. The so-called 'functional foods revolution' cannot be seen simply as an autonomous reflex by consumers, but as the outcome of a complex set of relationships between consumers, food companies, scientists and public relations firms. While the major companies have been careful to qualify their claims over the role of lycopene, nevertheless they have been extremely adept in building consumer awareness of the correlations between lycopene intake and cancer risk.

With this in mind, it is interesting to note that research into the role of lycopene in cancer prevention is still hotly debated. Heasman and Mellentin (2001, p.91) argue that the promotion of lycopene in reducing prostate cancer risk 'is based on interesting, but far from complete, peer-reviewed, published scientific research'. They cite the work of Clinton (1998) in arguing that 'if lycopene indeed contributes to lower risk of prostate cancer, the mechanisms remain speculative … much more evidence is necessary before a causal relationship can be established'. Moreover, the importance of lycopene to the totality of cancer prevention should not be over-stated: 'Put in context, nutrition accounts for only five per cent of the possible risk factors associated with developing prostate cancer' (Heasman and Mellentin, 2001, p.92).

Green tomatoes: organics and sustainability

Disparate elements of what is now known as the organics sector evolved in the early twentieth century and were spurred on by counter-cultural movements in the 1970s and 1980s, before being catapulted into a major force in the world food industry during the mid-1990s. With this growth, qualitative changes have occurred in the social and economic character of the industry: 'from sandals to suits' as one researcher has suggested (Lyons, 2001). The 1998 purchase of the 'Seeds of Change' organics brand by the Mars Corporation was a bell-wether for large-scale corporate engagement in this sector. Campbell and Coombes' (1999, p.69) research into the New Zealand organics sector points to the role of large corporate actors the transformation of this industry. In 1990 New Zealand exported approximately $0.5 million (US) of organics but by 1997, with the entry of corporations such as H.J. Heinz Co., organic exports had grown to approximately $10 million (US). The motivations of these firms for entering the organics market relate to the desire to capture lucrative niche markets plus, importantly, an attempt to enhance the environmental qualities of their foods generally to combat potential 'green protectionism' Campbell and Coombes (1999).

This raises the question of what role environmental considerations play in consumers' perceptions of processed tomato products. Unilever's *Growing for the Future* program, initiated in the late 1990s, provides an interesting example in this regard. The program has the aim of embedding sustainability principles within the company's core businesses, thereby promoting consumer trust and mitigating against a potential 'green' backlash. As purchaser of five per cent of the world's processing tomatoes (Unilever, 2001), Unilever identified tomatoes as one of five crops subjected to field-based sustainability assessment. Field research in Brazil, Australia and California identified a range of variables in the environmental and social sustainability of tomato production (Table 3.5). Oversight of the Unilever initiative is carried out by an advisory team comprising representatives from the World Wide Fund for Nature, the United Nation's Food and Agricultural Organization, the OECD, and the University of Essex in the UK, among others. One of its key aims is to establish within Unilever an environmental management system based on ISO standard 14001.

The development of sustainability initiatives by companies such as Heinz and Unilever challenges some theoretical accounts of corporate agriculture and the environment. To some observers, and in some instances, such initiatives represent 'green-washing', which is essentially a public relations strategy designed to establish a veneer of environmental respectability, while actually initiating little substantive change in corporate practice. Beder's (1997) book *Global Spin*, which was a highly influential study in some areas of the environmental movement, paints corporate initiatives ultimately as cynical and self-serving.

Table 3.5 Unilever's sustainable tomato initiatives

	Brazil: Unilever Bestfoods	Australia: Unilever Australasia	California: Unilever Bestfoods USA
Start	1999	2000	2001
Stake-holders	Farmer groups. Potential NGO partners identified.	Farmer groups. Potential NGO partners identified. State government participating as a project partner.	Preliminary discussions with a number of growers and with University of California.
Major challenges	Pest management and biodiversity.	Water & soil fertility management.	Tomato growing in California is a highly competitive, large-scale operation. Profit margins are small. Main challenges are Integrated Pest Management and Biodiversity.

Source: Adapted from Unilever (2001).

Clearly the disputed politics of the environment reaches into any analysis of the legitimacy or otherwise of corporate sustainability initiatives. As Campbell (1996, p.165) has commented: 'grafting sustainable production techniques onto the social structures of production currently in existence poses a new set of questions as to the viability or desirability of such a system of production'. Although companies may seek to source their food inputs in sustainable ways, other elements of the food production and distribution system (such as transport and packaging) may be less pervasively 'green'. Burch et al., (2001) coined the phrase 'first-phase' or 'partial' greening to describe the potential contradiction between efforts to make agriculture more sustainable, when set within global (perhaps unsustainable) systems of food distribution and consumption. 'Second phase' greening, in contrast, seeks to incorporate sustainability principles more widely within production-consumption systems. The well-documented initiatives by the UK retailer Sainsbury, involving changes to package and procurement strategies, are an example of second-phase greening (Burch et al., 2001).

The point about examples such as Heinz's involvement in organics and Unilever's sustainability initiatives, is that these entities are engaging actively with, and helping to re-compose, consumers' attitudes to 'green consumerism'. There is a dynamic inter-play whereby food producers attempt both to respond to and shape consumer attitudes towards foods' environmental values.

Genetically modified tomatoes

The past decade has witnessed significant changes in consumer attitudes towards genetically modified (GM) foods. Tomatoes have been one of the key food products caught up in these controversies, and illustrate the critical elements which have been a feature of this debate.

There have been eleven GM tomato products approved for sale in the US since the first GM tomato product was marketed in 1992 (Barach, 2000). Of these, the most influential have been Calgene's *Flavr Savr* fresh tomato (introduced in 1992) and Zeneca's GM tomato puree (introduced in 1996). The *Flavr Savr* tomato, in particular, was a landmark product within the wider GM foods debate. Because this commodity utilized DNA sequencing derived from fish to provide fresh tomatoes with longer shelf-life, it served as a marker to raise public awareness of the cross-species implications of GM. Furthermore, with the benefit of hindsight, this product seems to have embodied key attributes associated with the initial period of GM foods. The main attribute of the *Flavr Savr* tomato was not that it would generate higher yields or offer greater pest protection but, with a longer shelf-life, could facilitate a restructuring of the fresh tomato distribution business. Calgene assumed that these qualities would interest the Campbell Soup Company, which at the time was reportedly considering expanding from processing vegetables into the fresh sector. Calgene's strategy was to develop the *Flavr Savr* tomato for ultimate on-selling to Campbell Soup (Martineau, 2001).

In fact, as Harvey (1999, p.17) reports, Calgene was positioned within the GM fresh tomato market only because of an agreement in the 1980s to split the tomato market with rival biotechnology firm Zeneca. According to this agreement, which was struck because both companies were working on DNA sequencing of tomatoes and had over-lapping research interests, Calgene would focus on fresh product and Zeneca would focus on processing tomatoes. This episode clearly demonstrates how the biotechnology industry's evolutionary trajectory in the 1980s and 1990s rested on a complex set of industrial and corporate actions. Using a case study of Zeneca's development of GM tomato puree, Harvey (1999) documents the minutiae of intended and unintended events that created this end product. Harvey's interest is the sociology of knowledge. His theoretical interest in the GM tomato lies in its capacity to shed light on the ways in which the production of this commodity hinged on the complex causal relationship between individuals and institutions, coming together by accident at critical moments to further advance a project whose goals, at the time, were often unclear. Harvey's narrative documents the scientific investigations into tomatoes carried out during the 1970s, the work of a particular scientist (Don Grierson) with Zeneca in the early 1980s, the first patenting of a tomato gene sequence in 1987, and the deals that allowed Zeneca to sell GM tomato puree to the British public. The critical element in this progression from laboratory to supermarket was the establishment of DNA sequencing as a tradeable good, and hence a commodity with economic value.

The irony in Harvey's narrative, and indeed, of the GM foods story to date, is that these products failed to earn the trust of consumers and hence their lengthy development was followed by an extremely short commercial life. In 1992 Zeneca's development of GM tomato puree had progressed to the point where a supply chain was established in the US and discussions were taking place with UK retail chains. However, another four years elapsed before the product was made commercially available. During that period, collaboration between Zeneca and Sainsbury and Safeway supermarkets established that GM tomato puree would be sold under the two chains' private labels (rather than have Zeneca establish its own brand), at a price that was 15 per cent lower than non-GM competition (Harvey, 1999, p.43). The actual product was ultimately manufactured under a co-packing agreement by Hunt-Wesson in California.

Between its entry into the retail sector in 1996 and the end of 1997, Sainsbury and Safeway supermarkets had together sold over 1.6 million cans of GM tomato puree. Harvey reports evidence from retail sources from that time: '[W]here both products are in store, GM products outsell conventional cans of puree [paste] by two to one' (1999, p.21). In 1997 and 1998 however, consumer attitudes began to swing sharply against GM foods. A coalition of 'radical ecologists' and 'agricultural traditionalists' (to use Harvey's description) argued against the commercialization of GM foods. Coincident public health and food scares including Bovine Spongiform Encephalopathy (BSE) further raised consumers' fears of the role of technology in food production systems. In 1999, the UK's major retail chains abandoned their sale of GM tomato products as part of the adoption of a wide policy of not selling GM foods. A large number of transnational branded food companies, including Cadbury, Nestle and Unilever, joined the chains in similar actions, while at about this time Calgene's *Flavr Savr* tomato was sold to Monsanto, which in turn dropped the business (Martineau, 2001).

This book is not the place to attempt to explain the swings in consumer sentiment towards GM foods. This is a complex issue that has not yet run its full course. A few relevant points, however, can be made. The GM food narrative exposes the simplicities of essentialist explanations of consumer behaviour. The consumer swing against GM foods in the late 1990s did not arise autonomously, but was a political outcome from the publicity – good and bad – accorded to GM foods within a particular historical moment and involving specific institutional coalitions. This does not mean, as some GM apologists sometimes assert, that consumer oppositions to GM foods are illegitimate. Such arguments run counter to the axiom (cited in every business textbook) that 'the customer is always right'. The simple fact of the matter is that in the late 1990s, GM foods failed to win the trust of consumers and (many) public institutions, an outcome which points to the fact that a dynamic inter-play of consumer-producer relations helps construct consumption discourses and revealed practices.

Conclusion

The processing tomato system involves a set of activities by which a relatively standardized agri-commodity (the tomato) undergoes a range of further transformations which successively add value and profit. Because processing tomatoes are used for many, varied end products, and because these activities occur in a range of different spatial arenas, the approach of this and the previous chapter has been to highlight important components of system, rather than attempt to document it in its entirety. Key elements of seeds production, cultivation, first-tier processing, branded product development and consumption, have each been considered in some detail.

It is timely to bring together the key arguments presented here. First, the world's processing tomato sector is currently not *global*, in the sense that it exists as a seamless commodity system unhindered by geography. National and regional factors are critically important to the organization of this industry. Of course this is a dynamic system and it is changing as some elements of the industry are becoming globalized. The extent of this change, however, should not be overstated. In line with these processes, the ways in which value and profit is created, is also shifting. Some industry participants, in some regional contexts, are feeling the brunt of restructuring as international competition becomes more intense and margins are squeezed. The responses to these pressures vary, as will be discussed in the following chapters.

Note

1 In 2001, with the Woodland factory no longer needed by Del Monte, it was purchased by Pacific Coast Producers.

Chapter 4

The persistence of national differences: grower-processor bargaining in Australia and Canada

Examination of the conduct of grower-processor bargaining provides a concrete illustration of the central argument of this book, namely, that processes of global agri-food restructuring are heterogenous and fragmented, bounded in multiple ways by the separations of geography, culture, capital and knowledge. Regulatory arrangements organized at national and sub-national scales continue to exert strong influence over the force, character and direction of global change in the processing tomato industry. The activities of growers and processing firms are 'regulated' not only through the various laws and ordinances of economic life, but also by the wider assemblages of cultures and norms that construct commercial and social behavior. Therefore, consideration of these issues needs to take account of the socio-cultural processes by which industries are embedded in their particular economic spaces. In a broad sense, the persistence of national regulatory difference in the bargaining arrangements of growers and processing firms exemplifies the mediating role of nation states as 'containers' (Dicken, 1998) for processes of economic regulation and accumulation.

These issues have generated robust debate within the processing tomato industry over recent years. Most of the world's processing tomato production occurs within the context of formal, industry-wide arrangements governing the bargaining rules between growers and processors. Although the exact nature of arrangements varies, typically these provide for collective bargaining through the recognition of grower and processor associations as third party agents representing their constituencies. However, in recent years, structural changes in the processing tomato industry and the shifting landscapes of international trade and competition have encouraged some industry participants to seek changes to these arrangements. According to industry critics, collective bargaining represents an anachronistic manifestation of an era when competition occurred largely within, rather than across, nation states. The presumption that nation states should intervene in markets to enshrine industry-wide stability and 'fairness' – central justifications for collective bargaining regulation – has been seen by some industry participants and observers as an unjustifiable barrier to efficiency within a fast-moving, globally competitive, industrial sector. According to such views, heightened global competition demands the development of flexible and individualized contract negotiations, which are broadly incompatible with statutory-based, collective bargaining.

Few would disagree with the argument that businesses are more efficient when they are provided with enhanced flexibility in the sourcing of key inputs. The critical question, however, is whether the pursuit of this goal requires a shift away from collective bargaining. For advocates of *laissez faire* models, individualized contracting generates superior industry outcomes because it provides optimal conditions for processing firms to use price-based systems to reward risk, productivity and performance. In contrast, supporters of collective bargaining tend to argue that *laissez faire* models may impose social costs in the form of unequal bargaining relationships. Moreover, it might be argued that, in many cases, collective bargaining is not necessarily detrimental to industry efficiency because industry-wide negotiations can reduce the transaction costs of grower-processor relations (grower and processor associations can write binding agreements on behalf of potentially hundreds of constituents). Furthermore, industry-wide agreements can facilitate information flows and trust via their perceived capacity to enshrine principles of 'fairness'.

The claims and counter-claims in this debate provide a backdrop to the content of this chapter. Our main interest is not to arbitrate on the merits of collective bargaining versus liberalized grower-processor relations, but rather to emphasize the ongoing reproduction of diversity within this arena of the world's processing tomato industry. Critically, we interrogate assumptions that global competitive pressures will lead *inevitably* to the deregulation of collective bargaining arrangements between growers and processors.

This is important because the concept of a 'globalized' industry implies a diminution of national-based regulatory systems, in favor of *laissez faire* arrangements that enhance market-based processes at an international scale. Evidently, the presence of collective bargaining regulations implies continued mediation of industry arrangements at national or sub-national scales, and thus the incompleteness of global agri-food integration. Echoing the arguments in Chapter Two, where it was suggested that the 'footprints' of mechanized harvesting differ across the world's tomato growing regions because of diverse agricultural, ecological, political and cultural contexts, this chapter argues that globalized competition in the processing tomato industry is being produced within a variety of institutional frameworks governing grower-processor relations, which in turn are creating diverse pathways of restructuring and change. Although there are strong forces worldwide encouraging the deregulation of industry-wide rules covering grower-processor relations, these impulses are being negotiated within the confines of a range of different contexts and, to date, it is unclear whether there is an inevitable shift towards an individualized model of grower-processor bargaining.

To analyze these issues, this chapter is organized in four parts. First, we develop a set of general arguments about the concept of regulation, as applied to grower-processor relations. Drawing on recent insights from research in the fields of regulation theory, institutional economics and economic sociology, the behavior and attitudes of growers and processors needs to be understood as being framed by the historical contexts of national socio-political systems. Grower-processor relations and industry structures exist in a dynamic inter-relationship, whereby shifts in one influence the other. Regulatory structures, therefore, are political artefacts, representing the interplay of regional agricultural *cultures*, economic and political processes at different scales, and individual agency. In short, these perspectives underline the complexity of context, rather than depend upon rarefied assumptions about the 'rational economic behavior' of individual actors. Through this approach, we depart from a binary framework that contrasts 'regulation' with 'the market', and instead establish these concepts as intersecting institutions.

Our general discussion on these issues is then supplemented by the empirical assessment of two contrasting cases. During the 1990s, there was a complete restructuring of grower-processor relations in the Australian processing tomato industry. National and sub-national (State) governments underwent a *volte-face* on the critical issue of how to regulate agricultural markets and collective bargaining was deemed illegal, to be replaced by individualized contract negotiations. This experience is contrasted with that of the processing tomato industry in the Canadian Province of Ontario, where there has been a long history of extensive governmental involvement in collective agricultural bargaining. Finally, this comparative analysis then provides a framework for the consideration of contemporary debates in California about grower-processor regulation, and the strategies available to grower organizations there. Through discussing these issues, this chapter emphasizes the persistence of difference, at national and sub-national scales, as a central theme of global agri-food restructuring.

What is regulation? The importance of socio-spatial context

The analysis of grower-processor relationships has been subjected to inquiry from a range of disciplines. Within the social sciences, a broad distinction can be drawn between studies which approach this issue from the perspective of positivist economic theory, and those coming from the perspective of critical agrarian political economy (see Chapter One). An appreciation of these two broad areas of inquiry frames the discussion of this topic within this chapter.

Orthodox economic theory suggests that, *ceteris paribus*, efficiency outcomes are optimized when markets are organized as self-regulating institutions. Following on from this view, much orthodox economic analysis of grower-processor relations in agriculture has attempted to model the ways industry-wide regulations may cause outcomes that deviate from free market practices, and hence may generate inefficiencies. Minimum price regulations, for example, are generally understood by orthodox economists to inhibit market responsiveness to shifts in supply and demand.

The strength of orthodox economic theory lies in its capacity to identify and model the relationships between supply, demand and the market. By seeking to explain market behaviour and, by extension, facilitating a critical evaluation of impediments to market equilibria and efficiency, this approach can generate important insights into production, competitiveness and social outcomes. Orthodox economic theory therefore, provides a set of tools to evaluate the effects of industry-wide regulation in terms of broad market outcomes.

At the same time, the limitations of such an approach also need to be recognized. The modelling of supply and demand (whether in tomatoes or any other commodity) is only as accurate as the validity of the underlying assumptions. Measurements of the alleged social costs of regulation are relevant only inasmuch as they are able to accurately reflect what markets would look like in the absence of regulation. Because of the inherent lack of precision of such an exercise, considerable caution needs to be exercised when interpreting the conclusions of orthodox economic analyses of grower-processor regulation. In particular, a predilection within agricultural economics in favour of models which assume conditions of *perfect competition* – in which market prices neatly respond to shifts in supply and demand and there is a free flow of information among participants – may lead to erroneous conclusions about market outcomes. Although the conceptions and assumptions of perfectly competitive markets are standard tools within agricultural economics, actual market processes often tend to be more complex. Information flows do not move seamlessly among participants in a market; actors may define 'rational' responses to market information in different ways; and price responsiveness to market conditions may be significantly shaped by institutional and socio-cultural factors.

These considerations limit the utility of models based on the assumptions of perfect competition, and point to a need for more sophisticated analyses of the interactions between markets, information flows and the behavior of participants. One approach to this question has been to construct models of imperfect competition (Stiglitz, 1985) and bounded rationality (Simon, 1959). However, while such developments have improved the frameworks for the analysis of the regulation of grower-processor relations, they remain indebted to the generally positivist and predictive epistemology of agricultural economics. In other words, their primary concern rests squarely on questions concerning what will happen to prices and resource flows given certain assumptions of market regulation. This may generate useful insights for agricultural policy-makers debating regulatory legislation, but contributes little that is useful to industry participants seeking to understand the socially-grounded character of grower-processor relations; i.e., why growers and processors act the way they do.

For these reasons, it is necessary to look beyond the strict frames of orthodox agricultural economic theory in order to develop a deeper and more insightful understanding of grower-processor relations. Two related sets of literature help us here. First, since the late 1980s, researchers in the traditions of critical agrarian political economy have made a considerable effort to understand and document agri-food regulation, particularly as it pertains to grower-processor relations (Marsden, 1997; Pritchard, 1999c; Wilkinson, 1997). The key insight from this research is that the concept of 'regulation' refers not only to the legally binding rules that shape behaviour by economic actors, but also an ensemble of written and unwritten rules and cultural norms. Seen this way, the concepts of 'regulation' and 'the market' are identified most accurately as intersecting institutions, rather than alternatives. Second, following the lead of Granovetter and Swedberg (1992), researchers in economic sociology have argued that economic behavior is embedded in socio-spatial structures. Wolf, Hueth and Ligon (2001, p.365) suggest that these arguments, which can also be labelled heterodox approaches, reject

> what Storper (1999) has referred to as the new Hobbesianism and the 'economics of suspicion' which accompanies pure methodological individualism ... That is, rather than viewing contracts as defensive tools imposed on agents (farmers) by principals (contractors), heterodox approaches focus on the 'social labor of coordination', which gives rise to localized agreements and permits their diffusion.

Together, these approaches argue that markets are constructed from social, legal, political and cultural processes, and that the concretization of dichotomies between 'the free market' and 'a regulated market' do not capture the subtleties of how markets actually behave, and why.

The importance of these insights is that in policy parlance, 'regulation' tends to be linked to the stultifying effects of 'red tape' on individual agency, enabling 'deregulation' to be presented as a paragon of economic freedom and efficiency. In fact, there are myriad examples across the globe of poorly constructed administrative regulations that impair or contradict their stated goals, as indeed, there are of 'free markets' which are costly and inefficient. This, however, is beside the point. From the perspectives of critical agrarian political economy and economic sociology, the key issue is to identify the social, cultural, political and economic mechanisms of regulation in particular settings, and to assess their varied impacts. To this end, regulation needs to be understood explicitly in terms of scale. Production regions without industry-wide regulation are best described and understood *not* as 'deregulated' or even 'free market' systems, but as regulated through private mechanisms, at the scale of individual economic actors. In turn, recognition of this framework transforms the questions to be asked about the role of industry-wide regulation. Rather than asking 'what are the social costs of industry-wide regulation?' (as implied by orthodox agricultural economic analyses), the question becomes: 'what are the relative social costs and benefits of various regulatory approaches'. Such an approach focuses attention on the complex manifestations of economic behavior.

The role of production contracts

This reformulation of the debate on industry-wide regulation brings under scrutiny the social and economic effects of production contracts, as a regulatory form. From the perspective of agri-food processing companies, production contracts are a means to exercise control over on-farm activities and practices, without becoming directly involved in the ownership and management of land. Compared with the alternative of direct farm ownership, contract farming allows for processing companies to maintain influence over farm activities, but without the risks and capital associated with farming (Rickson and Burch, 1996).

Global trends towards tighter vertical coordination in agri-food chains have promoted more extensive use of production contracts in an array of territorial and commodity contexts. Much of the literature from critical agrarian political economy focusing on this phenomenon has sought to document connections between the implementation of production contracts on the one hand, and farm restructuring on the other, with particular attention to shifts in economic power (Burch, Rickson and Annels, 1992; Mabbett and Carter, 1999; Watts, 1992).

Recent research on these issues by Wolf, Hueth and Ligon (2001) extends this literature. By adopting an inter-disciplinary perspective (with insights from both agricultural economics and critical agrarian political economy), these researchers provide a new approach to the problem of assessing whether production contracts 'delocalize' power and authority away from farmers. Critical to the findings of Wolf, Hueth and Ligon (2001) is the way they invoke the notion of socio-spatial embeddedness to interpret survey findings on contract farming in Californian horticulture. Through this approach, the researchers 'unpeel' previous research in this area, highlighting the complex environmental contexts in which production contracts are situated. Wolf, Hueth and Ligon (2001) conclude that whereas contract farming *does* tend to be associated with a redistribution of power and authority to off-farm agents, these outcomes are generated in the context of (i) the wider dynamics of agri-industrial supply chain restructuring, and the 'increasingly fuzzy boundaries between farm and nonfarm agribusiness firms, the role of services and outsourcing in production networks, and emergent forms of conflict and cooperation in an economy of quality' (Wolf, Hueth and Ligon, 2001, p.377); and (ii) the micro-economics of relations, whereby processors have strong incentives to address the potential problem of farmers failing to meet their specified needs for regular throughput of raw materials (what economists call a 'moral hazard'). This approach allows Wolf, Hueth and Ligon (2001) to build a bridge between positivist scholarship in orthodox economics (which has tended to emphasize production contracts as a mechanism for improved coordination and efficiency among supply chain actors) and critical scholarship in other areas of the social sciences (which has tended to emphasize production contracts as a mechanism for the subsumption of on-farm interests).

The utility of this type of approach is exemplified in Raynolds' (2000) study of production contracts in the processing tomato sector in the Dominican Republic. This study is worth considering in detail because, apart from being one of the few recent research articles specifically examining production contracts in this sector, it represents an empirical case of regulation as constructed through complex, multi-scaled social and political processes. While the Dominican Republic produces only a small volume of processing tomatoes, the sector expanded over the 1990s and, by the late 1990s, there were 6,500 contract growers in the country (Raynolds, 2000, p.444). By and large, these growers were peasant farmers for whom contract tomato growing constituted one of a number of cash and subsistence farming activities. From Raynolds' (2000, p.444) research, it is apparent that the involvement of these farmers in the industry hinges on a series of intersecting regulatory institutions. First is the contract itself, the primary purpose of which is to provide a means by which processing companies can gain access to irrigated farmland. Dominican Republic law prohibits large agricultural land ownership, creating a dependency by processors on contract growers. Hence, production contracts are situated within a specific national-scale socio-political context. Second, the content of production contracts reflects the division of power between processing firms and peasant farmers. According to Raynolds, processing firms retain extensive control over the cultivation of tomatoes, based on assumptions of peasants as inexperienced and lazy farmers; the phrase *los colonos no son agricultores* (the contract growers are not farmers) is used by processing firm representatives to rationalize these attitudes (Raynolds, 2000, p.445). Contract growers, however, self-identify as 'farmers' and find the attitudes of processing firms a challenge to their self-esteem. Thus, divisions of responsibility orchestrated via the contract reflect not only economic preferences, but are shaped by cultural and political contestation over the roles of each party. Third, the (commercial) effectiveness of production contracts rests on a wide range of socio-cultural factors, not least of which is the role of the family. Raynolds (2000) details how for peasant farmers, contract tomato growing is attractive because it provides cash payment for farming tasks usually undertaken by household members, hence monetarizing peasant farming. In summary therefore, a full understanding of the regulation of grower-processor relations in the Dominican Republic's processing tomato sector is achieved via an approach that defines the concept of regulation in broad socio-cultural terms, rather than in the narrower terms of economic discourse.

The foregoing discussion emphasizes that analyzing grower-processor relations requires a broad definition of regulation, encompassing its written and unwritten, economic, legal, cultural and social aspects. In the processing tomato sector, grower-processor relations are regulated via private, cooperative and statutory institutions at individual and industry-wide scales. The challenge for researchers is to document and assess the diversity of these structures, in an attempt to shed light on the relationships between different regulatory processes and social outcomes. To this end, attention is now directed towards two contrasting frameworks for the regulation of grower-processor bargaining: the liberalized, or 'enforced individualism' of Australian bargaining relationships, and the statutory collective bargaining arrangements of Canada.

Enforced individualism: the Australian approach

The restructuring of Australian grower-processor relations during the 1990s provides an example, *par excellence*, the supposed inevitability and efficacy of grower-processor deregulation in an environment of intensified global competition. The early years of the 1990s saw the emergence of a severe competitive threat to the future of the industry, in the form of relatively low-cost imports of canned product from Thailand, China and the European Union. During a period of general deregulation of Australian agriculture, these developments provided an impetus to dismantle statutory, collective bargaining arrangements, for the avowed purpose of increasing the industry's 'flexibility'. Later in the decade, these directions of change were given further sustenance when, under the guise of Australia's National Competition Policy (NCP), there was an enforced shift to individualized contract bargaining. Australia's competition regulator advised growers and processors that even voluntary collective agreements (say, between a single processor and a group of growers) would be deemed to contravene the anti-collusion principles of Australia's *Trade Practices Act*.

The sections below document the Australian experience of regulatory change, and illustrate how this intersects with wider discourses concerning the necessity of introducing individualized and 'flexible' regulatory structures in response to global competition. As noted earlier, our approach is not to criticize *per se* the shift away from statutory, collective bargaining. In recent years Australian tomato growers have displayed considerable agency and nous in their dealings with processor firms, and it is by no means apparent that they are worse off. Nevertheless, it is still necessary to interrogate the assumptions that have underpinned arguments about the supposed necessity of deregulation. Accordingly, our key motivation is to question why these developments occurred, and to what extent they have contributed towards their anticipated, *ex ante*, outcomes. The first step in this task is to situate these developments historically.

Restructuring of the Australian industry during the early 1990s

The Australian processing tomato industry is situated in the Murray-Darling Basin of southeast Australia (Figure 4.1). Although Australia makes a relatively small contribution to the global industry, producing just 1.6 per cent of global processing tomato output in 2001 (see Table 1.1), nevertheless it has displayed characteristics that are typical of the changes occurring in other countries. These changes include a decline in grower numbers alongside increased yields, a shift towards a bifurcation of processing activities between paste marketers and re-manufacturers (see Chapter Three), and heightened exposure to international competitive pressures.

Prior to the 1990s, the Australian processing tomato crop was grown by a large number of diversified family farmers. In 1984, approximately 350 growers supplied 183,000 tonnes of processing tomatoes, indicating an average production level of 523 tonnes per grower (Burch and Pritchard, 1996, p.109). At this time, the cultivation of processing tomatoes would typically represent one of a number of activities sustaining family farms, which might include a range of processing fruits and vegetables under contract, fresh market horticulture, dairy, beef and sheep. The diversification of on-farm activities implied that there were few specialized processing tomato growers during this era, with many farmers 'opting into' the industry when prices increased, and 'dropping out' when prices fell.

The structure of the grower community changed radically during the late 1980s and the 1990s. According to industry sources, progressive reductions in the price per tonne of processing tomatoes during the period 1984 to 1988 (Table 4.1) encouraged many growers to leave the industry. Data on grower numbers is extremely unreliable before 1990 (Rendell McGuckian et al., 1996, p.37), but it is apparent that by the beginning of the 1990s, the number of Australian processing tomato growers had fallen to about 100. During this same period, annual industry production tonnage fluctuated around 200,000 tonnes, implying an average output per grower of about 2,000 tonnes (Table 4.2).

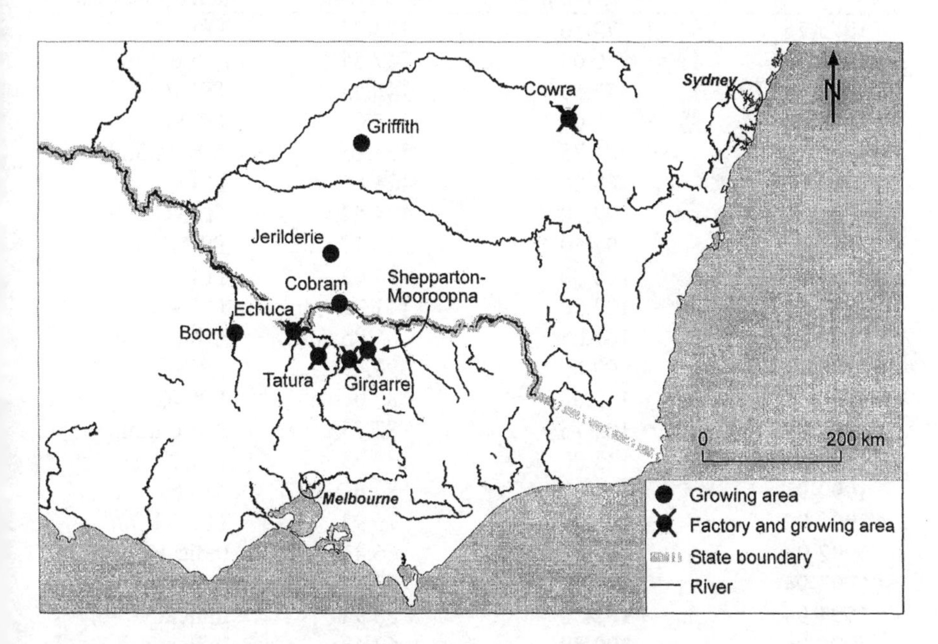

Figure 4.1 The Australian processing tomato industry, 2002

Until the mid-1990s, growers sold their fruit to a number of producer cooperatives and branch plant operations of transnational food corporations. All processors operated just one first-tier processing facility each, and all undertook in-house re-processing of first-tier products (Figure 4.2, Table 4.3). Although most processing firms initiated capital expenditure programs during the 1980s, these were insufficient to prevent an ageing of plant and equipment. Many of the industry's processing factories were originally constructed to meet the demands for processed foods generated by the Second World War, and over the 1980s, the Australian fruit and vegetable processing industry in general was characterized by relatively low rates of return to capital and the financial position of some operations were especially vulnerable (Pritchard, 1995b). In this environment, increased import competition during the late 1980s provided a catalyst for a major crisis within the industry.

Apart from some seasonal variation, import demand for *industrial tomato paste* during the late 1980s and early 1990s was relatively static, with Australian processors satisfying approximately 80 per cent of local needs (Rendell McGuickian et al., 1996, p.75). However, supply and demand conditions for

Table 4.1 Prices paid to growers in the Australian processing tomato industry, 1974-75 to 2001-02

Year	Nominal price (AUD)[1]	Real price[2]	Nature of price determinations
1975-76	72.50	173.75	TPINC
1976-77	70.00	147.55	Arbitration
1977-78	72.50	140.43	TPINC
1978-79	75.00	135.02	TPINC
1979-80	81.75	133.15	Arbitration
1980-81	85.00	125.66	TPINC
1981-82	92.50	124.52	TPINC
1982-83	97.50	117.77	TPINC
1983-84	100.00	114.02	TPINC
1984-85	102.50	111.95	TPINC
1985-86	101.25	101.25	Arbitration
1986-87	99.00	90.29	TPINC
1987-88	102.00	87.01	TPINC
1988-89	104.75	82.75	Arbitration
1989-90	115.00	87.75	TPINC
1990-91	118.00	84.61	Arbitration
1991-92	109.00	76.95	TPINC
1992-93	95.00	66.41	Indicative
1993-94	105.00	72.14	Indicative
1994-95	102.50	68.94	Indicative
1995-96	102.59	67.48	Indicative
1996-97	95.86	62.23	Industry average
1997-98	92.85	60.28	Industry average
1998-99	96.97	62.20	Industry average
1999-2000	99.47	62.27	Industry average
2000-01	99.04	58.28	Industry average
2001-02	97.47	55.69	Industry average

Notes: Arbitration = decision on statutory price determined by the Minister for Agriculture (or delegate), State of Victoria. TPINC = Tomato Processing Industry Negotiating Committee. 1. Nominal prices represent statutory prices until 1992-93; 'indicative prices' between 1992-95 and 1996-97; and industry average prices (as calculated and published by Horn [various years]) between 1997-98 and 2001-02. 2. Real prices are measured in 1986 values, deflated by the Australian Consumer Price Index. Calculations by the authors.

Sources: Data from 1974-75 to 1994-95: Burch and Pritchard (1996a, p.113); data from 1995-96 to 2000-02: Horn (various years).

Table 4.2 Restructuring of the Australian processing tomato growing sector, 1990-91 to 2001-02

Year[1]	Number of growers	Total production (tonnes)	Average output per grower (tonnes)	Number of growers with production exceeding 12,000 tonnes
1991	107	262,000	2,449	0
1992	94	205,272	2,184	0
1993	84	152,339	1,814	0
1994	76	247,025	3,250	0
1995	71	244,088	3,438	0
1996	64	287,740	4,496	0
1997	55	299,212	5,440	2
1998	48	334,176	6,962	6
1999	45	309,331	6,874	7
2000	45	367,607	8,169	9
2001	38	379,900	9,997	12
2002	33	374,314	11,343	12

Note: 1. 'Year' refers to the calendar year in which the harvest season finishes (i.e., 2002 refers to the season finishing in the first half of 2002).

Source: Horn (various years).

canned tomatoes shifted dramatically in the late 1980s. During the ten years until 1987-88 (Australian economic statistics are recorded from July to June), imports of canned tomatoes averaged approximately 6,000 tonnes per annum, with no significant growth. Accurate figures on local production of canned tomatoes during this period are not available, but industry sources interviewed during the early 1990s indicated that these imports represented a penetration rate of less than 25 per cent. Over the two years from 1988 to 1990 however, imports of retail canned tomatoes surged, reaching 16,900 tonnes in 1989-90 (Australian Bureau of Statistics, various years). These trends continued in the early years of the 1990s, representing an import penetration rate of approximately 50 per cent of total domestic consumption (Rendell McGuckian et al., 1996, p.75)[1].

Two vital issues are raised with respect to the import surge of the late 1980s and early 1990s. Firstly, the growth in imports reflected both an increase in supply from traditional source nations, as well as a number of new entrants. Throughout this period, Italy retained its traditional position as the major source country for canned tomato imports, accounting for over half of all Australian canned tomato imports over the late 1980s and early 1990s (Australian Bureau of Statistics,

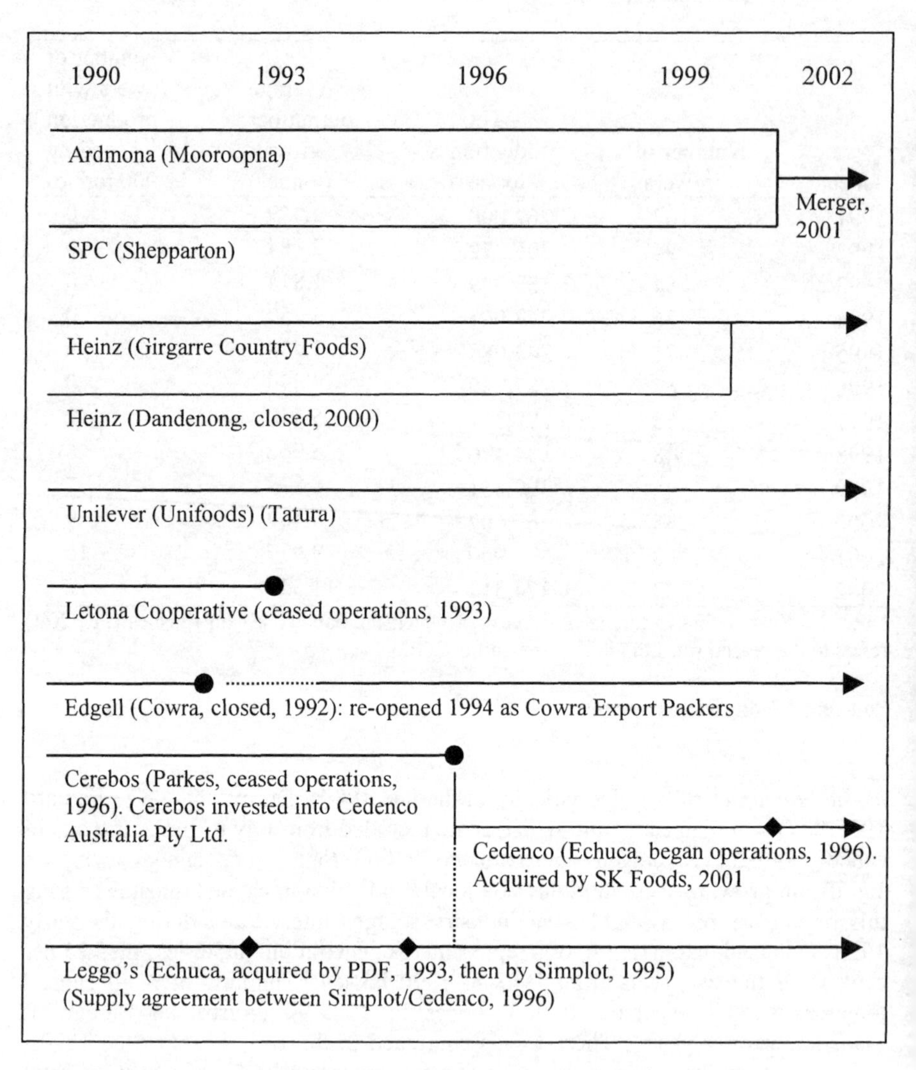

Figure 4.2 Corporate restructuring of Australian processing tomato firms, 1990-2002

Notes: SPC = Shepparton Preserving Company. PDF = Pacific Dunlop Foods.

various years). The competitiveness of the Italian product was closely connected to the extensive subsidization arrangements provided through the EU (discussed in detail in Chapter Five). In addition, Australia became an even more attractive export destination in the late 1980s following a decision by the US to impose countervailing duties on Italian canned tomatoes as a consequence of US-EU trade disputes during the Uruguay Round of the GATT (Pritchard, 1995a, p.280). However, in addition to Italy a number of other nations, notably China and Thailand, also became key import sources for retail canned tomatoes during this period. Thailand, in particular, proved an aggressive competitor, with its share of the canned tomato import sector increasing from 16.1 per cent in 1988-89 to 31.3 per cent in 1990-91 (Australian Bureau of Statistics, various years).

Table 4.3 Raw product throughput of major first-tier processing tomato firms, Australia, 1994-95 and 2000-01

Company (factory location)	Throughput 1994-95 (tonnes)	Throughput 2000-01 (tonnes)	Comments
Heinz (Girgarre)	58,000	95,000	Paste and re-processing facility.
Unilever (Tatura)	67,000	60,000	Paste and re-processing facility.
SPC (Shepparton)	34,000	58,000	Canned, diced and paste products.
Ardmona (Mooroopna)	19,000	26,000	Canned, diced and paste products.
Leggo's/Simplot (Echuca)	36,000	---	Since 1996, first-tier products supplied by Cedenco.
Cerebos (Parkes)	23,000	---	Closed 1996.
Cedenco (Echuca)	---	118,000	Paste, whole-peeled and diced producer for other users. Not operating in 1994-95.

Note: SPC = Shepparton Preserving Company.

Sources: 1994-95 data from Rendell McGuckian et al. (1996, p.73). 2000-01 data courtesy of individual processors, rounded to the nearest thousand.

Secondly, the surge in canned tomato imports represented a new articulation of power relations in the Australian processing tomato sector, with supermarket chains for the first time exercising their capacity for global sourcing and thus by-passing traditional suppliers. The initial impetus behind this increased global sourcing was partly the shortages of domestic raw materials caused by a poor season in 1988-89. But more importantly, underpinning these trends was an increasing share of generic and supermarket own brand (or private label) sales in the Australian canned tomato market. Between 1989 and 1990, the share of generic and private label canned tomatoes increased from 36.6 per cent to 48.8 per cent of total market share (Canned Food Information Service, 1991).

As one of Australia's largest food retailers, the actions of the 'Coles' supermarket chain illustrate the scope of these sourcing practices. Coles moved into the global sourcing of canned peeled tomatoes in the mid-1980s and, during the following decade, the company sourced canned tomatoes for its generic ('Savings') label from no fewer than three production sites – Australia, China and Thailand. In addition, its own-brand 'Farmland' label sourced whole peeled tomatoes from Spain and Australia; its 'Farmland' 'no added salt' whole peeled tomatoes line sourced product from Italy and Australia, and its 'Farmland' triple concentrated tomato paste was sourced from Hungary.

In order to engage in the global sourcing of these products, supermarkets established new supply chain linkages. In this context, the role of new actors in the chain of agri-food production and distribution, such as international trading companies and global food-broking firms, becomes important. When sourcing its 'Savings' brand of canned tomatoes from Thailand, for example, the Australian supermarket chain 'Coles' engaged the services of a Thai buying agent and broker, Champaca Ltd, to act on its behalf in the purchase of canned tomatoes. Champaca is a Thai-owned company that was established in 1975 with operations in timber trading. It diversified into the export of canned tuna and, subsequently, a wide range of food and non-food products on behalf of overseas buyers. In the case at hand, Champaca would provide Coles with a quote for delivery of a given volume of canned tomatoes, and if awarded the contract, would make purchases from local Thai processing tomato firms. These brokerage activities would also encompass responsibilities for packing, labelling, providing letters of credit and other consignment needs. In addition, in order to operate in the global agri-food system, and to meet the specifications and high product standards demanded by western retailers, Champaca would maintain a fully-equipped laboratory capable of conducting recognized tests (Business in Thailand, 1987, pp.99–100).

The role of brokerage agents such as Champaca however, provided just one model of supermarket sourcing structures. Japanese general trading companies (or *sogo shosha*) filled this role in the past, while in the early 1990s, Thai variants on the Japanese *sogo shosha* model, including companies such as CP Intertrade, and Siam Cement Trading Company, carried out these activities. During the period in question, six separate Thai-based processing tomato companies were involved in the sourcing of Thai canned tomatoes by Australian supermarkets (Australian Customs Service, 1991). In 1991, the largest exporters were the Sun Tech Group, SCT and the Royal Project (see Chapter Six). Sun Tech also exported large volumes of canned tomatoes to Canada for supermarket 'own brand' sales, but in this case the company assumed direct responsibility for sales rather than working through the existing trading companies and food brokerage firms (Anti-Dumping Authority, 1992). The significant number of Thai exporters involved in the Thai-Australian canned tomato trade of this era reflects the fact that Australia was the largest destination for Thai canned tomatoes, representing well over 50 per cent of exports by volume and value (Thailand: Department of Customs, various years).

The surge in canned tomato imports primarily affected the cooperatives and locally-owned firms specializing in these branded products. The case of Edgell Birds Eye illustrates the point. During the 1970s and 1980s, Edgell Birds Eye, which had extensive processing tomato operations at its Cowra (New South Wales) factory, was owned by a succession of parent companies. By the late 1980s, the business had become a division of the Adelaide Steamship Company Ltd ('Adsteam'), an investment vehicle managed by Sir Ron Brierley, a New Zealand entrepreneur and noted 'corporate raider', with a reputation for engaging in hostile takeovers with the aim of liquidating assets. In 1991, Adsteam sold Edgell Birds Eye to Pacific Dunlop Ltd, an Australian public company with extensive manufacturing investments, which had decided to move into food processing for the purposes of strategic diversification. Within a year of making this acquisition however, Pacific Dunlop announced the closure of the Edgell Cowra factory (which undertook the company's processing tomato operations) because it was allegedly not generating a sufficient rate of return. For the next few years, the 'Edgell' brand of canned tomatoes was imported from California and Italy by third parties under co-packing arrangements (Pritchard, 1995a, pp.286–300). Similar problems beset other Australian tomato canneries. For much of the 1980s, the Letona Cooperative remained afloat largely on the basis of financial support from the State Government of New South Wales, and when in 1993 its cash flow position deteriorated, it quickly collapsed (Burch and Pritchard, 1996). In 1990, the Shepparton Preserving Company (SPC), a grower-owned cooperative based in Shepparton, Victoria, narrowly avoided bankruptcy. A subsequent financial restructuring plan saw the entity's status converted from a cooperative to a public corporation. Clearly, the early 1990s was a time of considerable stress for processing firms within this industry.

The uneasy transition to globalization

The crisis of the late 1980s and early 1990s did not augur well for the future of the Australian processing tomato industry. Writing in 1995, in an attempt to forecast future industry structures, the current authors described the tumultuous period of the middle 1990s as 'the uneasy transition to globalization'. It was argued that:

> the kind of processing tomato industry that survives in Australia will be far removed from the one which developed through much of the post-war period (Burch and Pritchard, 1996a, p.124).

Developments during the second half of the 1990s validated this assessment. As detailed in Table 4.2, the number of growers fell consistently through the 1990s, at a time when industry output steadily increased. In 1996, almost half of Australian processing tomato growers cultivated less than 5,000 tonnes, and no growers cultivated over 15,000 tonnes. By 2001-02, only four growers cultivated less than

5,000 tonnes, and there were nine growers who each cultivated over 15,000 tonnes, and who supplied half of the industry's raw product (Horn, 2002, pp.4–5). According to industry sources, these larger growers expanded their tonnages incrementally, by taking over other growers' contracts as they left the industry. These changes coincided with the introduction of bulk handling arrangements, with the level of such handling increasing from five per cent to 52 per cent between 1995-96 and 1998-99 (Horn, 2002, p.14). Another important innovation of this time involved the rapid penetration of drip irrigation techniques. By 2002, there were only 33 growers, with an average output of over 11,000 tonnes per grower.

Significant restructuring also occurred among the processor companies during this period. Increasingly, the industry has moved towards a bifurcated structure (see Chapter Two), divided between specialist first-tier producers and branded food companies. The key event in this transition occurred with the arrival of a new entrant to the industry, Cedenco, in 1996. Cedenco was originally established in New Zealand in 1985 by two New Zealand processing tomato growers, Dean and Cedric Witters. They had built their tomato growing business through supplying Watties New Zealand (later taken over by Heinz), and in the mid-1980s saw an opportunity to develop their own first-tier processing tomato operation. In 1985, they commissioned a Rossi & Catelli processing tomato facility in Gisborne, on the east coast of New Zealand's North Island. The business was developed as a vertically integrated operation, combining tomato cultivation with first-tier processing. In this, it resembled Californian firms such as Ingomar and SK Foods. Initial success led to contracts to supply tomato paste to Australian branded food manufacturers, and the public listing of Cedenco in 1993.

In 1996, Cedenco decided to relocate its processing tomato operation to Australia. The company established a joint venture with the Japanese agri-food corporation Suntory (which owned the Cerebos tomato factory at Parkes, and was a significant downstream user of processed tomato products through its ownership of the 'Fountain' brands), and shipped its Gisborne tomato factory across the Tasman Sea to Echuca, Victoria. Cedenco Australia established its factory adjacent to a facility owned by the branded food company Simplot Australia Ltd, a subsidiary of the US J.R. Simplot Company which owned the 'Edgell's' and 'Leggo's' brands. Cedenco Australia signed a long-term supply agreement with Simplot for aseptic tomato paste, and diced and whole peeled tomatoes. Additionally, Cedenco established a large-scale tomato farming enterprise (near Griffith, New South Wales), and a contract harvesting and contract transporting business.

Cedenco's entry bolstered capacity at a time when other factories had been closing, and presented a new model for the operation of processing activities. By specializing in bulk industrial outputs, the company hoped to generate the kind of economies of scale that had been experienced by Morning Star in California (see Chapter Three). To the time of writing, however, these benefits have proved elusive, largely as a consequence of the fact that, while large by Australian standards, Cedenco's Echuca facility was small by world standards. The plant's regular throughput of approximately 130,000 tonnes of raw product was insufficient to generate the kind of cost efficiencies gained by international rivals. Furthermore, the company experienced a number of financial problems, and in

2001, Cedenco New Zealand advanced $797,000 (NZ), approximately $310,000 (US), to its Australian joint venture in an attempt to address these problems.

In 2001, Cedenco New Zealand's largest stockholder, Brierley Investments, sold its 48.5 per cent share of the business to California's SK Foods. The Maori Investment company, Mangatu Investment, also sold its 6.3 per cent stake, giving majority (54.8 per cent) equity to SK Foods. This latter company also purchased Cerebos Australia at about the same time, giving it approximately 75 per cent ownership of Cedenco Australia. At the time of writing, SK Foods' ambitions for this business have not been revealed, although its CEO was reported to be considering using Cedenco for further food acquisitions in Australia and New Zealand (Stride, 2001).

The Cedenco case illustrates the momentum of change that reverberated through the Australian processing tomato industry during the 1990s. In 2000, H.J. Heinz closed its landmark Australian processing factory at Dandenong, in outer Melbourne, with major implications for the ways it sources processing tomatoes within the Australian and New Zealand markets (see Chapter Three). Then in July 2001, reflecting the pressure for ongoing rationalization of the industry, a merger was announced between SPC and Ardmona, two grower-owned companies that were traditionally strong rivals. Growing pressure to remain globally competitive in a volatile international marketplace, has underscored an unfinished restructuring dynamic in the Australian processing tomato industry. For the substantive issues under consideration in this chapter however, the key question relates to the ways in which the industry responded to these challenges. It is to this question that we now turn.

The search for breathing space: industry complaints about dumping

One of the first responses by the Australian industry to the intensified import competition of the 1990s was to lobby for protectionist measures. The main weapon in this attempt to reduce imports was the application by the Canned Food Information Service (CFIS), an industry group representing food companies, metal manufacturers and supermarkets, for anti-dumping duties to be imposed on imported canned tomatoes from Thailand, China, Spain and Italy. The CFIS claimed that there was widespread dumping of canned tomatoes, as well as subsidies to production, processing and marketing, which represented unfair competition. The importation of such goods at subsidized or dumped prices, it was suggested, was causing significant material injury to the local industry (CFIS, 1991).

In its application to the Australian Customs Service for duties to protect the local industry from dumping, the CFIS acted on behalf of the seven major processing companies then in operation in Australia, namely Ardmona, Berrivale Orchards, Edgell Birds Eye, Letona Cooperative, H.J. Heinz, Plumrose and SPC. Relevant grower associations also supported the application. The essence of the case advanced by the CFIS was that canned tomatoes from Italy, Spain, Thailand and China were being exported to Australia at prices which were below the costs of production, with dumping margins ranging from three per cent to 61 per cent. Furthermore, it was argued that the EU directly subsidized exports from Italy and Spain, and that the Chinese and Thai Governments indirectly subsidized their product. Of the four countries against which dumping was alleged, the adverse material consequences were said to be largest in the cases of Italy and Thailand.

With respect to Italy, it was argued thatit was the EU export restitution payments which allowed Italian tomato canning firms to dump product on Australian markets. These payments meant that canned tomatoes were able to enter Australia at prices that were 25 to 50 per cent below prevailing Italian domestic prices (Pritchard, 1995a, p.280). In the case of Thailand, it was argued that canned tomato imports were being marketed in Australia at prices significantly below those of the local product. According to the example offered by the CFIS, in April 1990, the price of a 400/425g can of tomatoes from Thailand was $0.69 (AU), which was $0.12 (AU) cheaper than the comparable Australian-made product (CFIS, 1991, p.23). The CFIS argued that these lower prices reflected the production subsidies made available by the Thai government. The dumping margin on a carton containing 24 cans of 425g each was said to be between $2.30 (AU) and $3.60 (AU) (CFIS, 1991, p.13), which was approximately equivalent to 15-20 per cent of the retail price.

According to the CFIS, alleged dumping by Thailand was made possible through investment privileges made available under the Thai Board of Investment (BOI). In particular, the CFIS argued that the large tomato processing plants located in northeast Thailand were within 'Zone 3' of the BOI Investment Incentive Plan, and therefore received maximum privileges and subsidies (see Chapter Six). The CFIS also pointed to a number of other supposed 'hidden subsidies' to canned tomato exports, including the policy of the then military rulers of Thailand of suppressing labor groups and keeping wages low. Although these factors did not conform to the definition of what constituted a subsidy under international law, the CFIS nonetheless argued that they were effectively a form of government assistance.

Notwithstanding the scope of arguments made by the CFIS, the Australian Government's inquiry into these issues was obliged to confine itself to the relatively narrow definitions of what constituted dumping, as set out in the various codes and agreements of the General Agreement on Tariffs and Trade (GATT). In general terms, the GATT allows national governments to levy anti-dumping and countervailing duties only in circumstances where there is clear evidence that exports are being sold below comparable domestic prices, and there is a causal link between these outcomes and material injury experienced by local producers. In the case at hand, this required three areas of proof: (i) that the producers/processors in the country of origin received any official assistance which enabled them to undercut Australian producers; (ii) that the retail price in Australia reflected the 'normal value' of like goods when sold in the domestic or third markets; and (iii) that material injury was caused as a direct result of these actions. On these criteria, the Australian Government initiated action to determine whether dumping had occurred and, if so, what responses were appropriate.

The detail of the administrative processes by which the Australian Government sought to levy anti-dumping and countervailing duties highlights the contested, complex and litigious nature of international trade relations. After the CFIS had lodged its application for protection in July 1991, the Australian Customs Service conducted a review and issued a preliminary finding in December of that year, recommending the provisional imposition of duties (ACS, 1991). The ACS inquiry found evidence that canned tomatoes were exported to Australian at subsidized and dumped prices by Italy, Spain and Thailand, and at dumped prices only from China. Following this decision, the matter was referred to Australia's Anti-Dumping Authority (ADA) for final resolution. In 1992, consistent with the earlier ACS review, the ADA recommended countervailing action against subsidized exports of canned tomatoes from Italy, Spain and Thailand, and anti-dumping action against exports from Italy and China (Anti-Dumping Authority, 1992). On these grounds, the Australian Minister for Customs approved the levying of anti-dumping and countervailing duties in April 1992. However, in July 1993 following legal action by La Doria, one of the major Italian exporters of canned tomatoes, Justice Lee of the Federal Court of Australia set aside these levies. According to Justice Lee, the ADA had made errors of law in its recommendations to the Minister. However, on appeal in February 1994, the Full Bench of the Federal Court found in favor of the ADA, leading to the re-instatement of duties. In line with Australian anti-dumping legislation, the continuation of duties is reviewed every five years. In 1997 (five years after the initial 1992 decision), the imposition of duties was limited to Italian canned tomatoes, on the grounds that Thailand had, by this time, ceased to be a significant exporter. In March 2002, the ADA approved the levying of anti-dumping and countervailing duties for a further five years, following an application by Australian processors.

A sense of perspective is required when analyzing the effects and relevance of Australia's imposition of anti-dumping and countervailing duties on canned tomato imports from certain countries. First, in recent years, the local production of canned tomatoes for retail sale has represented only about 10 per cent of the utilization of Australian processing tomatoes (Horn, 2002, p.15).[2] Consequently, the imposition of duties had only indirect and marginal relevance for the competitive situation facing the majority of Australian production. Second, although import penetration rates in the Australian canned tomato industry have fallen in recent years, this does not seem to have resulted from the imposition of the duties. During the second half of the 1990s, import penetration remained relatively steady at approximately 40 per cent of the Australian canned tomato market, before falling to about 25 per cent. In 2001, Australian production of approximately 50,000 raw product tonnes of retail peeled tomatoes for the domestic market, was complemented by approximately 17,000 tonnes of imports (Horn, 2002, pp.15, 19–20). However, this reduction in import penetration occurred while duty levels remained unchanged, suggesting that the explanation for this trend seems to lie in factors such as exchange rate movements or the competitiveness of the Australian production. Third, the stated economic effect of anti-dumping and countervailing duties was not to impose penalties on imports, but to neutralize the impacts of EU subsidy payments in the Australian marketplace.

According to a complex set of mathematical equations developed by the ADA (ADA, 1994), duties were set with the aim of establishing what is called a 'non-injurious free-on-board' price, that is, an assumed price that excluded the effects of subsidy payments. Accordingly, it is not unreasonable to suggest that the *political relevance* of anti-dumping and countervailing duties being levied (representing a statement by the Australian Government against EU agricultural policies) outweighed their *economic effects*. Use of remedial provisions under international trade law did not, and was never intended to, provide a 'magic bullet' that would address the competitive problems of the Australian processing tomato industry. Although these measures provided some 'breathing space', there was widespread consensus that the survival of the industry would depend on its competitive merits, rather than government actions.

The philosophy of 'self-regulation' and the shift to enforced individualism

In line with these sentiments, a philosophy of 'self-regulation' has dominated the Australian processing tomato industry since the middle 1990s. These developments have been consistent with general ideological shifts in Australian agricultural policies during this period, in which leading policy-makers and many industry representatives have argued that government 'interventions' in areas such as supply management and the setting of industry priorities (say, in the area of research and development) should give way to market-oriented approaches (Pritchard, 1999b). In the context of the processing tomato industry, regulatory changes to the conduct of grower-processor price negotiations were the centerpiece of the shift towards market-oriented approaches.

At the time of the major import surge in the late 1980s, grower-processor relations were subject to industry-specific legislation. In the State of Victoria, this took the form of the *Tomato Processing Industry Act* (1964), amended in 1976 to establish formal rules on price setting negotiations. In the State of New South Wales, statutory provisions on the conduct of annual price negotiations were included in regulations attached to the *Marketing of Primary Products Act* (1983).[3] Under the Victorian and New South Wales legislation, annual price negotiations were specified to occur within the Tomato Processing Industry Negotiating Committee (TPINC), a body consisting of five representatives from processors and five from growers, and one (non-voting) representative from the Victorian Department of Agriculture. The TPINC was empowered to make an annual recommendation on minimum prices for processing tomatoes which, when endorsed by the Victorian Minister for Agriculture, became legally binding. In situations where the TPINC could not agree, the Minister was obliged to determine a minimum industry price through arbitration.

As detailed in Table 4.1, between 1975-76 and 1991-92, the TPINC reached consensus on 12 occasions, and resorted to arbitration only five times. In general, arbitration was required only at times when there were rapid or unexpected imbalances in supply and demand. In such cases, market uncertainty inclined grower and processor representatives to make 'broader-than-usual' claims in defence of their interests. As a case in point, in 1990-91, when the last arbitration took place, the Government-appointed arbitrator, Victorian Prices Commissioner, Professor Allan Fels, entered the fray reluctantly, and criticized negotiating tactics that diminished the chances of consensual outcomes:

> This arbitration comes at a time when there is an apparent excess of demand over supply. This is illustrated by the fact that growers did not meet the local demand for tomatoes for processing for the 1989-90 season, despite a significant increase in local output… I point out that when I last acted as arbitrator in 1988, a significant matter then was, as now, the potential for substantial replacement of domestic production with imports and a subsequent running down of processing capacity. The view of the processors was that any price more than $100 per tonne would open the potential for import substitution. There is little evidence that this eventuated, given last year's negotiated price of $115 per tonne and greatly increased crop under contract. Indeed, the doom and gloom scenarios painted by both parties at recent arbitrations have not become a reality (Fels, 1990, p.6).

These comments embrace a subtext that acknowledges the role of arbitration as a last resort in conditions of uncertainty. At the time Professor Fels made this determination, the Australian processing tomato industry was experiencing intensified import competition. With industry participants unsure how to interpret and respond to these pressures (the 'doom and gloom' scenarios noted by Fels), arbitration proved a means to commit participants to an agreed price, thus stabilizing industry relationships.

Despite the fact that the TPINC resolved price disputes by consensus in the majority of its determinations, the 1990s witnessed a complete reversal of policy and a shift away from statutory collective bargaining. The opening development in this transition occurred in 1990, when the State Government of Victoria decided to review all statutory regulations made between 1972 and 1988. This decision was prompted by concerns expressed by business lobby groups who wanted to expunge allegedly outdated statutes. As such, the impetus to reform was quite unrelated to the specific situation of the processing tomato industry. Under this decision, regulations on collective bargaining in the industry, established under the aegis of the *Tomato Processing Industry Act* (1976), were required for review prior to June 1992.

The Victorian Government review did not provide any substantive evidence on the shortcomings of statutory collective bargaining. Nevertheless, it decided to move towards liberalization. This apparent contradiction can be explained by the fact that key policy-makers in the Victorian Ministry of Agriculture were at the time ideologically committed to market reforms, and their agendas dominated over alternative regulatory discourses. In interviews undertaken by the authors in the early 1990s with senior Victorian bureaucrats responsible for these areas, it was made clear that the over-riding concern of policy was to 'get government out of agriculture'. This position was given force with the passing of the *Agricultural Industry Development (Tomato Processing) Act* (1992), in which the Victorian Parliament repealed the previous 1976 legislation, including its provisions on price determinations. In its place, the Government introduced the *Tomato Processing*

Industry Development Order (1992), which specified a more limited set of responsibilities for industry bodies. Critically, the TPINC was abolished and replaced by a new committee, with the (confusingly similar) acronym 'TINC', or the Tomato Industry Negotiating Committee. The TINC is best described as a 'half-way house' in the transition from collective bargaining to the free market. Its membership structure was identical to the previous TPINC (with five processor and grower members each, and one delegate from the Minister for Agriculture) but, unlike the prior committee, it held no statutory powers to set industry prices. Instead, the TINC was empowered to recommend 'indicative' prices which, while not legally enforceable, could be used as general guidelines during the conduct of price negotiations.

Coinciding with the shift from statutory collective bargaining, the TINC sponsored the establishment of an industry 'Code of Practice', which specified the broad parameters that should govern grower-processor relations. The development of this Code was critical in the period between statutory collective bargaining and the move towards liberalization. According to the first edition of the Code (which was intended to be updated annually), its purpose was to 'underwrite the process of self-regulation in the Victorian processing tomato industry in establishing agreed yet voluntary standards of conduct' (Tomato Industry Negotiating Committee, 1994, p.1). Hence, although having no formal legal status, the Code of Practice provided a framework for deregulation at a time of industry uncertainty. The Code provided a reassurance to growers that processors would still be expected to act in 'good faith', thus providing a discourse of continuity between the era of statutory collective bargaining and the 'brave new world' of liberalization. For this reason, during this period growers generally accepted the shift towards deregulation. Although the initiative for deregulation originated with the Victorian Government, its implementation was largely based on a process of agreement between government, processors and growers.

On the assumption that the TINC embodied a set of transitory arrangements on the road to full liberalization, the *Tomato Processing Industry Development Order* (1992) was scheduled to expire in 1996. Despite a mid-term review of arrangements that recommended a continuation of the TINC beyond its specified term (Fisher et. al, 1994), the Victorian Government revoked the Committee's mandate in 1996, and the TINC held its final meeting in June 1997. For the 1997-98 season, processors were expected to notify growers of price and other contractual matters on an individual basis. Henceforth, there was no formal role for industry bodies with respect to price-setting decisions, although the Australian Processing Tomato Industry Council (APTIC) has subsequently had an important function in industry representation and leadership.

In an attempt to manage the transition towards a liberalized system, the TINC sponsored a major benchmarking report into the industry in 1996 (Rendell McGuckian et al., 1996). The intention of this report was to help generate a 'common view' of the key issues facing the industry in this period of change. Importantly, the benchmarking report encapsulated an ideology of 'self-help', which was based on the view that the industry itself was responsible for identifying its problems and acting upon them. This reflected an important break from previous policies, which relied heavily on governments to identify and address industry concerns.

The significance of the benchmarking report rested with its agency as both a *process* and *product*. As a 'product', the benchmarking report set out the key issues facing the industry, and established a series of recommendations for further actions. However, of equal if not greater importance was the fact that the benchmarking report initiated a *process* of improved information exchange and strategic debate within the industry. Industry data collected during the course of the benchmarking study was updated in annual industry surveys (Horn, various years), which have been produced with the support of both processors and growers. Furthermore, in the years following the benchmarking study, growers, processors and other interested parties have met on a regular basis to discuss issues of mutual strategic relevance. The commissioning of the benchmarking study, and its wide acceptance throughout the industry, thus acted as a catalyst for improved modes of communication and cooperation by industry parties, across a range of areas.

Notwithstanding the ability of the benchmarking report to help create a sense of 'shared vision' at an industry level, the initial effect of the demise of the TINC was to produce much immediate confusion and friction within the industry. In interviews carried out in the late 1990s, we were told of extended contractual disputes between processors and growers on an individual-by-individual basis, during the first year after the TINC had been abolished. In the absence of formal mechanisms for dispute resolution, and with parties lacking experience in the new environment of liberalization, disagreements could quickly escalate into major problems. During the 1997-98 season, a number of growers sowed their tomato crops without a contract and, for some, contracts were not agreed upon until the fruit was actually harvested. In one case, described to us by a number of informants, a conflict emerged between the role of H.J. Heinz as the industry's largest provider of hybrid seeds (see Chapter Two), and its role as a major processing firm. Traditionally, Heinz had provided its contract growers with seeds free-of-payment at the time of sowing, with payment being deducted at the time of harvest. However, with growers reluctant to sign production contracts in the climate of uncertainty associated with the abolition of the TINC, Heinz questioned whether previous arrangements were appropriate, and suggested growers without contracts should pay 'up-front' for seeds. Although Heinz was clearly acting within its legal rights, this stance reverberated through the industry, because many growers used Heinz seed to fulfil production contracts with other processing firms. Following the intervention of industry leaders, this crisis was resolved; however, it underlined the frictions in the industry that accompanied the abolition of the TINC.

In a broader sense too, an erosion of trust between some growers and some processors helped fuel disputes in other areas. For example, a number of growers interviewed by the authors alleged that some processors were manipulating the results of quality grading, to the disadvantage of particular (often more vocal) growers. (However, it was also noted by some growers that in years of tight supply, grading decisions may have been made in their favor, as an 'unofficial' bonus payment.) Whatever the facts, it was apparent that, unlike the situation in California and Ontario, discussed later in this chapter, there was no independent grading regime in the Australian processing tomato industry. This meant that growers had to trust the processors, if disputes were to be avoided. The alleged failure of the Code of Practice to provide for effective intervention in this and other areas provided a lightning rod for criticisms. A number of growers were disgruntled by an alleged unwillingness of some processors to abide by the Code's clauses. In this environment, many growers' expectations as to what the Code could deliver diminished. In effect, the Code lost its currency as an institutional text of self-regulation because participants perceived it as divorced from the material practices of the industry. The accuracy or otherwise of growers' claims about processor behavior notwithstanding, the growers point to the dramatic deterioration in grower-processor relations that accompanied the initial introduction of liberalization. While there was broad acceptance within the grower community of the need for market-orientation in the industry (as noted earlier, growers generally accepted arguments for deregulation), there was considerable concern about the precise effects of regulatory changes. However, with liberalization deemed 'inevitable', there were few channels for growers to express these views, beyond occasional interviews with agri-food researchers and the traditional domain for complaint, the front bar of the local hotel. On the rare occasions when growers formally aired their opinions, they were keen to emphasize that they did not oppose market competition *per se*, but its manifestations in this industry. For example, according to one grower, in a letter to the Victorian Department of Agriculture:

> The people who are serious about tomato growing plan their crops up to three years in advance. Earthworks generally commence nine to 18 months prior to sowing. Fertilizer is applied and bed preparation is completed six months before sowing to ensure the growing of a quality crop. The monetary outlay to reach this stage is higher than for any other broadacre crop grown in this climate. The result [of deregulation] is that on one side a group of people [i.e., growers] has hundreds of thousands of dollars of their own money tied up, is trying to negotiate with a group of people [i.e., processors] with no personal outlay. The side effect of this is that growers who waited until four weeks after the beginning of the sowing season to know what price they would be growing for … are hesitant to commit themselves to forward preparation lest they leave themselves open to the same situation next year. Unless there is some mechanism put in place to set a fair price in the event of a deadlock, the grower basically has no bargaining power except if they decide to leave the industry…There is no harmony at present, whereas previously [name of processing company deleted] had a rapport with their growers that was the envy of the industry. Is it a coincidence that this rapport that has existed for decades has been completely destroyed in the few short years since deregulation? I do not propose that deregulation be discontinued. Howcver because of the disparity between the negotiating parties it is essential if deregulation is to work, that some form of arbitration is in place … Unless a repeat of this season's fiasco can be prevented in future, I don't believe the tomato industry is going to be in a position to grow and develop the way it was intended when the industry was deregulated.[4]

Similarly, in an opinion piece published in 1998 in an industry newsletter, *Tomato Topics*, the growers' organization argued:

> A major problem with the present negotiation process is that it does not set out the forward signals by which growers and processors can make informed decisions about their future – particularly for growers, as processors have access to management expertise to better determine these things. Where growers come to negotiations each year with no other basis than getting/taking whatever they can, then there is no mechanism to help them look at the next season … and, for many, that is as important as the one coming up (Tomato Growers Council Victoria - Victorian Farmers Federation, 1998, p.7).

Central to many of the growers' concerns was the fact that the abolition of the TINC coincided with the advent of new approaches to competition policy in Australian agriculture. Many industry participants assumed that the abolition of the TINC would lead to the situation whereby processors would negotiate prices with so-called 'grower groups', representing their contracted growers. This type of arrangement exists in California (discussed below) and was widely known within the Australian industry. However, such assumptions proved misleading. In 1993 the Australian Federal Government released a report with the aim of establishing a consistent national framework for competition in the Australian economy (Hilmer, 1993). The so-called 'Hilmer Report' and the subsequent agreement by Australia's Federal, State and Territory Governments on National Competition Policy (NCP) in 1995 led to a strengthening of Australia's *Trade Practices Act* (1974) and the establishment of a new agency to regulate its provisions, the Australian Consumer and Competition Commission (ACCC). These developments encouraged a more robust interpretation of what constituted anti-competitive behavior. When liberalization was introduced to the processing tomato sector, the ACCC alerted industry participants to the fact that collective negotiations over price – say, in the case where a processing firm negotiated with a 'grower group' – could be interpreted as collusion. To protect against potential breaches of the *Trade Practices Act*, the ACCC indicated a preference for individual contract negotiations, whereby a processing firm would be required to establish separate contracts with each of its growers individually, and that the terms of those contracts would remain confidential to the specific parties under contract. Under this regulatory regime, processors could meet with grower groups to discuss issues such as harvest schedules, preferred seed varieties or general commercial conditions, but were disallowed specifically from discussing prices.

This framework appears to give new meaning to the term 'deregulation'. Under current Australian competition policy, the lengthy and expensive processes of civil law provide the only legal remedy for growers claiming breaches of contractual conditions. In addition to the commercial dangers of being earmarked as a 'trouble-maker' and excluded from being offered future contracts, the recourse to the law means that growers face major disincentives in addressing alleged breaches of production contracts. The implementation of this *enforced individualism* is based on the erroneous assumption that because growers and processors are equal parties in legal terms, this also extends to their economic relationship. Under the laws of competition policy, group negotiations on price are sanctioned only through the awarding of 'authorization' by the ACCC, which requires proposals be assessed in line with a public benefit test. In practice, the ACCC is reluctant to award authorizations unless there is overwhelming evidence in favor of a public benefit. Accordingly, these policies represent the *re-regulation* of the industry, insofar as they have led to a set of government rules that heavily circumscribe the freedoms and activities of participants.

The inefficiency of this re-regulated system is evidenced by the fact that, following the disastrous experiences of the first year of liberalized bargaining in 1997-98, growers and processors both sought to undermine the spirit, if not the letter, of these laws. For example, industry sources suggest that a common strategy of some processors was to meet with a single influential grower, with the view to informally propose a price for the upcoming season. The reaction of the grower (who presumably would have previously consulted with fellow growers on what, hypothetically, would be an acceptable price) would then set the ground for a subsequent, formal price offer.

While liberalization exacerbated the concerns expressed by many growers over an imbalance of power relations in price negotiations, processors for their part were encouraged by the opportunities for contractual flexibility that were apparently opened up by liberalization. To gain an insight into the effects of such contractual flexibility during this period, the authors analyzed the production contracts used by four separate processing firms over the period 1993-94 to 1999-2000. Contracts are supposedly confidential to their immediate parties, although their contents tend to be discussed widely within the grower community. However, a range of contracts was made available to us as researchers, although because of their confidential status, the names of the processing firms have not be revealed.[5]

Production contracts in the Australian processing tomato industry are lengthy and comprehensive. Within the sample examined by the authors, contractual conditions varied across 52 separate themes, including issues such as tonnage, area to be sown, seed varieties, delivery scheduling, procedures for weighing and grading, margins of tolerance for MOT (materials other than tomato), penalty charges, definitions of standard grades, ownership of bins, ownership of fruit, mode of delivery, and time of day of harvest. In broad terms, there is considerable 'pattern similarity' in the specifications of these themes within the observed contracts. Although no two firm's contracts were exactly the same, there was a surprisingly high level of correspondence among them in terms of their structure and wording. In 1996-97, for example, 32 of the 52 contract themes contained conditions that were worded almost identically, for at least three of the four firms. This suggests contracts are written on the basis of the common set of industry principles articulated in the Code of Practice.

Crucially, however, there are significant differences among the contracts on the question of how prices are set. Using 1996-97 as an example, three of the major Australian processing tomato firms used highly divergent methods for establishing grower prices (Table 4.4). In particular, one processing firm made significantly greater use of pricing incentives, compared with its rivals. 'Firm A' in Table 4.4 provided growers with a lower base price per tonne, but paid a considerable price premium if a grower was able to satisfy the processor's demands for quality, scheduling and soluble solids. Over the 1990s, this processing firm increasingly came to utilize these types of price incentives. As Figure 4.3 shows, since the introduction of these arrangements in 1993-94, the margin steadily widened between what a 'good grower' and a 'poor grower' could be paid.

Table 4.4 Pricing and payment terms, three Australian processors, 1996-97

	Firm A	Firm B	Firm C
Base price per tonne	$90.00	$100.00	$98.50
Premium for solids/ tonne (per 0.1% either side of 5.2)	$0.87	—	$0.94
Premium for delivery target/tonne	$1.40	$2.00	$1.00
Tolerance for contracted quantity	95-100%	90-100%	95-100%
Additional premium for attaining 100% of contract	—	$5.00	—
Premium for preferred varieties	—	$2.00	—
Premium for early growers 1	$10.00	—	—
Premium for early growers 2	$5.00	—	—
Premium for quality (below 2% standard grade)	$7.00	—	—
Premium for quality (2-3% standard grade)	$6.00	—	—
Premium for quality (3.1-4% standard grade)	$4.50	—	—
Premium for quality (< 4% rejects of preferred varieties)	—	$1.00	—
Deductions for soil (0.6% to 1.5%)	$2.50	—	—
Deductions for soil (1.6% to 3.0%)	$5.00	—	—
Advance payment for early signing	10%	—	—
Limits for rejection	—	—	—
Soil & MOT	3%	3%	3%
Soil, MOT & culls	10%	10%	10%
Terms of payment	2-3 weeks	2-3 weeks	2-3 weeks

Note: All currency figures Australian dollars.

Source: Authors' field data.

These initiatives to introduce incentive and penalty payments need to be understood as a strategy by processors to more effectively align growers' rewards with their own requirements. During the first few years of liberalization, the most aggressive pursuit of this agenda occurred with Cedenco's proposal to establish a 'grower alliance'. Cedenco's attempt to implement this scheme followed the adoption of a variety of unsuccessful procurement strategies by the company. Upon entering the Australian industry in 1996, Cedenco attempted to replicate the vertically integrated strategy it developed in New Zealand, establishing a large-scale tomato farm near Griffith, in New South Wales. This initial strategy, however, was shelved after one season in favor of procuring tomatoes through standardized production contracts. After two seasons of using contract farmers, however, Cedenco then proposed a 'grower alliance' under which the processor and growers would share profits. Growers would be paid a guaranteed minimum price equivalent to their costs of production; the processing firm would be paid a minimum price relative to its costs of production, and any subsequent profits would be shared equitably between Cedenco and growers based on the relative proportions of capital employed by both. Accordingly, the proposal envisaged that through a sharing of profits between the parties, traditional adversarial relationships would be replaced by mutual cooperation. As part of this arrangement, growers would sign exclusive, three-year supply contracts with Cedenco. The 'grower alliance' model represented an innovative attempt by Cedenco to restructure the character of grower-processor relations, in a way that would expand the rewards for risk, productivity and performance. However, in the

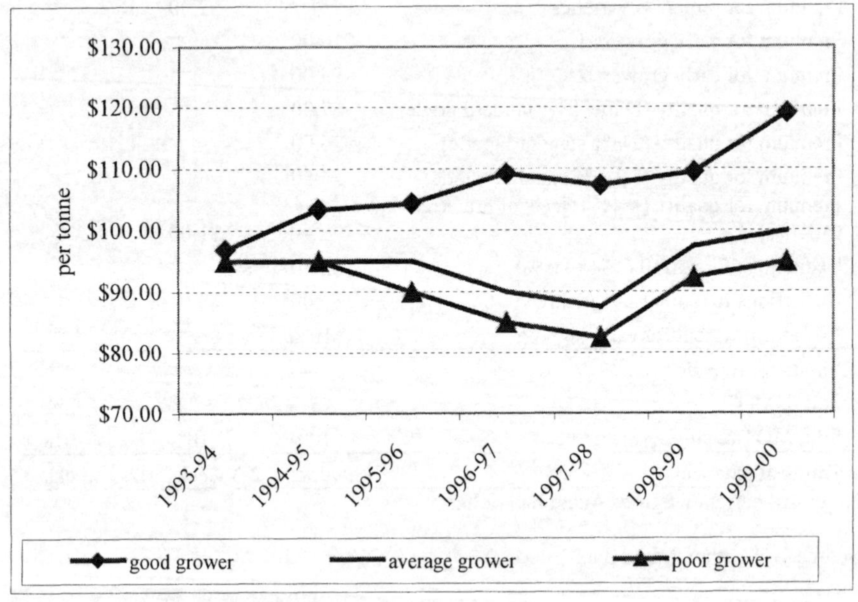

Figure 4.3 Changes to pricing payments by one Australian processing firm, 1993-94 to 1999-2000

Note: 'Good' grower is defined as obtaining premiums for solid content $\pm$ 0.1% of 5.2; within 10% of delivery target; delivering less than 2% of tonnage outside standard grade; and signing early. An 'average' grower receives base price only. A 'poor' grower receives deductions for soil content > 1.6%.

post-liberalization environment many growers remained skeptical of the intentions of processing companies, and Cedenco's proposal received only lukewarm support. Central to many growers' reaction to the Cedenco proposal was a fear that such an arrangement would institutionalize unequal bargaining. Although growers tended to support multi-year contracts because they provided for increased certainty, the concept of a multi-year *exclusive* contract, based on a low minimum price and a dependence on the processor to generate profits before any premiums were paid, appeared to hold little appeal. This was especially the case in light of concerns by some growers over the financial position of Cedenco. When the 'grower alliance' model failed to to garner support, Cedenco returned to standardized production contracts for tomato procurement.

The attempts by Australian processors to generate enhanced contractual flexibility continued in the following seasons, but added to this agenda were intensified demands for cost improvements. H.J. Heinz Co. led the call for improved cost competitiveness, giving effect to these views in the 2001-02 season, with an initial 'offer' price which was well below growers' expectations. At the time, some growers attributed this decision to the closure of Heinz's Dandenong factory, and the possibility of the company sourcing the Australian and New Zealand markets using imported Chinese tomato paste, both of which lessened the company's dependence on Australian tomato growers. For its part, at an industry workshop in May 2001, Heinz maintained the ambiguity of its commitment to Australian growers by stating: 'There are no long-term guarantees from Heinz about the future of paste at Girgarre' (Rendell McGuckian, 2001, p.6). This position was consistent with Heinz's strategic shifts away from first-tier processing, and towards branded product distribution and marketing, analyzed in Chapter Three. The perceived vulnerability of growers to calls for reduced prices was given further strength in 2000-01, when a company called AgReserves, owned by the Church of Latter-day Saints, established a processing vegetable farm of 46,000 hectares near the town of Griffith. AgReserves gained a large processing tomato production contract with H.J. Heinz and, while the terms of this contract were confidential to the parties, many growers believed that it contained a significant price discount compared to other suppliers. Rightly or wrongly, this development was perceived to be a major threat to other growers' livelihoods because, under Australian legislation, AgReserves was defined as a charitable institution, exempted from paying company income tax. According to one grower:

> Our profit margins are very fine and we've been made to become very efficient, and it's all come at a big cost and a lot of debt and the margins are very thin, and that tax exemption [to AgReserves] could top it [that is, could threaten viability] (Chirnside, G., cited in Upton, 2001, p.1).

Ironically, despite its tax-exempt status, AgReserves decided in 2002 that the profit margins in the processing tomato industry were insufficient to warrant its continued involvement, and the company left the sector.

Assessing Australian liberalization

During the early 1990s, when grower-processor regulations in the Australian processing tomato industry were liberalized, the industry was reeling from import competition and, in the view of many observers, faced an uncertain future. In the period since the middle 1990s the industry has expanded its output, albeit with a substantial loss in the number of tomato growers – from 71 in 1995, to 33 in 2002. Nevertheless, at an industry workshop in May 2001, participants concluded that:

> There are reasons to be optimistic about the future. The industry could experience a 400,000 tonne crop over the next two years, and there is potential to be processing 500,000 tonnes in five years (Rendell McGuckian, 2001, p.6).

For its advocates, the apparent turnaround in the industry's fortunes since the early 1990s could be interpreted as offering *prima facie* evidence of the effectiveness of liberalization. However, the coincidence of liberalization with evidence of improved performance within the Australian industry requires further consideration. The fact that certain events coincide (in this case, liberalization and industry performance) does not necessarily imply causality one way or another. To assess the effects of grower-processor liberalization requires an approach that identifies the *role* that liberalization has played in improving the industry's competitive performance since the early 1990s.

The starting point for such an assessment is to ascertain the drivers of international competitiveness during this period. As discussed earlier in this chapter, although anti-dumping duties levied on Italian canned tomatoes played a role in stabilizing import penetration in this component of the industry, the underlying issues of cost competitiveness have been the central factors in the restructuring of the processing tomato sector. In this regard, the industry has benefited greatly from the depreciation in the value of the Australian dollar against most major currencies since the middle 1990s. Movements in the exchange rate of the Australian dollar have been critical to the industry's competitiveness because, despite significant improvements in yields and haulage costs, the industry's overall cost structures expressed in local currency terms, were not reduced significantly over this period. Lower prices paid to growers and improved haulage costs were reflected in a fall in the 'factory gate' cost per tonne of tomatoes, from $124 (AU) in 1995-96 to $114 (AU) in 2001-02. However, because of declining soluble solid ratios during this period, the factory gate cost per tonne of solids (the critical input cost measure for paste production) remained relatively unchanged, at $2,476 (AU) per tonne of solids in 1995-96 and $2,458 (AU) per tonne of solids in 2001-02 (Horn, 2002, p.15). Nevertheless, because of the depreciation of the Australian dollar during this period, there was substantial improvement in the industry's international competitiveness in terms of factory gate prices, when measured in US dollars. Whereas in 1994-95 the factory gate price per tonne of Australian processing tomatoes was approximately $75 (US) compared with $62 (US) in California, by 2001-02 the exchange rate depreciation meant that this price differential had been eliminated, with the factory gate price per tonne of Australian processing tomatoes standing at approximately $63 (US), compared with $64 (US) in California (calculations based on data in Horn, 2002). These exchange rate effects are illustrated in Figure 4.3, which indicates that, when expressed in Australian dollar terms, the factory gate price of Californian processing tomatoes has risen considerably since the mid-1990s, providing an important competitive cushion for the Australian industry.

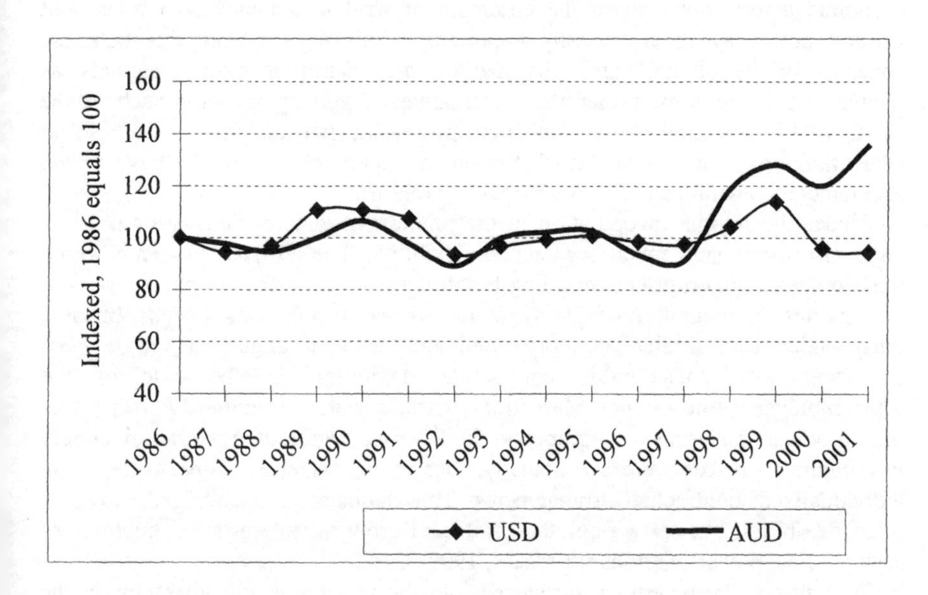

Figure 4.4 Factory gate cost of Californian processing tomatoes, in US dollar terms and Australian dollar terms, indexed 1986-2001

Source: Adapted from data in Horn (2002, p.29).

These issues frame the assessment of the role liberalization has played in the recent restructuring of the Australian processing tomato industry. For its proponents, an underlying assumption of liberalization is that, because it enables growers and processors to strike individually flexible arrangements, it helps generate lower factory gate prices. Yet, as the foregoing analysis has suggested, such developments have been only modest in scope since the mid-1990s. The price per tonne of soluble solids in the industry has remained largely unchanged, with improved international competitiveness coming mainly from exchange rate movements. Of course, there is no way of knowing what may have occurred in the absence of liberalization. However, the key point of this discussion concerns the association of liberalization with improved industry performance. Liberalization *may* have played a positive role in the Australian processing tomato industry since the mid-1990s, but it has *not* been the driving force of improved industry competitiveness.

Furthermore, although there is a clear trend towards increased contractual flexibility as the industry moved forwards under greater liberalization, this could well have eventuated in the presence of other, alternative regulatory approaches. Consistent with the approach discussed earlier in this chapter, the concept of 'regulation' refers not solely to the legally binding rules that shape behaviour by economic actors, but also to the ensemble of written and unwritten rules, and cultural norms which are equally important in this regard. Seen this way, the concepts of 'regulation' and 'the market' are identified most accurately as intersecting institutions, rather than alternatives. Applying this approach to the policy of liberalization of the Australian processing tomato industry leads to an evaluation of the ways market-oriented approaches could have been accommodated within a collective bargaining regime.

Under the regime in operation in the processing tomato industry until 1992, there already existed much contractual flexibility. The TPINC was empowered only to set minimum prices, meaning that processors were free to pay growers at higher rates in recognition of quality, in accordance with their own needs. In some other industries, similar statutory minimum price arrangements have been consistent with considerable contractual flexibility. In the case of the Murrumbidgee Wine Grapes Marketing Board, a statutory authority with price-setting powers covering wine grapes in the Riverina region of Australia, relatively low minimum prices were set annually, thus giving wineries scope to negotiate individualized, contractual arrangements. This deliberately encouraged (upward) price flexibility, but set a floor that protected growers' interests in situations of asymmetrical power relations (Pritchard, 1999b).

In contrast, the measures introduced into the processing tomato sector in the mid-1990s prohibited 'grower groups' from negotiating on price with processing firms. This was the most troubling element of the liberalization reforms, from the perspective of growers and indeed, some processors. The enforcement of this provision of Australian competition law was perceived by a number of industry informants as an unnecessary intervention within the sector, in an environment in which prices had already been deregulated and processors were keenly developing their own pricing models. There was a strong perception that this placed growers in a structurally weaker bargaining position, with little justification in terms of industry efficiency. In conclusion, although Australian grower-processor liberalization has been associated with industry expansion, its contribution to that outcome remains unclear.

Enshrining the rules of the game: grower-processor relations in Canada

Like the industry everywhere, the processing tomato sector in Canada has experienced major restructuring in recent years. The response to the pressures for change, however, did not result in wholesale liberalization, as was the case in Australia, although there were sections of the industry which called for these steps.

Indeed, the Canadian industry has not only been able to maintain a high level of regulation in production and marketing, but has also responded in effective ways to its competitive challenges. Why is this the case? What is it about the Canadian experience that allows it to operate effectively on the basis of an industry model that has been so strongly challenged in other production regions? These questions are of particular significance, given the proximity of the highly competitive US production sites of Ohio and California, and the closer integration of the US and Canadian economies via the Canada-US Trade Agreement (CUSTA) of 1988 and the North American Free Trade Agreement (NAFTA) of 1994. For these reasons, the Canadian experience is particularly interesting, and reveals much about the nature of the responses and strategies of adaptation that may be put in place in order to ensure survival in a globalizing industry.

Canada's historical experience is of considerable significance in the context of such questions, and it is proposed to analyze the local processing tomato industry in terms of a number of stages. The early history of the processing tomato sector in Canada was very closely bound up with the development of the vegetable processing sector in general, and much of the regulatory framework governing the processing tomato sector emerged from this wider framework. As a result, the first stage of this analysis focuses on the origins and growth of the vegetable processing industry as a whole, and the emerging framework of regulation that was established in the 1920s. The second stage of the analysis focuses more directly on the processing tomato sector up to the late 1980s and early 1990s, when the pressures for deregulation and trade liberalization (culminating in the CUSTA and the NAFTA) began to emerge. The third stage of the analysis discusses the impacts of these agreements, with particular emphasis to the institutional responses of Canadian growers, processors and governments. This examination reveals the vital role of statutory, industry-wide regulation, in the context of rapid shifts in the industry's competitive environment.

The processing vegetable industry in Canada

Most of Canada's vegetable production is located in the eastern provinces, but with its milder climate, the southwest region of Ontario Province emerged as a particularly important production area, and remains so to this day (Canadian International Trade Tribunal, 1991b, p.62). The modern processing industry emerged out of the operations of the large number of small, locally owned vegetable processing companies, established in this region in the nineteenth century. For example, Nabisco Ltd., one of Canada's largest producers of canned tomato products, soups, sauces and ketchups (as well as other vegetable lines), traces its origins to the establishment of the Dresden Canning and Pickling Company in 1891, and the merger between it and fifteen other small companies to form Canadian Canners Consolidated Companies Ltd. in 1903 (Nabisco, 2001). In 1956, Canadian Canners, by this time the largest fruit and vegetable canning company in the British Commonwealth, was sold to the California Packing Corporation (better known as Del Monte), and again changed hands when purchased by R.J. Reynolds in 1979. Nabisco acquired the Dresden facility in 1985, and then it was sold to Kraft (owned by Philip Morris) in 2000. Over this whole period, similar changes were occurring throughout the vegetable processing sector in Canada, resulting in a high level of concentration of ownership – especially foreign ownership – in processing, and the emergence of a few, larger firms (Winson, 1993, pp.129–54).

The high level of regulation that characterizes the modern vegetable processing industry in Canada also goes back many years, and can be traced to the historical conditions of the 1920s, and to the developments in the agricultural cooperative movement at that time (McMurchy, 1990). Producers' cooperatives had been introduced into the grain-handling sector in 1911 for the purchase of production inputs (Winson, 1993, p.24). However, whilst collective purchasing and handling was a very effective way to reduce costs, collective action to improve prices to growers was more problematical, since cooperative membership was voluntary and those who remained outside the system could undermine farmer solidarity over prices. This was a major problem during the Great Depression of the 1920s and 1930s, when falling farm prices and a series of mergers and acquisitions in the processing sector reduced the number of buyers and put strong downward pressure on farm incomes (McMurchy, 1990). These conditions gave rise to 'compulsory cooperatives', or Marketing Boards, which were an Australian innovation, first established in the Queensland *Primary Products Act* of 1922. Marketing Boards covering dairy products, tobacco, beans and fruit and vegetables were first introduced in Canada in 1927, under provincial (state) legislation, with the aim of strengthening the bargaining positions of farmers at a time of increasing concentration, and to bring stability in supplies, prices and incomes. In Ontario, Marketing Boards were introduced under the *Farm Products Control Act* of 1937, although these were significantly different from the 'compulsory cooperatives' which were established in other provinces, and which handled the sale of a commodity under quasi-monopolistic conditions. Instead, the Marketing Boards established in Ontario adopted a negotiation-based system of regulated marketing which provided for a 'negotiating agency' comprising representatives of both the growers and the buyers of commodities (mostly the processing companies). As its name suggests, this agency had the authority to negotiate prices for farm products, but also had a provision for binding arbitration where a negotiated price could not be agreed.

The *Farm Products Control Act* of 1937 was repealed in 1946 and replaced by the *Farm Products Marketing Act*, and the legislative arrangements governing the marketing of a wide range of agri-food products have remained largely unchanged to the present day. Under the 1946 legislation, Marketing Boards were made answerable to the Farm Products Marketing Commission, a supervisory body operating as a branch of the Ontario Ministry of Agriculture and Food, and composed of civil servants and private citizens appointed in order to provide wide geographical and sectoral coverage. The Farm Products Marketing Commission was also made the licensing authority for processing companies, and was given wide regulatory authority, including the power to establish negotiating agencies and conciliation and arbitration boards, with the authority to settle disputes over prices, terms and conditions of production, and more (McMurchy, 1990).

Since 1946, the key role of many of the Marketing Boards has been to negotiate prices of regulated commodities. The Ontario Vegetable Growers' Marketing Board (OVGMB, renamed the Ontario Processing Vegetable Growers, OPVG, in January 2000) was established as the statutory body representing the processing vegetable sector. The OVGMB's remit covers 12 processing vegetable crops, including tomatoes, sweet corn, carrots, peas, beans, cucumbers and peppers. Membership of the OVGMB is compulsory for primary producers wishing to sell these regulated commodities to processing companies. The OVGMB allocates licenses to growers and, similarly, processing companies are required to have a license in order to process vegetable products.

In conjunction with extensive industry-wide regulation of the Ontario processing vegetable sector, tariffs had long been in place in order to provide protection from US competition. In the late 1980s, the tariff rates reflected the levels recommended in 1979 by the (then) Tariff Board, but such protection was gradually eliminated under the CUSTA from 1989, and by the NAFTA from 1994. So while the Canadian vegetable processing industry has remained highly regulated, it has also been subject to pressures for change as a result of the influences of the CUSTA and the NAFTA, as well as the actions of transnational food companies in production and retailing. In recent years, such factors have resulted in a high level of concentration in the processing sector, and a significant reduction in the numbers of growers. In 1980, there were 232 vegetable processing establishments in Canada, 190 of which were canning companies, and the remainder in freezing. Of the 232 establishments in total, 108 (47 per cent) were located in Ontario. By 2000, as a result of mergers and acquisitions, there were only 24 companies in Ontario. Over the same period, the number of contract farmers also declined. In 1981, there were approximately 2,600 contract growers producing vegetables for processing, a number which had declined to 875 by 2000 (OPVG, 2001).

By way of contrast, between 1981 and 1990, total volume of output from the processing vegetables sector increased from 810,000 tons to 931,000 tons, and the value of the commodities increased from $86 million (CA) to $115 million (CA) (Canadian International Trade Tribunal, 1991b, p.91). Farm sizes also increased. In aggregate terms, the number of farms in Canada declined by over 18 per cent between 1976 and 1996, from some 277,000 to 227,000, a loss of about 50,000 farms, while the area under cultivation remained constant (Statistics Canada, 1996). Within this framework of regulation, organization and transformation, the processing tomato sector in Canada has flourished.

The processing tomato industry in Canada

The Canadian processing tomato industry has long been dominated by US ownership, and the influence of significantly larger production levels in the United States. In 2002, 11 companies were processing tomatoes in Ontario, but the three largest of these – H.J. Heinz Co., Nabisco, and Sun-Brite – accounted for 85 per cent of output. Heinz and Nabisco are, of course, owned and controlled by US parents, and decisions about what and how much to produce are made in the context of these companies' international operations, in which Canada plays only a minor part. In terms of production levels, Canadian output of processed tomato products has been small by world standards, and is a fraction of US output. The scale of production and the range of processed tomato lines produced in Canada from 1973 to 1995 are indicated in Table 4.5.

Table 4.5 Processing tomato supply and disposition, Canada, 1973-1995 (thousand tonnes, finished product volumes)

Year	Tomatoes canned	Tomato juice	Tomato pulp/ paste/puree
1973	38.96	96.18	9.93
1974	32.83	77.06	n/p
1975	34.78	82.86	1.09
1976	36.06	98.84	1.62
1977	40.83	121.10	n/p
1978	42.39	115.38	3.21
1979	50.66	97.29	5.22
1980	44.40	82.62	5.82
1981	53.86	96.01	9.92
1982	75.55	92.58	16.28
1983	50.88	77.61	12.03
1984	73.61	95.01	16.60
1985	56.84	70.19	20.97
1986	48.88	67.98	25.68
1987	56.02	73.40	28.09
1988	67.54	79.50	31.81
1989	70.13	67.27	31.17
1990	64.18	63.71	41.77
1991	63.13	N/p	29.27
1992	43.01	N/p	22.84
1993	61.89	62.67	29.59
1994	68.61	52.88	28.16
1995	79.64	N/p	31.62

Note: n/p = not published because data are secure or confidential.

Source: Statistics Canada (2002).

Until 1995, the industry displayed fluctuating levels of output, but with a long-term trend towards increased production of canned tomatoes and tomato paste, and declining production of tomato juice. Tomato paste output was not of great significance until 1982, when H.J. Heinz Co. commenced local production. Nabisco followed Heinz, commencing tomato paste production in 1986. Winson (1993, pp.131-2) attributes Heinz's entry into Canadian tomato paste production (and the effect of marginalizing small local paste producers) to the pricing practices of US processing companies and the policies of the Ontario provincial government, which subsidized the entry of foreign companies into the Canadian market:

> The late 1970s had witnessed what effectively amounted to the dumping of cheap tomato paste from offshore producers into the Canadian market. This had driven ten or twelve canners out of business in the province [Ontario]. However, GATT negotiations in 1981 established what was hoped would be a more stable price for tomato paste for provincial processors. At the same time, the provincial government announced a new program, ostensibly to aid all processors of concentrated tomato products in the province. The program made considerable funding available for companies to expand facilities, but much of the funding was allocated to the largest tomato processor in the country, H.J. Heinz. (Local) tomato processors who were looking to benefit from a newly stabilized price were to be victims of this government policy...The problem went beyond the world of tomato paste, however, to whole canned tomatoes ...The provincial government funding had helped Heinz establish a 'pilot plant' for whole canned tomatoes, the mainstay of many small packers.

Of course, the entry of Heinz into Canadian tomato paste production also had the effect of offsetting imports, increasing the market share held by domestic (albeit US-owned) producers (Table 4.6). Hence, increased Canadian self-sufficiency in production came at the cost of increased penetration of foreign capital. What is beyond dispute is that this period saw a significant increase in the volume of processed tomatoes grown in Ontario (Table 4.7), which has largely been sustained over the 1990s.

Table 4.6 Canadian supply and disposition of processing tomatoes, 1980-89 (thousand tonnes, raw product)

	1980	1985	1989	Average 1980-84	Average 1985-89
Production	380	492	539	445	500
Imports	255	212	285	242	237
Exports	0.5	9	13	3	10
Domestic producers' market share	60%	69%	65%	65%	67%

Source: Canadian International Trade Tribunal (1991b, p.259).

Table 4.7 Processing tomato production in Ontario, 1983-90

Year	Number of growers	Harvest area (hectares)	Production (tonnes)	Average output per grower (tonnes)
1983	n/p	10,712	401,518	n/p
1984	n/p	12,032	538,445	n/p
1985	n/p	10,465	474,771	n/p
1986	637	11,010	489,199	768
1987	620	11,530	498,337	804
1988	533	12,596	517,507	971
1989	477	11,673	519,722	1,090
1990	400	11,318	568,693	1,264

Note: n/p = data not published.

Source: Canadian International Trade Tribunal (1991b).

Significantly, these shifts in the corporate and industrial structures of the Ontario processing tomato sector occurred in the context of an unchanged environment of statutory regulation, in which growers, via their representation on the OVGMB, retained substantial influence over price determinations, marketing and production:

> The marketing of processing tomatoes in Ontario is controlled and regulated by the Ontario Vegetable Growers' Marketing Board (OVGMB). All growers are required to have a license in order to market processing product to processing companies. Likewise, all processing companies must have a license to process vegetable products. The OVGMB negotiates tomato processor prices on behalf of growers through a negotiation agency which is made up of six members; three members representing processors and three members representing growers. The main thrust of the OVGMB is to increase the bargaining power of growers when dealing with processors (CITT, 1991a, p.6).

Clearly, throughout the 1980s the Canadian processing tomato industry maintained a high level of statutory regulation that accorded considerable powers to growers. These statutory institutions formalized the rules of price bargaining with the effect of maintaining a balance of power between these growers and the large companies that had come to dominate the industry. The critical question for the industry in the future was – could it continue to operate in this way, particularly following the introduction of the CUSTA and the NAFTA?

Prior to 1988, imports of tomato paste into Canada carried a duty of 13.6 per cent, while canned tomato products and finished goods carried a duty of 15 per cent. Under the CUSTA, tariffs on fresh and processed fruits and vegetables imported from the US were to be phased out over a ten-year period (CITT, 1991b, p.10). The adoption of the NAFTA in 1994 did not affect these arrangements between Canada and the US, and, to date, Mexico's impact on the Canadian processing tomato industry has been limited, although the same cannot be said for fresh tomatoes (United States Department of Agriculture Foreign Agricultural Service, 2001, pp.4–5; Barndt, 2002).

The suggestion by some authorities that the processing tomato sector would be a 'loser' under the CUSTA (Canadian Horticultural Council, 1990, p.9) prompted the Government of Canada to conduct a study into the competitive factors affecting the production and marketing of fresh and processed fruits and vegetables in the Canadian and US markets. The study was carried out by the Canadian International Trade Tribunal (CITT), and was supported by a considerable amount of research. The CITT report (CITT, 1991b) suggested that in comparison to the US Midwest and California, the Canadian industry was in a weak competitive position, partly as a result of the higher costs that were incurred by growers and processors. For example, the cost of raw product for tomato paste was some 10 per cent higher for Canadian processors compared to their counterparts in the US, as a consequence of higher costs for overheads, labor and energy inputs. Canadian production was cost competitive only on the basis of the additional transport and tariff costs borne by US competitors (Table 4.8). Closer examination of these data revealed that the significantly lower yields that Canadian growers were achieving accounted for a major component of reduced competitiveness (Table 4.9), and it was this issue that the industry focussed on when US competition emerged as a significant threat in the late 1980s.

Table 4.8 Factory gate cost of processing tomatoes, Ontario and Ohio, 1989

		Adjustments		
	Average cost	For transport	For tariffs	Adjusted cost
Ontario	113	0	0	113
Ohio	98	18	11	127

Note: All figures Canadian dollars.

Source: Canadian International Trade Tribunal (1991b, p.194).

Table 4.9 Average cost of processing tomatoes, Canada and USA, 1989

	Average cost of production ($/hectare)	Yield (tonne/hectare)	Average cost ($/tonne)
Ontario	4.57	40.5	113
Ohio	4.42	45.0	98
California	3.96	63.0	63

Note: All figures Canadian dollars.

Source: Canadian International Trade Tribunal (1991b, p.192).

The Ontario industry's response to the competitive threat of US imports in the late 1980s and early 1990s involved significant reform, in particular focusing on the concept of 'productivity pricing' as a way of providing an incentive for both the grower and the processor to improve yields. The idea behind the concept of productivity pricing was that the growers and the processors would work together in order to improve the industry's overall competitive position relative to other major production regions, and to establish a formula for sharing the gains achieved through increases in yields. According to the OPVG, productivity pricing meant that as the 'collective average yield per acre of tomatoes that each processor accepted from his entire group of contract growers increased over a predetermined base yield, the processor unit costs would decrease incrementally' (OPVG, 2001). In other words, higher yields translated into both improved gross returns per acre to the grower and lower unit costs to the processing company. As illustrated in Figure 4.5, yields did increase significantly over the 1990s, and it is reasonable to assume that productivity pricing played a role in achieving this outcome (although, as noted in Chapter Two, tomato yields increased everywhere in the 1990s, due mainly to improved cultivars).

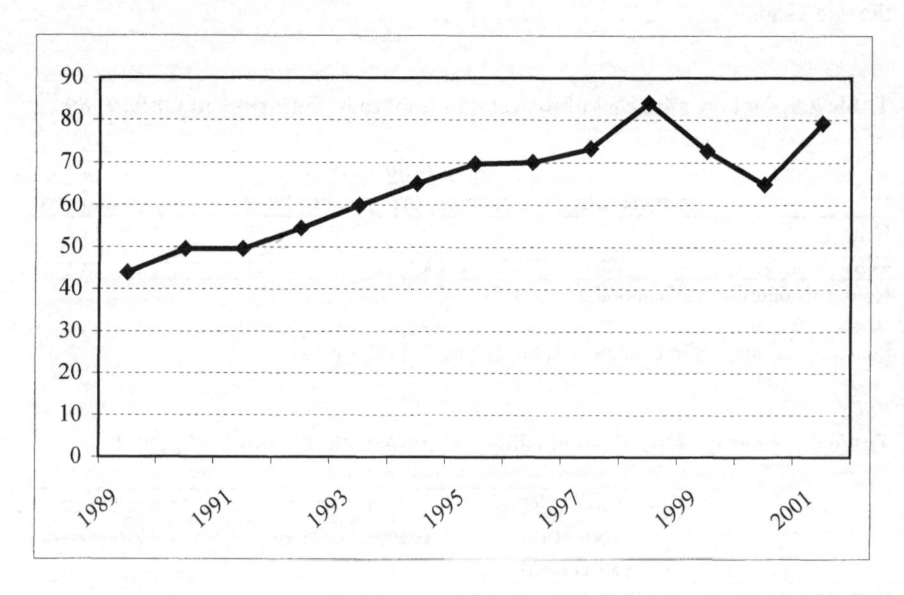

Figure 4.5 Canadian processing tomato yields, 1990-2001 (tonnes per hectare)

Source: Ontario Processing Vegetable Growers (2001).

Structure and regulation in Canadian processing tomato cultivation

During the 1990s, there was some growth in Canadian processing tomato output, which as with other growing regions, came from a steadily falling number of growers (Table 4.10). Prices have also been under pressure, although in many parts of Ontario, tomatoes still provide the highest returns of any processing vegetable. The price paid for paste in 2000 varied between $96.12-$101.57 (CA) per ton, depending on the contract option. Tomatoes used for the production of tomato paste commanded the lowest price to farmers, with an average of $86.85 (CA) per ton, compared to $97.35 (CA) for whole pack and $96.34 (CA) for juice. The price paid in 2001 was some four per cent lower than the price in 2000, although prices increased in 2002, by five per cent for paste, three per cent for whole peel and two per cent for juice (OPVG, 2001).

Table 4.10 Canadian processing tomato production, key statistics, 1996-2001

	1996	1997	1998	1999	2000	2001
Number of growers	222	192	183	187	173	173
Measured hectares	7,084	6,582	6,693	7,496	7,030	6,729
Tonnes harvested ('000)	502	497	562	545	456	533
Average yield (tonnes per hectare)	71.07	75.76	84.25	72.97	65.09	79.20
Gross farm value ($m)[1]	52.5	50.3	58.9	57.3	44.2	50.9

Note: 1. Canadian dollars.

Source: Ontario Processing Vegetable Growers (2001).

While the experiences of Ontario's industry broadly parallel those of other growing regions, Ontario is distinctive for several reasons. These unique characteristics, in turn, have relevance for debates on the evolution of Ontario's grower-processor regulations. The short growing season is the most important factor that distinguishes the Ontario industry from other production sites. Because of its climate, the Ontario tomato processing industry has a growing season which runs for about 55 days for paste and 43 days for diced tomato products (compared for 85 and 73 days for California). Planting of the tomato crop begins in early May, as soon as the risk of overnight frosts has disappeared. Again, because of the short season, all processing tomatoes in Ontario are grown from transplants (or 'seedlings'), raised in nurseries for about six weeks before being established in the open fields (Tomato News, 1997). This has not only resulted in a substantial increase in productivity in recent years – germination rates in the greenhouse are 10 per cent better than those occurring in the field – but has also given rise to a substantial transplant industry in Ontario. The growth of the local transplant tomato seedling industry is testament to the persistence of the role of government regulation in the Canadian processing tomato industry, and of the importance attached to such regulation by influential actors. For despite the fact that it did not fully evolve until the early 1990s, it is very significant that the local transplant industry was still accommodated within the organizational model of industry-wide regulation that has its roots in the era of the Great Depression.

The framework of regulation which governs the negotiations over contracts between the OPVG and the tomato processing companies, stipulates the procedures to be followed in annual negotiations over prices and the procedures which are brought into play when problems or disputes occur. Every year, negotiations between the OPVG and representatives of the Ontario Food Processors Association are held in accordance with an agreed and predetermined schedule. For the crop year 2001, for example, the regulations booklet – a detailed document prepared annually by the OPVG – specified that the negotiation agency for tomatoes would hold an initial round of negotiations which would end at 4 p.m. on 1 March 2000. By 13 March 2000, the OPVG would send to processors a copy of a proposed agreement (if concluded) or a notice of intention to negotiate, and by 20 March 2000, a processing company would respond either by returning a signed agreement or indicating an intention to negotiate. If no agreement has been reached, negotiations would continue until the deadline set at 4 p.m. on 1 April 2000, and if the parties had still not agreed on contract conditions, the matter would normally follow a prescribed course and go to mediation, followed if necessary by arbitration. Under these circumstances, the contract terms and conditions would be decided by a conciliator acceptable to both sides and appointed by the Farm Commission established under the *Farm Products Marketing Act* (OPVG, 1999). In the ten years 1991-2000, there were 612 agreements reached between the OPVG and processors, covering all crops, with arbitration being utilized on only 27 occasions. Seven of these arbitrations were in the processing tomato sector. The result of the 27 arbitrations favored the growers in 17 cases, the processors in seven, and three went to further arbitration (OPVG, 2000).

Beyond the terms and conditions established during sector-wide negotiations, there is the scope for processor specific negotiations, which provide an opportunity for agreement on the particular requirements of processors, and which build some flexibility into these highly regulated institutional structures. This combination of industry-wide regulation with flexibility at the scale of individual processor-grower relations clearly attempts to deliver the 'best of both worlds' to interested parties, and has parallels with other situations. For example, describing the comparable regulatory system in the Murrumbidgee wine grapes industry, in eastern Australia, Pritchard (1999b, pp.435–36) observes:

> Th[e] network of social relations existing between grape growers and wineries – what can be called the 'informal' tier of market relations above and beyond the legal minimum pricing arrangements enforced by the [Wine Grape Marketing] Board – nevertheless benefits from the existence of statutory marketing. Through the Board's minimum pricing and vesting powers, a safety net is established which facilitates the construction of trust between growers and wineries on a range of grape supply issues. In the words of one grape grower interviewed as part of the research for this paper, 'the Board is my union': the implication being that the Board's ability to assure fair and timely grape supply payments enables him to adopt a more cooperative approach with wineries generally.

The relationship between industry-wide regulations backed by statutory provisions and the individual relationships between growers and processors is mediated by the content of production contracts. In the Ontario processing tomato industry, production contracts typically specify tonnages, prices, and grading standards. Volumes are assigned to growers for a three-year term, on the basis of total demand by the processor with whom the grower has a contract. For example, if in a year in which Heinz's total volumes of processing tomatoes stand at 100,000 tonnes (say Year 1), and a grower signs a contract with Heinz to deliver 5,000 tonnes (or five per cent of Heinz's total take), then in the two following years (Years 2 and 3), the same grower has a right to five per cent of Heinz's total volume, up to a limit of 5,000 tonnes. So if Heinz reduces its volumes to 90,000 tonnes in Year 2 (a 10 per cent reduction), the growers can also expect a 10 per cent reduction and will get a contract for 4,500 tonnes. However, if in Year 3, Heinz increases its volumes to 110,000 tonnes, then the grower cannot routinely expect more than the 5,000 tonnes contracted for in Year 1 (although he/she *may* get some of the extra tonnage). Any additional tonnages required by Heinz (or any other company) can be allocated to any of its existing growers, or to a new grower, or a combination of both.

One implication of the adoption of three-year contracts based around these provisions is that it encourages grower loyalty to particular processing firms. Unlike many other tomato growing regions, it is unusual for growers in Ontario to have tomato production contracts with more than one processing company. Growers contracted to Nabisco have no tomato contracts with other processors; Heinz can veto any grower entering into a tomato contract with a direct competitor, but may allow some flexibility where a grower seeks to enter into a contract with another company which is not in direct competition with Heinz. However, growers typically do not solely grow processing tomatoes, but diversify their operations among a combination of those thirteen crops covered by the OPVG, and may well have contracts with a number of processors for different commodities. Thus, several farmers interviewed in the course of research indicated that while holding a tomato contract with Heinz, for example, they may also grow peas and/or sweet corn for Nabisco. For most farmers growing tomatoes though, this commodity accounted for anything between 50-80 per cent of gross income, and was usually their largest single commodity line. These issues are important because, in places without industry-wide statutory regulation, such as Australia, anecdotal evidence suggests that a heightened sense of vulnerability encourages growers to enter into production contracts with a number of processing firms, to defray risk. What appear to be higher levels of grower loyalty in Ontario then, may be related to the 'safety net' provisions of Provincial legislation and the empowerment of Marketing Boards.

There seems little doubt that despite the decline in grower numbers in recent years, the prevailing system of regulation has wide grower support, and has enhanced security in production for those farmers who remain. More importantly perhaps, the 173 Ontario processing tomato growers in 2001 saw themselves as part of a larger organization comprising over 800 vegetable growers, and with such numbers, grower support for the continuation of the current system translates into significant political influence at the Provincial level.

Another important element of the regulatory regime in Ontario relates to crop insurance, the monitoring of production, and the grading of produce. Up until 2001, these services were provided by Agricorp, a private crop insurance and consultancy organization. In 2002, another company, Upgrade Consulting, was contracted to perform these functions. Agricorp monitored crop acreages using a global positioning system (GPS) and reported the results to grower and processor groups. Additionally, Agricorp acted as an independent agent in the grading of processing tomatoes. Tomato growers are responsible for harvesting their own fruit and arranging for delivery to the processing plant, where grading is carried out. In 2000, for example, Agricorp graded 456,000 tonnes of processing tomatoes, at 11 processor sites, and employed 57 graders and four group leaders over a 68-day period. The cost of grading this crop was $380,000 (CA), shared equally between the processors and the growers. Grading is performed by selecting samples of delivered crop and processors can choose from a range of grading options which are mainly determined by the processor tolerance for green tomatoes. A measure known as the Agtron color level test (measured by an instrument developed by the Agtron Corporation of Sparks, Nevada) determines the price paid per tonne. Other quality and grading measures, including the presence of extraneous matter and crop defects of various kinds, can result in penalties. As discussed in the section on Australia above, the existence of independent fruit grading is a vital component of the regulation of grower-processor relations, as disputes over grading may act as a lightning rod for wider antagonisms between these parties.

The final element in this analysis of Canada's regulated system of production concerns the emergence of a local seed and transplant industry. As noted earlier, because of its climate, the industry in Ontario has long relied on transplanted seedlings, but until the mid-1980s, most of these were imported from Georgia, in the southern United States. At the peak of production in 1988, Georgia growers produced some 684 million tomato plants valued at $7 million (US). One of the largest individual growers produced some 35 varieties on 40 hectares, mainly for H.J. Heinz Co. (Harrell, 1988). The move to produce local transplants (also known as 'plug plants') in Canada began in 1988, around the time that the CUSTA was imminent. There were several reasons for this shift in the production of transplants from Georgia to Ontario. First, there was considerable inflexibility in delivery from Georgia, with a minimum three-day lag between ordering plants and their delivery to farmers in Ontario. There could be even longer delays since Georgia suppliers would not 'pull' transplants on a Sunday, or if it was raining. This meant that farmers in Ontario were not in a position to take advantage of optimal conditions for planting. In addition, imported seedlings did not always travel well and, as was discovered subsequently, were of inferior quality to those that were to be produced locally (La Prise, 2001).

The transition to the full use of local transplants took seven years, and contributed greatly towards Ontario's ability to compete with US producers. Whereas nearly 700 million seedlings were imported from Georgia in 1988, the local industry is now able to maintain the Canadian industry's output on the basis of annual production of some 200 million transplants. This figure in part reflects the fact that the local product is healthier and more productive, but is also evidence of the reduced need for 'replants', when seedlings fail to grow. 'Replants' may result from frost, wind damage, disease and poor plant quality, and the rate of replant in 2001 was around 1.6 per cent, compared to 15 per cent ten years ago.

The local transplant local industry was originally based on traditional greenhouse production but in recent years has moved towards large-scale production with the construction of huge steel-frame structures spanned by an opaque flexible PVC cover. Production within these greenhouses is highly automated, with seeds individually dropped into trays with 200, 288 or 406 compartments ('cells') moving along conveyor belts. These trays are then stacked and stored while germination takes place. Germination occurs within a few days, but transplants are grown for a period of about six weeks before being taken to the field. In operational terms, a processing tomato company will supply the greenhouse companies with the types and volumes of seeds it will need for that year and inform its contract tomato growers when they should collect and plant the seedlings that are specified. Growers collect the plants at the appointed time, and transplant them, using specialist mechanized equipment produced in Canada. However, this process still requires some labor in extracting the plug plants from their trays and positioning them in a 'carousel' for automatic planting.

At the peak of industry development, there were 34 companies producing tomato seedlings, but currently there are about 18. The five largest suppliers provide about 50 per cent of the Ontario industry's requirements, with one company alone supplying 20 per cent. For many of the largest plug plant companies, tomato seedlings are just one of a number of product lines, and in the case of the largest companies, tomato seedlings account for only 55 per cent of business revenue (La Prise, 2001). Transplant producers are not tied to any one processing tomato company, although some processors, most notably Nabisco, has in the past tried to encourage exclusive arrangements with preferred suppliers. On the other hand, some processors such as Heinz favor diversifying their sources of supply, thereby spreading their risk. Most of the varieties grown in Ontario are hybrids, with the exception of those purchased by Nabisco, which utilizes open pollinated varieties exclusively.

The producers of tomato seedlings are represented by Tomato Seedling Producers Marketing Board (TSPMB), which serves for them many of the same functions that the OPVG carries out for farmers. For example, seedling producers first agree amongst themselves the target price that they would like to receive for seedlings, and the Board then negotiates with the tomato processing companies on price and other conditions of production. Contracts between the seedling growers and the tomato processors extend over a three-year period, and specify the volumes to be supplied (up to the level of the base volume allocated to each tomato farmer – see below), the varieties to be supplied to contract tomato growers, and the time at which growers collect seedlings from the transplant supplier. The three-year contract was introduced in the early 1990s as the local transplant industry became more established and the processing tomato companies sought some guarantees on supply as their dependence on the local industry grew. At the same time, the emerging transplant industry was seeking the security of longer-term contracts as it became larger and more capital-intensive. In a trade-off with the tomato processors, the transplant producers agreed to a five per cent reduction in the price for transplants in return for the three-year contract introduced in the early 1990s. Despite the lower returns this initial agreement implied, the stability of a longer-term contract enabled the transplant industry to address efficiency issues, such as achieving greater economies of scale rather than simply seeking an annual increase in price. This led to enhanced productivity which was to contribute to the economies and productivity gains made elsewhere in the industry, and which enabled Ontario producers to better meet the challenges from US producers which were to emerge over the 1990s. In the early 1990s, the average sale price of processing tomato seeds was $0.30-$0.50 (CA) per thousand, compared to a current price for hybrids of between $3.00-$9.40 (CA) per thousand. Currently, the processing companies usually provide the seed and deduct the cost from the prices paid to transplant growers.

As a major contractor of processing tomatoes in Canada and the proprietary owner of many of the most successful hybrid seed types in the processing tomato industry, Heinz has had a long-term interest in the organization and operations of the Ontario tomato seedling industry. The company has about ten suppliers of seedlings who not only supply the Heinz tomato growers but also service those other companies which use Heinz seed. Having calculated the volumes of Heinz seedlings that will be required, negotiations between Heinz and the transplant producers on the price and timing of production begin in April each year. The contracts that Heinz enters into with its tomato growers are typically twice the size of other processor contracts, and the company has on occasions argued that it should be given a price discount by transplant producers on account of the higher volumes of seedlings it purchases. However, such approaches directly challenge the industry-wide regulatory system orchestrated via the TSPMB, and at the time of writing, the pricing system remains regulated according to industry-wide standards.

The above discussion on the conduct of negotiations over production contracts, on the importance of independent grading, and on the relevance of the seedling industry as a marker for the continued application of a high level of formal regulation, demonstrates that the state still has a major say in how the processing tomato sector in Canada operates. However, there *are* pressures for liberalization, most notably emanating from the processing companies. How these pressures have been identified and dealt with are the key issues in the following section.

Canadian processing tomato firms and industry-wide regulation

As noted earlier, since the 1980s the Ontario processing tomato industry has become highly concentrated. In 2001 there were eleven companies processing tomatoes in Ontario, but Heinz, Nabisco and Sun-Brite accounted for 85 per cent of output. According to industry sources, Nabisco and Heinz, which are both US-owned, accounted for 67 per cent of production (Heinz 40 per cent and Nabisco 27 per cent), and Sun-Brite, a locally owned company, accounted for 19 per cent. The corporate strategies and structures of each of these firms vary on account of differing histories and management goals, with the effect that each intersects in unique ways with the Ontario system of industry-wide regulation.

H.J. Heinz Co., Ontario's largest processing tomato firm, provides the appropriate starting point for this discussion. Heinz began its Canadian operations in 1909 in Leamington, Ontario, where it operates to this day. The processing plant at Leamington is one of the largest Heinz factories in the world, and produces branded ketchup, soups, tomato juice, baked beans, baby food, pasta sauces and vinegars, and some private supermarket and generic brands. H.J. Heinz Co. accounts for over 40 per cent of the output of Canada's processing tomato sector, although it contracts with only 57 growers. Heinz provides an interesting case for the analysis of the substantive issues under discussion, because it has aggressively expanded its output from Leamington, despite being publicly critical of the system of industry-wide regulation. In order to understand such behavior, it is necessary to locate the company's Canadian operations within its evolving global strategy.

Heinz's Canadian expansion began in the early 1980s, when it introduced tomato paste production to its Leamington plant, following an agreement between the company, the Ontario provincial government and the OVGMB. Until that time, tomato paste was imported into Canada, and amounted to approximately 36,000 tonnes of finished product, valued at $30 million (CA). In order to reduce import dependence and encourage local production of manufactured goods, the Ontario provincial government introduced the Board of Industrial Leadership and Development (BILD) which, among other things, allocated grants to strengthen Ontario's industrial base. Despite protests from some local Canadian-owned companies who complained that their requests for similar support had been turned down, in 1982 the Ontario Ministry of Agriculture and Food made available a BILD grant of $3 million (CA) to Heinz, which in turn agreed to invest $15 million (CA) to expand its tomato paste operation. The growers agreed to supply tomatoes at a discount for five years to help establish the plant. According to the Ontario Government, the result was to reduce tomato paste imports by $12 million (CA) per annum. Heinz increased its tomato intake by 26 per cent, leading to $5 million (CA) in new income for tomato growers.

This investment proved vital for Heinz Canada in the 1990s, when the parent company undertook a global restructuring of its operations. In the wake of the CUSTA and the NAFTA, Heinz implemented a series of factory restructurings in the USA and Canada. Then, through 'Project Millennia' of 1997 and 'Project Excel' of 1999 (see Chapter Three), the company sought to focus on a small number of core business activities, each with global scope and utilizing world 'best practice' facilities. These developments impacted significantly on Heinz Canada and, in particular, the Leamington factory, which was identified as a critical cog in Heinz's emergent global structure. However, this did not eventuate without considerable conflict between the company, its workers and contract growers, stretching over several years. In 1993, Heinz announced Leamington would be the sole manufacturer of baked beans for the whole of North America. At the same time however, Leamington's production of soups, sauces and sandwich spreads was shifted to the United States, with the effect of making 200 workers redundant (Financial Post, 1994, p.4). Soon afterwards, in 1994, a major conflict occurred between the company and its factory employees. In January 1994, not long after the decision to relocate production of soups and other products, Heinz moved to reduce labor costs by offering its plant workers a new contract which involved a reduction in pay rates. The company announced that it needed to find some $10 million (CA) of savings from its Leamington plant, and $3.8 million (CA) of this was sought through reductions in wages paid to factory workers in the collective agreement then coming up for renewal. When employees rejected this move, the company threatened to move the production of baby food and baked beans to the United States, at the cost of 450 jobs (out of a total of 850, down from the peak of 1,400 in 1988). With no resolution to this problem in sight, Heinz sent lay-off notices to these 450 workers (including long-term employees) and initiated plans to close the Leamington plant and move all production to the United States. At this point, the processing employees agreed to the terms laid down by the company, and in return, Heinz agreed to pay stiff penalties if they closed the Leamington plant during the four years of the new agreement (Craig, 1994).

Having succeeded in reducing its processing costs at Leamington, Heinz then followed this up with an attempt to reduce its raw material costs. 1996 proved to be something of a watershed in the system of regulation then prevailing in Ontario, and it was a difficult year for processing negotiations. A change of government at the provincial level, which brought the Progressive Conservative Party to power in a landslide vote, reportedly led some processing companies to provoke multiple arbitrations in an attempt to persuade the new government that the system of regulated marketing had broken down. 1996 saw negotiations over contract prices for vegetable production collapse on 12 occasions, with each breakdown requiring arbitration (although none of these was in tomatoes). In this climate, Heinz exerted considerable pressure on growers to reduce prices. A North American reorganization at this time made Leamington more directly responsible to company headquarters in Pittsburgh and, according to some growers, the changed managerial structures encouraged heightened pressure for cost reductions. Ultimately however, Heinz relented in its attempt to reduce raw material costs. This may be because at this time, all Canada-based processing companies were developing the capacity to reduce their input costs and improve efficiency via the mechanism of productivity pricing, i.e. through cooperation with the growers and the OPVG, rather than confrontation. The company worked with some 70 growers at this time, and its future in Canada clearly has depended on continued cooperation on a range of activities which Heinz has undertaken as part of its global restructuring. Subsequently, Heinz Canada has reaffirmed its commitment to the Ontario processing tomato sector, including major new capital investment decisions at Leamington in 2001 and 2002.

Nabisco's recent experiences in the Ontario processing tomato sector provide a number of themes which parallel the experience of Heinz. In 2000, Nabisco was acquired by Kraft for $18.9 billion (US), in what was (at the time) the world's largest food sector acquisition. Kraft itself at the time was owned by Philip Morris, one of the world's largest tobacco companies. In 2001, Philip Morris floated Kraft via an Initial Public Offering, formally separating its tobacco and food businesses.

Nabisco's factory at Dresden, Ontario, is the second largest processing tomato facility in Canada (after Heinz, Leamington). It was built in 1948 to cater for the expansion of Canadian Canners. This company passed out of Canadian ownership when it was taken over by Del Monte in 1956, and was subsequently acquired in 1979 by R.J. Reynolds, before being sold to Nabisco in 1985. Nabisco's core business is now in biscuits, baked products and confectionery, and its processing tomato operations are essentially carried over from the period when Del Monte acquired Canadian Canners. Currently, Nabisco markets canned peeled tomatoes, and diced and stewed tomatoes, which account for 50 per cent of the company's tomato purchases; the remaining 50 per cent is turned into paste and used for pasta and pizza sauces, soup, and ketchup. These products are marketed under the 'Primo' and 'Del Monte' brands, as well as supermarket private labels. Nabisco also produces and markets Del Monte branded products in Canada under a licensing agreement. In 2000, the Dresden plant processed 125,000 tonnes of tomatoes (as well as 3,000 tonnes of red beets, 3,000 tonnes of potatoes and between 200 to 300 tonnes of sweet potatoes). The company has purchased paste from California for the production of soups, while using its own high viscosity paste output for ketchup, and has purchased tomatoes from non-contract growers in years when supplies were low. At the peak of the season, the Dresden plant operates seven days of the week, for 24 hours per day. In this 24-hour period, there are two nine-hour shifts engaged in tomato peeling and further processing, with six hours set aside for the cleaning of equipment. At this peak period, the plant employs 500 people, compared to 150 at other times. Peak labor demands are met by migrant workers, many of whom are from Jamaica. In the 1970s, when tomatoes were hand-peeled, the company employed some 300 migrant workers, but in the year 2000 only about 90 were recruited.

The company currently contracts with 40 growers and of all processing tmato firms, is the least flexible when it comes to allowing its growers to contract with other processors. The reason for this is partly related to the practice of utilizing its own seed varieties, which are grown to transplants by the local seedling industry. By requiring growers to produce exclusively for Nabisco, the company is able to maintain control of its seed and ensure that other companies do not use its varieties.

According to industry sources, the Dresden plant could not survive solely as a paste producer, even though Nabisco is paying less for paste than it would if purchased from California. Canned peeled products give higher returns, and are essential to the future of the plant. The company's ability to source raw materials at competitive prices was in some measure due to the role played by the OPVG, which was focussed on keeping processing tomato capability in Canada. Through the introduction of productivity pricing, the OPVG was able to negotiate agreements that reflected competitive prices for raw tomatoes. In other words, the OPVG agreed to price reductions in order to guarantee longer-term security by ensuring that Ontario growers were competitive with the alternative sources of supply, and able to deliver a degree of flexibility so that companies like Nabisco would remain in Canada.

However, competitiveness in production does not guarantee survival, particularly in the face of the newly emerging relationships with the retail and food service sectors, which increasingly determine the parameters of agri-food production. About 40 per cent of Nabisco's processed tomato production is for supermarket private labels, mainly in canned peeled tomatoes, and diced and stewed tomatoes. The Dresden plant deals with two major supermarket chains, Sobey's and the Weston Group (including Loblaw's, owner of Canada's largest private label, 'President's Choice'). Both of these supermarket chains have clear marketing strategies that involve them carrying only two lines of any particular food product; a branded premium product and a private label version of the same commodity. In the case of tomato and related products, the supermarkets that Nabisco deals with prefer to carry Heinz lines as their branded products, insisting that Nabisco supply their private label products. As noted in Chapter Three, processors are in a difficult situation because supermarkets control access to consumers via their control of shelf space, product positioning, special offers, marketing and so on. Furthermore, in recent years the supermarket chains have themselves entered into a series of mergers and acquisitions at the national and the global level, which have strengthened their position *vis-à-vis* food processing companies.

This newly emerging role of the retail sector and its influence over the production sectors has been a major driver of the high level of mergers and acquisitions in agri-food processing in recent years, as food manufacturers seek to generate a countervailing power and greater leverage over the retail chains. Such developments certainly underpinned Kraft's decision to acquire Nabisco in 2000. However, processing tomatoes have always been a small part of the Nabisco operations, and even before the Kraft takeover, Nabisco wanted to divest its tomato-based operations and concentrate on its core business in biscuits and baked products. Following the merger with Kraft Foods, Nabisco's tomato-based products were an even smaller part of the total food operations of the merged company, and did not fit with Kraft's ambition of holding either the first or second market position in all of its food and beverage offerings. As a consequence, in 2001, Kraft proposed the sale of the Del Monte and Primo pasta and tomato-based brands, with the intention of concentrating on its core activities (Biswas, 2001). What this suggests is that while Nabisco's Ontario operations have been made competitive through the cooperation of the OPVG and the company's contract growers, factors relating to the restructuring of retail channels continue to raise questions about the factory's ongoing operations.

A similar set of issues faces the third major processor in the Ontario processing tomato industry, Sun-Brite. Unlike the two previous companies examined here,

Sun-Brite is locally owned, having been established in 1973 by Henry Iacobelli, an Italian migrant who originally worked as an engineer in the food processing industry. In its first year of operations, the company produced 35,000 cases of 24 oz. and 28 oz. cans of whole peeled tomato. Apart from the processing of some red beans and other minor vegetable crops, the company is mainly a processor of tomatoes. Currently, the company produces three million cases of product annually at two plants; the main factory at Ruthven and a secondary facility nearby. The company produces tomato products under its own UNICO label, and for a range of private supermarket labels, including Canada's largest supermarket brand, Loblaw's 'President's Choice'. Some 30 per cent of its peeled tomato products is used for supermarket private labels but the company also produces large drums of tomato products which are used for re-processing. Sun-Brite processes about 130,000 tonnes of tomato products and purchases large volumes of paste from California – 2-4 million tonnes in most years and 5 million tonnes in 2001. The main product lines are peeled tomatoes (whole, diced, stewed, and flavored with herbs and spices), sauces (pasta and pizza sauces), and tomato paste. All peeled product is processed from local supplies made available over the growing season, while purchased paste is used to produce other lines and at other times of the year. The company prefers seed varieties that display good tomato taste, and mainly uses Heinz seeds, supplemented by some Italian varieties. However, one industry observer suggested that Sun-Brite sometimes sources seedlings from Georgia, presumably in order to maintain some leverage over the terms and conditions upon which it is able to acquire its inputs.

Sun-Brite contracts with about 25 growers, and although relations are generally harmonious, Sun-Brite management has argued strongly in favor of the deregulation of Ontario's grower-processor bargaining arrangements. According to Sun-Brite management, the current system provides insufficient flexibility in production and marketing, particularly when it comes to the issue of the price paid to growers. In 2001, Sun-Brite management indicated that Ontario processors should be paying $80-85 (CA) per ton in order to stay competitive with California, rather than the agreed $95 (CA) per ton (Iacobelli, 2001). Moreover, it was alleged the three-year contract system was not fair to processors because the grower could quit the contract at any time, and although the processor could sue a defaulting farmer, this was not usually a realistic proposition.

There are obvious commercial pressures relating to Sun-Brite's desires for lower-priced raw product. With its reliance on supermarket private label production, Sun-Brite is a price-taker in a highly price-sensitive industry segment. Both Sun-Brite and Nabisco have a lot of capacity tied up in the supply of supermarket private labels. The sourcing policies of the retail outlets, and the flexibility which they have in shifting suppliers at relatively short notice, is a major problem which creates great uncertainty and generates pressures for price reductions all the way back down the supply chain.

Ontario's system of industry-wide regulation: an assessment

The years since the introduction of the CUSTA have seen significant restructuring in the Canadian processing tomato industry, which has been characterized by periodic high levels of conflict at numerous points in the commodity chain. Processing companies, for example, have seen their traditional market dominance erode in the face of demands from supermarkets for private label products which compete with the processor's own premium brands, and the demands by fast food outlets for raw materials which meet specific quality or preparation requirements. Even a company the size of Heinz is not always able to resist the demands of retail chains, and while it may be willing to participate in the supply of private labels in some commodities (e.g., in soups in the US market, where Campbell Soup is traditionally dominant), it is not always able to do this on terms that it finds attractive. Under these circumstances, processing companies inevitably bear down on the growers to reduce costs and increase efficiencies, while at the same time looking to other options – for example, to get bigger through mergers and acquisitions or to relocate production and/or source raw materials from cheaper production sites – to maintain or improve market share.

In considering these issues, the Canadian experience is of particular interest, because quite clearly, growers in Ontario have been able to retain the system of regulated marketing that has enabled them to exert some control over the terms and conditions by which they engage with production. Why is it that, at a time when neo-liberal policies have resulted in the deregulation of markets almost everywhere, Canadian growers have been able to retain a highly regulated system?

In attempting to answer this question, there are a number of factors that should be taken into account. There is, firstly, the fact that no Canadian government ever fully embraced the type of neo-liberal philosophy that led to deregulation in Australia. A legacy of economic nationalism, developed as a consequence of Canada's proximity to the world's largest economy and ever-expanding domination by US corporations, tempered the neo-liberal impulse and gave support to the idea that it was important for Canada to maintain some vestiges of economic independence. Thus, while the conservative Federal Government of Brian Mulroney advocated measures for deregulating some sectors of the Canadian economy, the main policy initiative with regard to the liberalization of markets and trade involved the introduction of the CUSTA and the NAFTA which, significantly, did not require Canada to dismantle the regulatory framework which governed the marketing of much agricultural produce.

As a result, by the end of the 1990s the Ontario processing tomato sector had averted the potential competitive threats posed by the CUSTA and the NAFTA. According to Heinz Canada CEO, James Krushelniski:

> The Canada/US Free Trade Agreement ensured that those companies that survived the shakeout significantly improved their cost structures and the quality of their products. As a result, they have become more effective and efficient producers ... Our Leamington facility manufactures for the Canadian and US marketplaces. We've

> rationalized our production in accordance with the production expertise in different plants (Food in Canada, 2001).

Canada's processing vegetable sector has scored some considerable successes since the mid-1990s. Exports (mainly to the US) have more than doubled while imports have grown by just 33 per cent. Canada's processed fruit and vegetable exports of $960 million (CA) in 2000 represent 27 per cent of all shipments by fruit and vegetable processors, compared to only 16 per cent in 1996 (Food in Canada, 2001). Like the situation in Australia, the significant depreciation of the Canadian dollar against the US dollar over the second half of the 1990s and beyond obviously provided a key element of the improved competitive position of Canadian tomato production. Yet, the relative success of Ontario vegetable processing, including processing tomato production, has not only relied on exchange rate factors. Over the 1990s Ontario's biophysical attributes (good climate and soils and cheap land) combined with advances in transportation technology and significant improvements in grower efficiency, worked to ensure that Canadian producers remained competitive.

Ontario's extensive, industry-wide regulation of grower-processor relations provides a constant backdrop to these outcomes. Defining its precise role is obviously problematic, although it seems reasonable to conclude that the certainty engendered by these structures has played a beneficial role in an otherwise turbulent period of change. Not only did this regulatory framework remain intact during the implementation of the CUSTA and the NAFTA, but in addition, the processing vegetable industry used it to meet and deal with more intense competition. The high level of regulation within the industry created a stable environment that enabled the growers' representative body, the OPVG, to negotiate changes to the pricing system that could deliver benefits to both growers and processors. Moreover, the growers, having long operated in a disciplined and controlled environment, were able to deliver on their commitments. They were able and willing to invest in the new technologies which could deliver improved yields and lower prices, whilst maintaining income levels over the longer term. In addition, the OPVG is clearly an organization that carries political influence, by virtue of the fact that it represents *all* processing vegetable growers, who can combine to exert pressure on governments when necessary.

Interestingly, evidence for the positive association between industry-wide regulation and sectoral growth comes from Heinz Canada. In October 2001 Heinz's Leamington facility benefited from a $4 million (CA) investment in a new production line, funded through a joint venture partnership comprising Heinz Canada, the Ontario Ministry of Agriculture, Food and Rural Affairs, and the Ontario Processing Vegetable Growers. Five months later, in March 2002, the plant received a further $3 million (CA) investment. These three-way initiatives (involving a transnational food company, the Provincial Government, and a growers' association) would be difficult to imagine in an industry without extensive statutory regulation. Heinz Canada, moreover, openly admitted as much when it noted that:

> A partnership of employees, management, Ontario tomato growers and the provincial government has played an important role in Leamington's growth and strong standing for the future. It is their collective contributions that have allowed Heinz Canada to expand our business prospects in North America and ultimately benefit the local economy of southwestern Ontario (Heinz Canada, 2002).

Undoubtedly, debate will continue on the future of Ontario's system of industry-wide regulation. However, in spite of recent industry restructuring it has remained intact, has been associated with industry growth, and (although criticized by some processors) is generally accommodated with relatively few tensions.

Which way to turn? Grower-processor regulation in California

The recent history of grower-processor regulation in Australia and Canada provides contrasting models of the ways in which two distinct production sites have responded to international competition in the processing tomato industry. Fundamentally, these two cases emphasize the argument that the analysis of grower-processor regulation should be sensitive to the nuances and particularities of an industry's geographical and socio-historical contexts. A 'one size fits all' set of prescriptions about grower-processor regulation will be likely to misinterpret these processes, with potentially adverse implications for industry outcomes.

Although these insights would seem to have application for the processing tomato industries of many production sites worldwide, their relevance to California, in particular, stands out. In recent years there has been considerable debate about Californian grower-processor regulation, especially in the context of arguments about the alleged merits of liberalization.

One of the leading advocates of liberalized grower-processor relations is Chris Rufer, CEO of the Morning Star Packing Company (Rufer, 1997; Rufer, Evans and Gashaw, 1999). As the principal of California's largest processing tomato firm, Rufer's views carry considerable weight within the industry. Hence, his views deserve close consideration.

According to Rufer, there is a natural evolution in trading systems towards arrangements where market arrangements are transparent and flexible:

> I think market pricing will eventually develop [in place of fixed price negotiations at an industry-wide level] because business drives to more efficient business models and flexible pricing throughout business has proven to work better than fixed pricing (cited in Tomato Land, 2002, p.8).

In the Californian processing tomato industry, Rufer identifies the system of industry-wide forward contracting as inhibiting market signals among participants, and thus encouraging inefficiency. This argument is defended through recourse to evidence on historical production and price trends within the industry, which suggests that when compared to either the fresh tomato sector or to other horticultural sectors, the Californian processing tomato industry has possessed significant short-term variability. Such variability is identified as a major impediment to cost reductions in the industry, because it encourages over-capitalization by processing firms, and problems of under- or over-production compared with market conditions (Rufer et al., 1999, p.ii).

According to Rufer, the reasons for production and price variability can be traced to the system of industry-wide forward contracting, because market conditions can and do change considerably in the period between raw tomato contracts being signed and processing firms disposing of product: 'the general principle… [is that] the earlier in the year that raw tomato prices are set, the less likely they will reflect final-product market conditions that prevail at harvest' (Rufer et al., 1999, p.6). Therefore, both growers and processing firms can be locked into contractual conditions that may bear little resemblance to the evolving conditions in processing tomato markets:

> Forward price contracts insulate growers' profits from sudden changes in the prices of finished tomato products. In some years, such as 1979, growers were protected from sudden drops in the price of tomato paste. In others, such as 1981, 1982, 1988 and 1989, fixed price contracts prevented growers from benefiting from an unanticipated rise in the price of tomato products. In 1998, tomato paste prices increased approximately 50 per cent between the time that the price for raw tomatoes was negotiated, and the time of harvest (Rufer et al., 1999, p.4).

To address these perceived problems, Rufer proposes that raw tomato prices be set in accordance with real-time, open market systems (Rufer et al., 1999, p.11). On the one hand, this implies a need to improve the transparency of industry information, and the establishment of new trading channels such as futures and options markets for raw tomatoes. In large part, these proposals are not controversial. However, on the other hand this also suggests a need to liberalize the industry, so that growers and processors are not obliged to use forward contracting as a means of defining intentions and securing product. This latter proposal is more controversial. In order to assess its significance, it is necessary to review the institutional framework of grower-processor relations in California.

Grower-processor relations in the Californian processing tomato industry have important differences from those in either Australia or Canada, because they operate under the principle of *cooperative industry bargaining*, which in essence provides a mechanism for industry-wide negotiations but without the forced, statutory basis of support that underpinned the Australian system until the early 1990s, and the Ontario system. An understanding of these principles is central to a wider analysis of the choices facing California.

Cooperative agricultural bargaining is the product of efforts by US legislators, over the course of more than one hundred years, to design statutes which both facilitate competitive marketplaces, and provide scope for individual farmers to negotiate fairly with (usually much larger) processing firms. These legislative aspirations are in continual evolution as political priorities change, and as the courts interpret legislative provisions in line with new cases. Furthermore, legal frameworks in this area comprise laws passed at both Federal and State levels, often creating a complex jurisdictional web.

Federal activities in this area originated with the *Sherman Act* of 1890, which declared illegal any attempt to restrain trade. Imprecise language in the *Sherman Act* however, led almost immediately to a need for the US Supreme Court to arbitrate on what comprised a 'trade restraint', which it did comprehensively in 1911 in Standard Oil Co. v. United States. According to the Supreme Court ruling, a 'rule of reason' needs to be applied to the *Sherman Act*, whereby trade restraints are deemed illegal only when they 'unreasonably' limit or monopolize commerce (Frederick, 1993, p.437). Included within this are activities such as price-fixing, which were labelled illegal *per se*.

This decision impacted directly on the ability of farmers to negotiate collectively over prices. If price-fixing was defined contrary *per se* to the *Sherman Act*, the collective actions of farmer groups could be construed as illegal. These implications led the US Congress in 1914 to pass the *Clayton Act*, which exempted agricultural organizations (along with labor unions) from antitrust scrutiny. The intent to uphold the ability of farmers to negotiate collectively was further enshrined with the *Capper-Volstead Act* of 1922, which has subsequently become the anchor for US agricultural policies in this field. Under the Act, farmers organized as a 'cooperative' (effectively defined as a non-profit organization representing more than half of an industry's output where voting rights are more-or-less 'one-member, one-vote') were shielded from antitrust prosecution, so long as their actions were not predatory. Accordingly, under the *Capper-Volstead Act*:

> producer-members of a cooperative may agree among themselves not only on the prices they will receive for their products but also on reasonable terms of sale (Frederick, 1993, p.440).

Notwithstanding the statutory authority of these provisions, farmers' rights to fair collective bargaining were found wanting under the *Capper-Volstead Act*. In the early 1960s, farmer cooperatives in a range of industries lobbied for improved legislation on the grounds that powerful processor interests allegedly discriminated against collectivized suppliers, *vis-à-vis* non-collective ones. Interestingly, the processing tomato industry provided one of the key disputes that prompted reconsideration of the *Capper-Volstead Act*. In 1959, Ohio tomato growers filed a complaint with the Federal Trade Commission (FTC) after processing firms refused to purchase fruit from members of the growers' association. In 1964 the FTC dismissed the growers' complaint, ruling that it could not arbitrate on this dispute. When, following this decision, the Ohio legislature enacted laws in 1965 protecting the rights of grower associations, growers lobbied for a national legislative umbrella (Frederick, 1990, p.681). After extensive debate in Congress, President Johnson signed the *Agricultural Fair Practices Act* (AFPA) into law in 1967. Effectively, the AFPA maintained key provisions of the *Capper-Volstead Act*, and added to them legislative prohibitions on the ability of buyers of agricultural products to discriminate against individual farmers or groups of farmers on grounds of price, quantity, quality or whether they were members of an agricultural marketing organization (Frederick, 1990, p.684).

In an analysis of the AFPA published in 1990, Frederick argued that the legislation had only partially achieved the goal of enshrining farmers' rights to collectively bargain. On the one hand, the AFPA seemed to limit unfair practices by processors. On the other hand however, the AFPA's powers were generally weak. Under the 'Disclaimer clause' of the Act, processors were not obligated to negotiate with grower associations and retained powers to choose which suppliers they used, so long as the provisions of non-discrimination were met. According to Frederick:

> The phrase [in the disclaimer clause] stating processors and other handlers are not required to deal with associations of producers gives processors justification to totally disregard a producer [farmer] association or to go through the motions of bargaining and then, as planting time, harvest, or some other critical period nears when the producers are under the greatest pressure, walk away from the table and offer growers take-it-or-leave-it contracts. Efforts to develop effective producer bargaining associations are also undermined by the language permitting handlers to refuse to do business with a producer for any reason other than membership in an association. The threat of reprisals against elected leaders, disguised as legitimate reasons to refuse to deal, can be a forceful weapon for someone who wishes to discourage association activity (Frederick, 1990, pp.690-91).

These limitations in the AFPA (from the perspective of farmer associations, at least) have encouraged a number of Congressional attempts to amend the legislation over a period of over thirty years. However, opposition from processor firms has effectively killed proposed amendments. Most recently, an attempt to amend the AFPA was initiated in 2001 by Representative Marcy Kaptur of Ohio, the ranking Democrat on the House of Representatives Agriculture Appropriations Committee. In January 2001, Kaptur introduced the *Family Farming Cooperative Marketing Bill* (HR 230) and then, in October of the same year, re-introduced the legislation as an amendment to the proposed *Farm Security Act*. Kaptur explained her motivations in introducing this legislation in terms of her belief that it was necessary to:

> restore farmers' separate and equal station in agriculture today. Most farmers aren't negotiating with their buyers. The farmer is faced with a 'take it or leave it' situation. Some contracts contain confidentiality clauses that prohibit farmers from revealing their terms (Kaptur, 2001, p.3).

The proposed legislation would amend the AFPA in four key ways:

(i) require the Secretary of Agriculture to establish a system to accredit voluntary cooperative associations of agricultural producers;
(ii) provide for good faith bargaining between processors or handlers and cooperative associations of agricultural producers;
(iii) allow for mediation by USDA to resolve impasses in bargaining; and

(iv) provide investigative and enforcement authority for the Secretary of Agriculture.

The requirement for parties to engage in 'good faith' bargaining is the central element of the proposed amendments. Accompanied by legislative sanctions allowing the Secretary of Agriculture to intervene where 'good faith' is not shown, the legislation would underwrite a significant shift in the character of farmer-processor relations in US agriculture. To the time of writing, the future of these proposed amendments remains unresolved, with Kaptur agreeing in October 2001 to withdraw them in exchange for Hearings to be held on these matters by the House Agriculture Appropriations Committee.

The Federal laws enshrined in the *Capper-Volstead Act* and the AFPA provide the overarching legislative framework governing grower-processor relations in the US processing tomato industry. However, they are augmented by requirements enacted by State legislatures. In California, provisions in the California *Food and Agriculture Code* govern grower-processor relations in the processing tomato industry. These structures provide the framework for an industry-wide system of cooperative bargaining. The broad objectives of the Californian regime are to marry principles of 'fair bargaining' with the competitive needs of processing firms. At the crux of the system is an annual 'Master Agreement' that sets out the terms and conditions for commercial relations between growers and processors.

The basis of law for the Master Agreement resides with the status of the California Tomato Growers' Association (CTGA) as an authorized third party agent. As a third party agent, the CTGA is empowered to negotiate on behalf of its members. The Master Agreement therefore, has the effect of a commercial contract between the CTGA membership on the one hand, and individual processors on the other. As such, the Agreement treats the growers and processors as equals; although an individual processor may deal with dozens of separate growers, the terms and conditions of those relations are formalized in one agreement. From the growers' perspective, this framework addresses the need for countervailing powers, namely, the fact that in a commercial relationship between one large processor and dozens of relatively smaller growers, economic power usually lies with the larger party. Furthermore, the Master Agreement supports industry-wide initiatives such as independent grading. Unlike the Australian situation, factory-gate tomato grading is undertaken by an independent agency, according to agreed industry protocols.

As evidenced by the comments of Chris Rufer, cited above, there are differences of opinion as to the effectiveness of these arrangements. Although the Californian system allows for some price flexibility (that is, there is no statutory fixed price within the industry), some processors oppose the principle by which they are required to treat all growers similarly. In contrast, grower organizations defend the current system on grounds of equity and efficiency. Notwithstanding these positions however, the Californian system needs to be understood as lying between the neo-liberal Australian model, and the statutory rules-based Canadian model. On the one hand it gives third party authority to the CTGA, but on the other it does not specifically prescribe the pricing and other conditions within tomato contracts. The future course taken by the Californian industry will depend on how these apparently differing objectives are reconciled, in the context of emerging pressures of international competition and restructuring.

The internationalization of grower representation

Although processing tomato growers in Australia, Canada and the United States face considerably different institutional contexts for grower-processor relations, they also possess common problems in terms of having to negotiate with processing firms, which over time, are increasing in size and geographical scope. In this regard, it is noteworthy that grower organizations have sought to defend their positions through international strategies designed to disseminate knowledge and to provide a forum for the discussion of matters of common concern. For over thirty years, tomato growers from California, the US Mid-West and Canada have met at the North American Tomato Conference, where issues under discussion have included market supply and demand, new seed varieties, pesticides and a variety of topics that affect the North American industry and its growers (OVGMB, 1996, pp.1-2). More generally, Canadian processing vegetable growers have also participated in the Pacific Coast Bargaining Conference, a meeting organized by US and Canadian growers at which vegetable growers' organizations exchange information about bargaining experiences across a wide range of commodities. This conference is mainly concerned with horticultural crops but has included discussions on commodities such as poultry and catfish.

In 1996, such organizational initiatives by growers were extended beyond North America by the creation of the Tomato Growers Alliance, which involved the creation of a loose association of grower groups from Canada (the OPVG), Australia (the Processing Tomato Growers Section of the Victorian Farmers' Federation) and California (the California Tomato Growers Association). In July 1996, these organizations signed a memorandum of understanding which, following a preamble which pointed to the significance of the globalization of the processing tomato industry through multinational corporations, agreed to:

(i) promote the interests of processing tomato growers in the international market arena;
(ii) promote the exchange of information and data between growers and organizations in relation to the tomato processing industry;
(iii) promote the exchange of information as it relates to market outlook reporting, price discovery, contractual arrangements and legislative and trade activities, and
(iv) promote the exchange of technical information as it may relate to the sustainable future of processing tomato growers in Australia, the US and Canada (OPVG, 1996, p.2).

Members of the three organizations exchange technical and commercial information, and come together at meetings of the World Processing Tomato

Congress. However, there are no formal agreements between participants on issues of production and prices, and the Alliance does not attempt to generate grower solidarity across different production sites as a way of reducing the capacity of processing companies to play one group off against another. However, while the Alliance largely serves as a clearinghouse for information, its importance should not be under-estimated in an environment in which some major processors flexibly source processing tomato products (notably, aseptically-packaged bulk tomato paste) from multiple sites across the world. Nor should the possibility be dismissed that growers, like the companies with which they contract, will come to organize globally. Indeed, attempts by growers to enhance their bargaining power recently led North American processing tomato grower organizations to participate in the so-called CLOUT (Cooperation and Linking Offer Unbeatable Transactions) Conference, held at the University of Iowa, Ames, in 2001. The conference involved a large number of North American farmer groups, covering a broad range of commodities, who wanted to express their concern at the prevailing marketing system and the low prices it delivered to producers. At this conference, the experience of the OPVG and Canada's system of regulated marketing was the subject of a presentation by John Lugtigheid, a past chairman of the OPVG and a tomato grower himself.

Conclusion

The commercial relations that enable tomatoes to pass from farm to factory represent a highly contentious field of activity. Negotiations between grower and processor groups can consume enormous amounts of energy, with major ramifications for industry competitiveness, growth and sustainability. The comparative analysis in this chapter of grower-processor bargaining in the Australian and Canadian processing tomato industries, illustrates how regulatory arrangements, organized at national and sub-national scales, continue to exert strong influence over these activities. Moreover, the case study material in this chapter supports the central argument of this book, namely, that processes of global agri-food restructuring are heterogenous and fragmented, bounded in multiple ways by the separations of geography, culture, capital and knowledge.

At a more specific level, several key conclusions arise from the analysis presented in this chapter. First, the evidence presented here challenges those accounts of global agri-food restructuring that posit the supposed *inevitability* of the liberalization of grower-processor bargaining relations. Seen through the positivist lens of orthodox agricultural economics, government activities in this area may appear as 'interventions' which represent an impediment to market efficiency. However, detailed empirical investigation into the social construction of regulatory arrangements reveals a more complex reality, in which the subtle effects of history and geography influence the relationships between regulation, efficiency and competitiveness.

Second, growers must be understood as possessing agency. With intensified global competition and the continued rise of large, transnational corporations, it may be tempting to discount or ignore the possibilities for growers to actively shape bargaining relations. However, the recent experiences of both Australia and Canada suggest otherwise, notwithstanding the wider constraints of the global market. In Australia, despite the limitations of enforced individualism, growers rejected the Cedenco 'grower alliance' proposed in the late 1990s, and moved towards multi-year contracts in 2002. In Canada, grower support for the OPVG has underwritten industry-wide regulation. The creation of international grower initiatives, described above, points to further evidence of grower agency.

Finally, the analysis of this chapter emphasizes the importance of regulatory forms for the social outcomes of restructuring. Comparing Canada and Australia over the 1990s, it is apparent that the processing tomato industries of both nations experienced substantial declines in grower numbers, although in Australia the fall was sharper. While there is no unequivocal way to assess how regulatory changes were implicated in these outcomes, it is not unreasonable to suggest that the Canadian model has played a positive role in maintaining grower numbers, whereas the greater uncertainties associated with the Australian model accelerated grower departures from the industry.

Notes

1. By the end of the 1990s however, there had been a considerable easing in the level of import penetration for retail canned tomatoes.
2. In 1999-2000, retail peeled tomatoes represented nine per cent of the processing tomato crop; in 2001-01, they represented 14 per cent (Horn, 2002, p.15).
3. The New South Wales legislation was written to support the Victorian industry structures so that, for example, the statutory prices received by New South Wales growers was defined as to be determined by the Victorian TPINC. For this reason, the discussion in this chapter concentrates on Victorian legislative changes.
4. Because this was a private correspondence, the author of this letter is not disclosed.
5. Production contracts were provided to the authors by a number of growers, on the condition that specific names of processing firms and their contracted growers would remain anonymous.

Plate 4 Processing tomato companies in San Antonio Abate, Italy

Photo: Michael Kay, reproduced with permission.

Chapter 5

The distinctly European tomato

The hand of Government, in one form or another, is evident in every tomato production region in the world. The Californian processing tomato industry, for example, has relied heavily on state interventions in the form of irrigation infrastructure development, water pricing, highway construction and the activities of the land-grant universities. The emergence of the processing tomato sector in Xinjiang Province, China, has depended on extensive state interventions for reasons of regional geopolitics, whilst in Thailand, Government subsidies to the tomato industry were made available as a way of alleviating poverty (see Chapter Six).

Yet in Europe, national governments and the supra-national European Union (EU) have played a more extensive role in the processing tomato industry than any other major producing region. Since 1978, the European processing tomato sector has been the recipient of significant direct and indirect industry support, and a complex array of rules and regulations has been developed to administer this support. Through the 1990s, the future of the EU's tomato support regime intersected with the multilateral politics of world trade reform and pan-European agri-politics, and although the EU substantially restructured its processing tomato policies in 2000, this has not significantly reduced overall levels of support.

This chapter analyzes how these policies have helped to construct specific and distinctive patterns of capital accumulation and restructuring in the European processing tomato industry. Contemporary changes in this sector are symptomatic of tensions between the impetus for, and limitations of, the globalization of economy and culture. The 'distinctly European tomato', outlined in this chapter, is the result of the separations of geography, culture, capital and knowledge that shape contemporary global agri-food restructuring.

To elaborate upon these issues, this chapter details key elements in the political economy of the EU processing tomato sector. First, EU processing tomato subsidy arrangements are placed in their world-historical contexts. These policies need to be understood as a strategic intervention by the state within the wider frames of rural development policy, agricultural trade negotiation and European politics. The chapter then examines the detail of the 1978 EU processing tomato arrangements and the decision to restructure this support in 2000. Finally, the implications of current EU regulations for this industry are assessed. Restructuring of the EU processing tomato regime reveals an uneven picture of winners and losers; specified in terms of regions, sizes of firm, and firm strategies. To summarize the key argument of this chapter, an understanding of the European processing tomato industry requires an appreciation of the ways it is embedded within various national and regional spaces, in terms of regulatory arrangements, commercial relations, and consumption environments. The analysis of these issues builds upon the basic theme of this book, namely the fractured and uneven character of global agri-food restructuring.

The European processing tomato industry in world-historical context

European Union regulation of the processing tomato sector exemplifies a particular manifestation of late-twentieth century European agri-politics. Over the past half-century, the EU has attempted to maintain its rural landscapes through a large, expensive and complex institutional structure of agricultural support payments. The direct and indirect costs and benefits of maintaining these structures are widely debated, and have formed a prominent 'sticking point' within recent multilateral agricultural trade negotiations under the auspices of the General Agreement on Tariffs and Trade (GATT) and its successor, the World Trade Organization (WTO).

In a world of globalizing capital and freer trade, with its concomitant Ricardian promise of improved living standards, the maintenance of Europe's social and economic structures, as well as its rural landscape, becomes problematic. Globalization renders economic costs more visible, and in recent years policy debates at national, supra-national and multilateral scales have increasingly been influenced by neo-liberal economic discourse. Whereas the world would be poorer place without the French village market, the Dutch cheese-maker, or the Spanish olive grove, expressing the costs and benefits of these social and cultural assets in economic terms can be done only with great difficulty and imprecision (indeed, some would argue they *cannot* be measured using economic concepts). Furthermore, European cultures and traditions are not static, but are shaped by an ongoing interplay of processes occurring at different scales. Whereas many European consumers maintain strong loyalties towards local food traditions, large supermarket chains and franchised fast food outlets now dot Europe's consumer landscape, encouraging European and imported foods to be insinuated into consumption arenas in new ways, and challenging traditional culinary cultures.

Market theory, which counterpoints 'state regulation' in markets against 'free market' alternatives, says little about why, at specific moments in history, industries are regulated and develop in particular ways. An answer to this question requires the deployment of agrarian political economy, in particular the concept of *food regimes* which, as the material in this chapter indicates, provides a framework from which to observe and assess the historical significance of contemporary European agricultural restructuring.

This concept suggests that the past century has seen two extended periods of relative stability in the global agri-food system. The first of these (dubbed the *first food regime*) coincided with the Colonial era before World War I. With the passage of Britain's 'corn laws' in the mid-nineteenth century, fundamental shifts occurred in world agricultures. By opening the British market to agricultural imports, the interests of domestic agricultural producers were sacrificed for the goal of cheaper food. Rapidly rising urban populations employed in manufacturing industries made the provision of cheap food a British national priority. Britain's evolution as 'workshop to the world' was mirrored in deepening food import dependence, and political control over colonies producing food and raw materials was vital to the economic wellbeing of the motherland.

New agri-technologies in the 1920s and 1930s presaged key shifts in the global food economy, and created the basis for an *intensive* regime of agri-food production dependent upon plant genetics, nitrogen-based fertilisers and farm machinery (Goodman and Redclift, 1991, pp.95–100). President Roosevelt's New Deal responses to the rural crises of the Great Depression accelerated direct state interventions in agriculture through orderly marketing arrangements (du Puis, 1993). These responses were emulated widely outside the United States. The creation of orderly markets enshrined price floors and state disposal of surplus production, placing agricultural producers within an institutional treadmill whereby they were less involved in decisions over the conditions by which their output was sold. These structures, in combination with the post-1945 *Pax Americana* and the collapse of the colonial regimes of the European powers, forged an international agricultural regulatory order that Friedmann and McMichael (1989) label the *second food regime*.

The longevity of the second food regime was assisted by the fact that agriculture was excluded from the mainstream of multilateral trade negotiations from the 1940s to the 1980s. The 1947 formation of the General Agreement on Tariffs and Trade (GATT) did not cover agriculture, since too many vested interests were affected by its inclusion. Hence, whereas trade barriers for manufacturing fell consistently through the post-War decades, similar outcomes were not apparent for agricultural trade. Instead, international trade in agriculture was organized in the context of extensive 'orderly marketing' frameworks at national scales.

By 1965, concerns expressed by developing nation exporters about market access and the terms of trade (especially with respect to agriculture) led to the creation of the United Nations Conference on Trade and Development (UNCTAD) (Dicken, 1998, p.95). This facilitated the creation of the General System of Preferences (GSP) for developing nations' exports, but did little to implement change within the developed nation economies of Europe and North America. Indeed during the 1960s and 1970s the European Common Agricultural Policy (CAP) expanded, with adverse implications for developing country agricultural exporters. Agricultural issues were discussed in the Kennedy (1962-67) and Tokyo (1973-79) GATT Rounds, but these discussions were on the understanding that outcomes were exempted from GATT disciplines and hence subject only to separate, loosely-framed agreements (Coote and LeQuesne, 1996, p.114). Step-by-step however, these developments encouraged the incorporation of agriculture within GATT's multilateral framework, which was achieved with the Uruguay Round of the GATT beginning in 1986.

The Uruguay Round commenced at an important juncture for global agriculture. State interventions in agricultural markets, an abiding feature of the second food regime, were escalating rapidly. Nominal rates of OECD agricultural protection rose from 40 per cent in 1979-81, to 68 per cent in 1986-88 (Roberts et al., 2001, p.1). The stabilizing mechanisms of the second food regime (i.e., orderly market interventions) were becoming key agents in the global destabilization of agriculture.

Through the Uruguay Round negotiations governments attempted to establish new structures for the global regulation of agriculture. After seven years of negotiations, the Uruguay Round culminated in the World Trade Organization Agreement on Agriculture (WTOAA), signed by 114 national participants in Marrakech, Morocco, in 1994. A key outcome from the Agreement involved the codification and limitation of market access barriers for agriculture. Under the Agreement, non-tariff barriers (such as quotas) were to be expressed as tariff equivalents and reduced progressively. Developed nations were obliged to reduce market access barriers by 36 per cent and by at least 15 per cent for any one item between 1995-2000 (based on 1986-88 baselines). Developing nations were obliged to reduce market access barriers by two-thirds of these amounts, over the period to 2004 (Podbury and Roberts, 1999, p.13). In the WTOAA, the global economy had, for the first time, entertained the prospect of a multilateral framework for agricultural trade.

The WTOAA represents a historically significant intervention in world agricultural trade, although its quantitative impact remains modest. Because the baseline period used to measure tariff-quota reductions was a high point in agricultural protectionism, much of the required action by member nations was minimal. Indeed, by the time the WTOAA had been ratified, a number of nations had already met WTOAA-mandated requirements. Even ardent advocates of freer agricultural trade expressed the view that the WTOAA did not result in significant change:

> In the light of the long history of agricultural protection growth in industrial countries, even achieving a standstill in agricultural protection via the Uruguay Round could have to be described as progress … As it turned out, though, only a little more than a standstill was agreed to (Anderson, 1998, p.25).

Nevertheless, the WTOAA's relatively modest outcomes (measured in quantitative terms) have provided the foundations for significant qualitative change in the scope and integration of global agriculture. Through agriculture's inclusion in multilateral trade negotiations, an array of factors central to the production and consumption of foods – rural support policies, food aid and security, quarantine policies, food labeling, food standards, land reform – are inserted into policy debates as potential 'barrier to trade'. Governments still retain wide powers to intervene on behalf of national agriculture on these (and other) matters, but interventions need to be consistent with provisions of the WTOAA. For example, food import restrictions based on quarantine concerns need to be justified and able to be publicly defended on rigorous scientific grounds, within WTO expert panels. This places the onus of regulation on its proponents; free trade must not be curtailed *except* within the provisions available through the WTOAA.

These changes execute qualitative shifts in the scope of national decision-making over agriculture. They benefit actors such as large food companies, traders and globally competitive agri-exporters, who are able to take advantage of greater openness and certainty within the global agri-industrial system. Hence, the significance of the WTOAA cannot be considered solely within quantitative debates about its effects on mandated maximum subsidy levels. By bringing agriculture under the umbrella of the multilateral trading environment, it encourages structural transformation within this sector. Thus, the WTOAA cannot be considered in isolation, but has to be seen as an essential component of an emerging global food system, involving at its core liberalized capital flows, globally integrative strategies by large corporations, and enabling information technologies that expedite international knowledge and commodity exchange. These perspectives provide the context for the observation and assessment of contemporary debates on the future of the EU agricultural support, including the processing tomato sector.

The view from Brussels: recent reforms to European Union agricultural regulation

The political project of European unification provides the broad context for EU agricultural policies. The formation of the EU was underpinned by the explicitly *political* imperative to create a sovereign superstructure that would negate armed conflict in Western Europe. In the wake of the Cold War, *economic* imperatives were accorded primacy.

European economic integration has resided within two inter-related processes of economic governance. First, integration has required the establishment of processes and institutions that lead the separate nations of Europe to act as if they were one. Under the 1992 Maastricht Agreement, EU members have been required to adopt uniform commercial regulatory arrangements, mutual standards recognition, free labor mobility, and to cede certain functions to the European Commission. Second, the principles of European economic integration have depended upon the implementation of economic policies which aim to address social and economic problems at a European scale. This has meant the establishment of support payments to disadvantaged regions as European-wide strategies designed to achieve social justice. This has resulted in significant monies being allocated to European agriculture under broadly defined terms of 'regional development'.

The allocation of financial resources to European agriculture is undertaken via the Common Agricultural Policy (CAP), an overarching framework for pan-European agricultural management. By the mid-1980s the CAP had developed into a massive funnel producing large agricultural production surpluses (Le Heron, 1993). Sectoral policies underpinning these payments were often opaque in terms of logic and accountability, and appeared to be driven largely by the need for countries with sizeable rural populations (in particular, France) to meet the economic demands of their large rural constituencies. Over time, growing pressures were placed on the EU to reform these arrangements, both by European taxpayers and external parties such as the Cairns Group of agricultural exporter nations.

The commencement of the Uruguay Round in 1986 established a window through which external pressure could be applied to EU agricultural policy. Led by the Australian Bureau of Agricultural and Resource Economics (ABARE), an armoury of economic data and modelling was developed, illustrating the alleged economically injurious effects of the CAP. These arguments were given expression through the Blair House Accord of the Uruguay Round and the WTOAA, and contributed to the restructuring of the CAP in the 1990s. Although Brussels continues to provide high levels of support for agriculture, the EU has attempted to improve the CAP's accountability, transparency and efficiency, as well as placing a curb on its massive cost. These aims were brought together within the EU's *Agenda 2000: For a Stronger and Wider Union* (European Commission, 2000) conceived in 1995. Following a series of negotiations and refinements over a period of four years, *Agenda 2000* was ratified in 1999 at a meeting of the European Council in Berlin.

In so far as agriculture is concerned, *Agenda 2000* treads a fine line between restructuring and maintaining the CAP. Structural reform to the CAP, especially the reduction of budgetary outlays for agricultural support, is a cornerstone of *Agenda 2000*. At the same time, the document reaffirms the EU's need to maintain the policy of economic and social cohesion, thus providing a commitment to ongoing funding in the name of regional and social justice. European agricultural policy is caught at the current time between these two opposing objectives.

Pressures to reform the CAP in conjunction with a continuing commitment to allocate significant resources for rural social justice have encouraged structural change in the operation of EU agricultural programs, and the creation of new concepts and discourses with which to position and justify them. In defence of their programs, European policymakers and trade negotiators now talk of the 'multifunctionality of agriculture' (MFA), which argues that outputs from agriculture include not production alone, but environmental and social amenity, and the maintenance of tradition and heritage. According to the EU's Director-General of Trade:

> Agriculture, apart from its production function, encompasses also other functions such as the preservation, the management and enhancement of the rural landscape, the protection of the environment, and a contribution to the viability of rural areas (EU Director-General of Trade, 1999, p.19).

The EU has argued that these 'spillover benefits' would be endangered by intensified international competition. Such competition could lead to the abandonment of some farm activities in Europe, or a shift to productivist agriculture involving fewer and larger farm enterprises. Deployment of the MFA transforms EU agricultural policies, and shifts the focus from a pre-eminent concern with food production and continental self-sufficiency, to a primary concern with the maintenance of European rurality.

These arguments are expressed in terms of the post-productivist transition (PPT) in European agriculture, which implies the political re-evaluation of agriculture within Europe. Since the early 1990s, European agricultural support mechanisms have been restructured to give lesser priority to quantitative output, and greater priority to the maintenance and promotion of the social and natural agricultural environment (Ilbery and Kneafsey, 1999; Lowe et al., 2002).

In a general conceptual sense, economists accept arguments that agricultural practices may have positive or negative spillovers (also known as externalities) and that there may be legitimate reasons for Governments to address these. The EU's deployment of the MFA to justify existing subsidies however, is seen as opportunist. One of the most vociferous critics of the MFA has been ABARE, broadly representing the interests of the Cairns Group nations supporting freer agricultural trade. ABARE suggests two faults with the MFA. First, addressing externalities through subsidies to agricultural production is seen as indirect and therefore 'messy'. ABARE argues that positive spillovers of European agriculture, such as maintenance of rural culture and the rural landscape, would be addressed more effectively through direct funding rather than via the indirect support of farm production. It uses the example of hedgerow protection in the UK to make its point. EU policymakers identify the protection of hedgerows as a key activity in maintaining a rural aesthetic in the UK countryside. However, ABARE disputes that this objective is best achieved through the general subsidization of UK farming. It argues that paying farmers directly for maintaining hedgerows (i.e., targeted and direct assistance) generates superior outcomes compared with subsidizing farm commodity production generally (with the implied assumed outcome that subsidized farmers will continue to protect hedgerows) (Freeman and Roberts, 1999). Second, ABARE argues that subsidizing agriculture may also generate negative externalities (such as environmental damage) because it encourages over-production. In July 2002, as part of the 'Mid-Term Review' of *Agenda 2000* arrangements, the EU addressed such issues, in part at least, by the *de-coupling* of farm assistance from production levels.

ABARE also contends that EU agricultural subsidies contradict the stated aims of policy which are used to justify providing assistance to small farmers. ABARE calculates that in 1996, 50 per cent of EU agricultural support payments were received by only 17 per cent largest farms, and that these were within the largest farm size. Thus, 'half the agricultural support in the EU results in a wealth transfer from EU citizens to the minority group of farmers with [higher than average] incomes' (Podbury, 2000, p.3). This occurs because much CAP funding is allocated on the basis of unit of output, meaning that the distribution of EU farm subsidies is more-or-less proportional to the distribution of farm size. In an environment in which farm output is skewed towards larger producers, larger producers clearly receive higher gross subsidy amounts than small farmers. However, the question of whether the CAP *further* skews farm-size distribution in Europe is not addressed by ABARE. The reality may be that although small farms receive lesser gross subsidy support than larger farms, the receipt of these funds may be critically important for their survival. The issue here relates to the impacts of subsidies on farm enterprises' marginal costs and revenues. Despite ABARE's implicit assertions to the contrary, the CAP may play a key role for small farm survival, notwithstanding the fact that larger farms may receive the bulk of subsidy payments.

In summary, advocates of agricultural trade liberalization contend that the MFA is a rationalizing vehicle for protectionism in EU agriculture. For the Australian Agriculture Minister, the MFA is conceived of as encouraging 'a Disneyland amusement park role for farming' (Truss, 1999). Many Europeans, however, do not share these views. Though subsidy reductions within EU agriculture have been marginal, according to well-publicized forward timetables, they have generated entrenched opposition within farming communities. In addition, opposition to reduced subsidies in some situations has combined with a more general opposition to liberalization, globalization and (particularly American) transnational food companies. Jose Bove, a farmer from the tiny village of St Pierre-de-Trivisy in France, gave substance to these arguments by driving his tractor through the plate-glass windows of the local McDonald's store. Through this action Bove became a figurehead for anti-globalization protesters in Europe and elsewhere. His book *Monde n'est pas une Marchandise* (*The World is Not For Sale*) (2001) has become a call-to-arms for new social movements based around the production and consumption of food. Bove's calls for the protection of local distinctiveness have gained currency because they echo broader socio-cultural trends in Europe. The slow food movement and organic agriculture have become important consumption niches within the European foodscape. These trends have been given further sustenance by the emergence of Bovine Spongiform Encephalopathy (BSE), the foot and mouth crisis of 2001, and continued consumer skepticism and distrust over genetically modified organisms. The long-term implications of these political and agri-ecological events now shape a commitment for further restructuring of the CAP. Whereas some continuing reform of the CAP is inevitable, it is not clear at all whether these policies will lead Europe down a free market path.

Recent restructuring of subsidy payments and regulations covering the EU processing tomato sector typify current debates on the future of European agriculture. Until 2001, the EU processing tomato industry was situated within an administratively complex web of minimum price frameworks, tariffs, export subsidies, payments to processors, and quotas. Sweeping changes to these arrangements for the 2001 season underscore a qualitative shift to the EU's role in this industry. The knock-on effects of these regulatory changes will cast a shadow over the industry for years to come.

European Union processing tomato policy

Over the late 1990s, between eight and nine million tonnes of processing tomatoes were grown annually in the EU. This constituted approximately one-third of world output. Processing tomatoes are grown in five EU member states (Italy, Spain, Greece, Portugal and France), with Italy routinely accounting for over half of all EU output (Table 5.1, Figure 5.1).

There are three elements to EU processing tomato policy. First, tariffs are levied on processing tomato products imported into Europe, although in accordance with the WTOAA, these are being reduced progressively. Second, an export restitution scheme operates to allocate funds to EU processing tomato exporters. Third, the EU administers an extensive domestic support program that has had the aim of maintaining farm gate tomato prices at levels well above those indicated by prices paid elsewhere in the developed world. Of these three elements, the latter is the most important. It has a significant budgetary impact on the EU: in 2000, the domestic support program cost Euro 300 million, or $270 million (US), and plays a critical role in the regional geography of EU processing tomato production. For reasons discussed below, domestic support policies have tended to discourage new investment in the industry, and helped keep viable a large number of small and relatively inefficient factories.

The EU's reliance on domestic support payments as a means of assisting the sector dates from 1978. The development of these arrangements was the outcome of a compromise between northern European (importer) nations and southern European (producer) nations. In the 1970s, calls by Italy for the provision of processing tomato subsidies were contested by northern EU nations, especially the United Kingdom, which was reluctant to encourage measures that would have price-increasing effects. Consequently, the EU formulated a policy framework dependent mainly on (price-depressing) domestic support payments rather than (price-increasing) tariffs. The total policy package was developed with the view that payments should be sufficient to enable local producers to meet Californian competition.

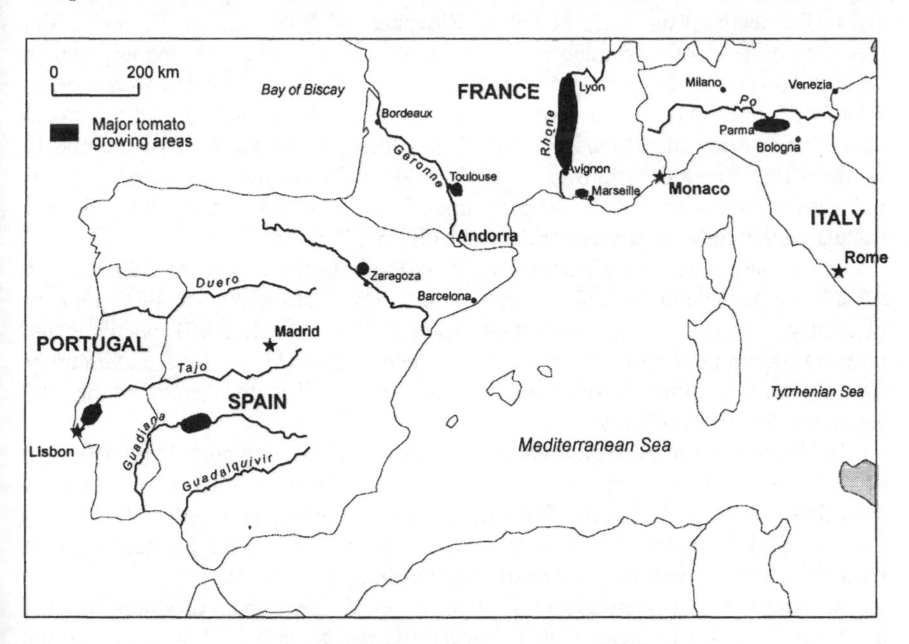

Figure 5.1 Major processing tomato regions, southwest Europe

Table 5.1 Estimated EU processing tomato output by major growing region, 1999 (tonnes, raw product equivalent)

Country	Region	Tomato paste	Canned tomatoes	Other	Total tonnage
Italy	North	2,160	0	250	2,410
	South	240	2,110	250	2,600
Spain	Extramadura	940	90	25	1,055
	Other areas	150	150	150	450
Portugal	Tajo River	955	8	32	995
Greece	Various	1,180	40	23	1,243
France	Various	280	55	30	365
Total		5,905	2,453	760	9,118

Note: Data for sub-national regions are the authors' estimates based on published information and industry sources.

Source: AMITOM (2000).

European Union export subsidies and tariffs for processing tomatoes

In general, the EU provides only limited *export subsidy* arrangements for the processing tomato sector. Under current regulations, subsidies of between Euro 45-50 per tonne for canned tomato exports are provided (excluding exports to the US and to the neighboring states of Latvia, Slovakia and Bulgaria). In the year 2000, these subsidies were available for a maximum of 135,000 tonnes, which constituted roughly 54 per cent of the EU's total exports of canned tomatoes. A subsidy of Euro 50 per tonne represents the approximate equivalent of 8.5 per cent of the f.o.b. price of a tonne of canned tomatoes leaving the Port of Salerno in southern Italy (according to Sumner et al., 2001, p.36, the average export price of EU canned tomatoes is Euro 530 per tonne). In the year to February 2001, the EU calculated that these subsidies would cost of Euro 6.3 million.

Export subsidies have a minor impact on the industry as a whole. Since these subsidies are available for only 135,000 tonnes, they impact upon just 1.6 per cent of total processing tomato production. Sumner et al. (2001, p.40) calculate that removal of this policy would reduce EU canned tomato exports by a maximum of four per cent. In other words, this is a policy intervention that generates minimal assistance for EU producers.

Tariffs provide a more substantial element of industry support. Until 1995, an *ad valorem* 18 per cent tariff applied to processing tomato imports, although in accordance with the WTOAA, this tariff has been progressively reduced. By early 2002, it had been reduced to 13.4 per cent with provisions for 0.5 per cent ongoing reductions each six months until eventually reaching 11 per cent.

Because this tariff increases local prices of processing tomato products, Sumner et al. (2001, p.47) estimate that it reduces the net welfare of Europeans by the equivalent of Euro five million annually. The elimination of the tariff would increase EU processing tomato imports by 108 per cent (raising tomato paste import penetration rates from approximately 10 per cent to 21 per cent), and lower EU domestic production by five per cent. These estimates however, are highly sensitive to assumptions concerning the level of processing tomato imports, and the uses to which imports are put. According to official statistics, in 2000 approximately 0.6 million tonnes of processing tomatoes (raw product equivalent) were imported into the EU. However, not all of these imports were consumed within the EU, because duty drawback schemes allow for the tariff-free import of processing tomato products subject to their re-export. Tomato processing companies in southern Italy, in particular, have developed considerable commercial activity through the re-processing of imported tomato products (mainly paste) from sources such as Turkey and, more recently, China. However, these re-processing operations sometimes take the form of simple repackaging of paste from drums to glass jars.

The problem with estimating the impact of the tariff is that it is virtually impossible to obtain reliable hard data on the volume of imported product destined for re-export. Because of the opaqueness of production statistics in southern Italy (discussed below), Sumner et al. (2001) were forced to make a 'best guess' estimate of the proportion of imports that were subsequently re-exported. They suggested at least 20 per cent of processing tomato imports entered the EU duty-free for re-export (Sumner et al., 2001, p.45). This assumption gives rise to the trade and production flows model presented in Figure 5.2. In volume terms this indicates that of the 0.6 million tonnes of raw product equivalent imported into Europe (boxes 'b' and 'c'), 0.12 million tonnes is re-exported.

Interviews with industry informants suggest that this estimate understates the proportion of imports which are re-exported. According to industry sources, in the year 2000, between 50,000 and 60,000 tonnes of Chinese tomato paste (at 28 brix) were imported to southern Italy, and used entirely for re-export. Moreover, these purchases were supplemented by an estimated 10,000 tonnes of paste purchased from non-EU Mediterranean nations. If it is assumed that 60,000 tonnes of paste (from all sources) were imported under duty-drawback provisions during 2000, this represents a raw product equivalent of approximately 0.35 million tonnes, or 58 per cent of total imports. This is well in excess of the assumption by Sumner et al. (2001) of 20 per cent of imports being re-exported (Figure 5.3).

The 1978 domestic support arrangements for processing tomatoes

Domestic support payments represent the most important element of EU assistance to the processing tomato industry, not merely in terms of their size but because they deeply influence the structural characteristics of this industry. In 2000, these payments redistributed approximately $270 million (US) from EU taxpayers to the industry. The introduction of these arrangements in 1978 provided a major boost to the EU's processing tomato nations (at that time, Italy and France). With the accession to the EU of Greece (in 1980) and Spain and Portugal (in 1985), the size of processing tomato support grew considerably. Production subsidies for Greece, Spain and Portugal were phased in over five years as a strategy to assist these nations to adjust to EU membership. As a result of a series of major problems with the 1978 arrangements, the domestic support regime was radically restructured for the 2001 season, and is discussed later in this chapter.

The 1978 domestic support arrangements were executed through a system of commodity, national and factory-based quotas. The allocation of quotas was based

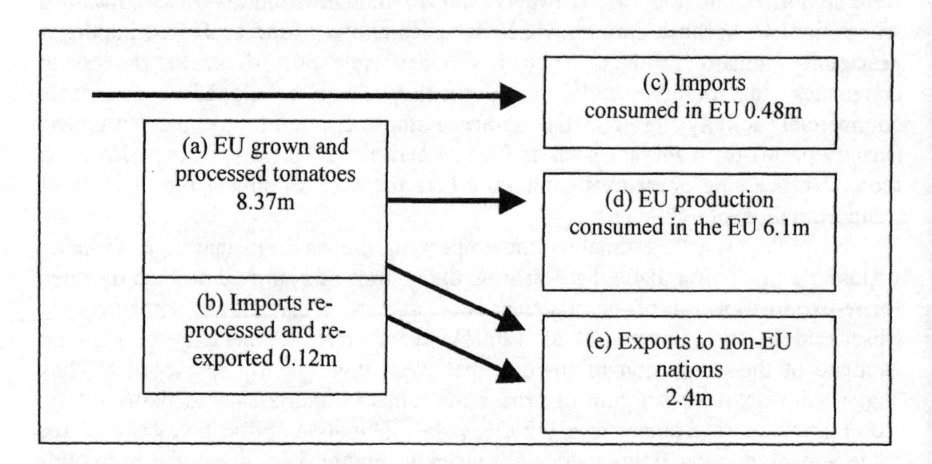

Figure 5.2 Sumner, Rickard and Hart's (2001) model of trade and production flows in the EU processing tomato sector, 2000 (raw tomato equivalents, millions of tonnes)

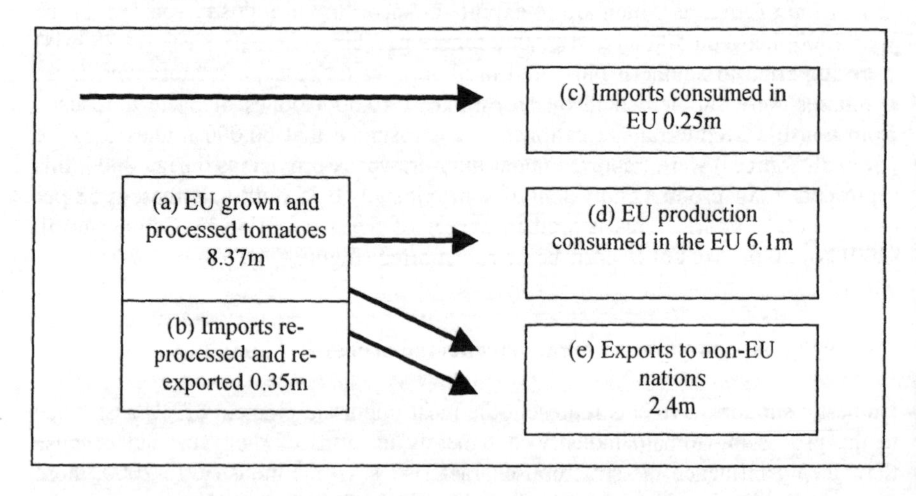

Figure 5.3 Authors' model of trade and production flows in the EU processing tomato sector, 2000 (raw tomato equivalents, millions of tonnes)

on historical records of the volume of raw tomatoes purchased at EU minimum prices, and quotas allocated to individual factories entitled these factories to apply for EU support payments (processor aid). For example, a facility with a quota for 20,000 tonnes of raw tomatoes to be used in paste production would receive quota payments for this amount. Additional (over-quota) production was not eligible for quota payment, but if tomatoes were purchased at EU minimum prices, this production could be used to argue for an increased quota in following years. Quotas were adjusted among nations based on an average of shifts in production over the previous three years, although the quota for an individual nation could not be changed by more than 10 per cent in any particular year. Through this system, industry outcomes could be manipulated by changing three variables: (i) the quota size; (ii) the level of processor aid per tonne of product within quota; and (iii) the minimum price to be paid to growers for in-quota tomatoes.

EU determinations of the total quota size and its apportionment between producer nations set the parameters for the processing tomato domestic support regime. When first initiated, the 1978 regime was intended to allocate production aid to the entire EU processing tomato industry. However, the allocation of support provided strong incentives for the industry to increase production levels and, as this occurred, the EU was faced with pressures to raise quotas, with obvious budgetary consequences. Quotas were increased in 1984, 1990 and 1992. In 1996 the EU began to resist arguments for quota increases, amending the regulations for setting quotas so that national and product allocations remained fixed for the years 1997-98 and 1998-99. For 1999-2000, quota allocations were based on average quantities produced (in compliance with minimum price regulations) for the two previous years.[1] As a consequence of this decision an increasing gap emerged between EU quota limits and actual production. As seen in Table 5.2, from 1994 to 2000 EU processing tomato production exceeded annual quota levels. By 1999, actual production was 133 per cent above quota.

The second mechanism which the EU could manipulate under the 1978 arrangements was the level of processor aid. As detailed in Table 5.3, processor aid fell steadily from the early 1980s until the demise of these arrangements in 2000. Between 1990 and 2000, the rate of processor aid was reduced by 55 per cent, expressed in European Currency Units. Until the late 1990s, processor aid per tonne of paste represented between 35-45 per cent of paste prices, expressed in US dollar terms. The incidence of processor aid fell significantly from 1997. These reductions were the most direct manifestation of EU attempts to rein in the cost of its processing tomato support arrangements. Because of reduced rates of processor aid, the budgetary cost to the EU of domestic support fell from ECU 470 million in 1992, to Euro 269 million in 2000 (Sumner et al., 2001, p.30). Reductions in the rate of processor aid were translated into a lower overall incidence of subsidy to the final price of tomato paste.

Third, under the 1978 arrangements, the EU could adjust the minimum prices set for processing tomato products. EU minimum prices were set for seven different tomato outputs. By raising or lowering minimum prices, the EU could influence both the competitiveness of the European industry and the supply intentions of growers. Under this mechanism, for example, Roma and San Marzano tomatoes used for whole peeled canning received a considerably higher minimum price than tomatoes used for paste.

Obviously, minimum prices applied only to 'in-quota' tomato purchases. For purchases outside quota, processors had two options. One option was to pay the statutory minimum price for the outside-quota tonnage. These purchases would then be used as the basis for the allocation of a higher factory quota allocation the following year. The second option was to pay a (generally lower than minimum price) market rate for outside-quota purchases, and thus provide growers with a 'blend' price based on proportions of in-quota and outside-quota tomato purchases. During the 22 years in which the EU set minimum prices (they were abolished for the 2001 season), they varied from ECU 122.26/tonne (in 1983) to ECU 88.05/tonne in 1999-2000 (Table 5.4). In ECU terms, minimum prices fell throughout the 1990s. Furthermore, because of significant over-quota production in the second half of the 1990s, the average price paid to growers fell by an even greater amount. In 2000 for example, growers in Italy received the statutory minimum price of 170 lire per kilogram for quota tomatoes, but only 130 lire per kilogram for over-quota production.

Table 5.2 Comparison of EU quota levels and actual production of processing tomatoes, 1991-2000 (thousand tonnes, raw product)

	Quotas				Actual production			
Year	Paste	Whole peeled	Other	Total	Paste	Whole peeled	Other	Total
1991	4284	1544	735	6561	4371	1215	840	6426
1992	4284	1544	735	6561	3636	1114	870	5620
1993	4284	1544	735	6561	3969	1377	882	6228
1994	4284	1544	735	6561	4798	1463	937	7198
1995	4284	1544	735	6561	4504	1278	909	6691
1996	4284	1544	735	6561	5304	1320	1245	7869
1997	4585	1336	915	6836	4661	1163	1061	6885
1998	4585	1336	915	6836	5413	1418	1414	8245
1999	4586	1323	930	6837	5912	1505	1707	9124
2000	4614	1321	929	6866	n/a	n/a	n/a	8379

Note: n/a = not available.

Source: AMITOM (2000, various pages).

Table 5.3 EU processor aid for tomato paste, 1978-2000

	Processor aid per tonne of paste				
	ECU	USD exchange rate	USD	Price for paste (ITL)/tonne (28° brix)	Processor aid as a % of paste price
1978	474.55	0.690	327.44	n/a	---
1979	461.51	0.636	293.52	n/a	---
1980	447.99	0.585	262.07	n/a	---
1981	486.63	0.822	400.01	n/a	---
1982	549.78	0.779	428.28	n/a	---
1983	567.53	0.739	419.40	n/a	---
1984	482.76	0.650	313.79	n/a	---
1985	326.03	0.658	214.53	n/a	---
1986	341.22	0.815	278.09	n/a	---
1987	358.96	0.942	338.14	952,000	---
1988	394.83	0.945	373.11	980,000	---
1989	383.65	0.869	333.39	1,288,000	---
1990	377.42	1.091	411.77	1,344,000	37.7
1991	374.09	1.007	376.71	1,288,000	37.9
1992	353.67	1.140	403.18	1,148,000	43.7
1993	332.39	0.927	308.13	1,036,000	45.0
1994	312.49	1.036	323.74	1,344,000	38.0
1995	301.55	1.345	405.58	1,546,000	44.5
1996	296.12	1.302	385.55	1,540,000	38.9
1997	267.83	1.107	296.49	1,260,000	40.2
1998	244.12	1.109	270.73	1,428,000	33.6
1999	216.19	1.042	225.27	1,400,000	29.3
2000	171.78	0.928	159.41	1,190,000	28.5

Note: Price for paste (hot break) is expressed in Italian Lire/brix/tonne. Italian paste is supplied in 200 litre drums and is packed 28 to 30 brix. Prices as at January each year. Processor aid per tonne of paste is based on US dollar equivalents. n/a = not available.

Source: AMITOM (2000, p.15).

Table 5.4 EU minimum prices for tomatoes used for paste, 1978-2000

	ECU/tonne	USD/ECU exchange rate	Minimum price expressed in USD
1978	95.76	0.690	138.78
1979	100.83	0.636	158.54
1980	105.66	0.585	180.62
1981	112.00	0.822	136.25
1982	119.28	0.779	153.12
1983	122.26	0.739	165.44
1984	121.04	0.650	186.22
1985	117.37	0.658	178.37
1986	111.50	0.815	136.81
1987	107.60	0.942	114.23
1988	107.60	0.945	113.86
1989	107.60	0.869	123.82
1990	107.42	1.091	98.46
1991	107.42	1.007	106.67
1992	107.42	1.140	94.23
1993	100.97	0.927	108.92
1994	96.94	1.036	93.57
1995	95.49	1.345	71.00
1996	95.49	1.302	73.34
1997	93.58	1.107	84.53
1998	90.77	1.109	81.85
1999	88.05	1.042	84.50
2000	88.05	0.928	94.88

Source: AMITOM (2000, p.15).

Table 5.5 EU factories and quota, 1994 and 2000

	Number of factories with quota			Total quota (million tonnes)		
	1994	2000	change	1994	2000	change
Spain	135	123	-8.9%	0.9670	1.0121	+4.7%
France	28	13	-53.6%	0.3795	0.2992	-21.1%
Greece	42	36	-14.3%	1.0136	1.0782	+6.4%
Italy	231	232	+0.4%	3.2940	3.6163	+9.8%
Portugal	18	16	-11.1%	0.8576	0.8581	0.0%
Total	454	420	-7.5%	6.5117	6.8638	+5.4%

Source: Unpublished data obtained from AMITOM.

The effects of the 1978 domestic support arrangements on European production efficiency

The 1978 domestic support regime provided a major brake on production flexibility and investment within the European processing tomato sector. Quota allocations could not be transferred between factories, thereby providing processors with huge incentives to retain (and increase) production within existing facilities.[2] With the possession of a (non-transferable) entitlement quota being a key element in production competitiveness, small and otherwise uneconomic factories across Europe were kept afloat, to the detriment of the overall industry productivity.

Tables 5.5 to 5.7 illustrate the role played by quotas in the industrial structure of the EU processing tomato sector. In 1994 there were 454 factories in the EU that held a processing tomato quota. By 2000 this number had fallen marginally, to 420. Hence the average production level per factory (raw product equivalent) was 15,858 tonnes in 1994 and 19,950 tonnes in 2000. By any stretch of the imagination, this reflects an extremely fragmented production system. By way of comparison, Morning Star's facility in Williams, California has an annual throughout of over one million tonnes of raw product. An analysis of the distribution of quotas by size of quota (Table 5.7) indicates that in 1994 and 2000 respectively, 20 per cent and 21.8 per cent of EU processing tomato factories held quotas of less than 1,000 tonnes and, in both periods, over 60 per cent of all tomato factories held quotas of less than 10,000 tonnes.

These data also highlight the way that quotas discouraged structural change in the EU processing tomato sector. At a time of rapid change in the global processing tomato sector, the EU's industrial structure remained relatively constant. Total factory numbers fell by only 7.5 per cent during the period 1994-2000. Factory closures in France and Greece accounted for about half of this reduction, highlighting relative stability in factory numbers throughout the rest of the EU, and especially in Italy.

Table 5.6 Average EU quota and output per factory, 1994 and 2000 (tonnes)

	Average quota per factory			Average production per factory		
	1994	2000	change	1994	2000	change
Spain	7,163	8,228	14.5%	9,475	10,715	13.1%
France	13,553	23,015	69.8%	9,882	24,154	144.4%
Greece	24,133	29,950	24.1%	26,450	29,500	11.5%
Italy	14,260	15,588	9.3%	15,877	20,866	31.4%
Portugal	47,644	53,631	12.6%	48,067	53,375	11.0%
Total	14,343	16,432	11.4%	15,858	17,593	10.9%

Source: Unpublished data obtained from AMITOM.

Table 5.7 Distribution of EU factories by size of quota, 1994 and 2000

	Distribution of factories by size of quota (thousand tonnes, raw product)						
	0.01–0.99	0.1–0.999	1–9.999	10–49.999	50–99.999	Over 100	Total
Spain: number of factories (and per cent of total)							
1994	1	47	70	10	7	0	135
	(0.7)	(34.8)	(51.9)	(7.4)	(5.2)	(0.0)	(100.0)
2000	10	51	44	10	6	2	123
	(8.1)	(41.5)	(35.8)	(8.1)	(4.9)	(1.6)	(100.0)
France: number of factories (and per cent of total)							
1994	5	3	13	6	0	1	28
	(17.9)	(10.7)	(46.4)	(21.4)	(0.0)	(3.6)	(100.0)
2000	0	3	5	3	1	1	13
	(0.0)	(23.1)	(38.4)	(23.1)	(7.7)	(7.7)	(100.0)
Greece: number of factories (and per cent of total)							
1994	2	7	9	17	5	2	42
	(4.8)	(16.7)	(21.4)	(40.5)	(11.8)	(4.8)	(100.0)
2000	1	3	10	14	5	3	36
	(2.8)	(8.3)	(27.8)	(38.9)	(13.9)	(8.3)	(100.0)
Italy: number of factories (and per cent of total)							
1994	4	22	124	66	10	5	231
	(1.7)	(9.5)	(53.7)	(28.6)	(4.3)	(2.2)	(100.0)
2000	3	21	110	85	9	4	232
	(1.3)	(9.1)	(47.4)	(36.6)	(3.9)	(1.7)	(100.0)
Portugal: number of factories (and per cent of total)							
1994	0	0	3	8	6	1	18
	(0.0)	(0.0)	(16.7)	(44.4)	(33.3)	(5.6)	(100.0)
2000	0	0	3	7	4	2	16
	(0.0)	(0.0)	(18.7)	(43.8)	(25.0)	(12.5)	(100.0)
Total: number of factories (and per cent of total)							
1994	12	79	219	107	28	9	454
	(2.6)	(17.4)	(48.2)	(23.6)	(6.2)	(2.0)	(100.0)
2000	14	78	172	119	25	12	420
	(3.3)	(18.5)	(41.0)	(28.3)	(6.0)	(2.9)	(100.0)

Source: Unpublished data obtained from AMITOM.

The 1978 domestic support arrangements and 'paper tomatoes'

In addition to restricting production flexibility, the 1978 arrangements were associated with problems of transparency and accountability. This lack of transparency gave rise to widespread allegations of financial misallocation and fraud. In the course of this research, many industry informants suggested that fraudulent activity, resulting in the over-payment of EU subsidies, was 'common knowledge', although many also added that inherent flaws in the arrangements encouraged these outcomes.

Not surprisingly, there are difficulties in generating hard evidence on the extent of alleged fraud of EU subsidy payments. Over recent years, official attention has been focused on the situation in southern Italy, but little evidence has been presented and few prosecutions have ensued. Even the USDA is cautious in estimating the extent of fraudulent practice within southern Italy. In its 2000 report on the Italian tomato industry, the USDA cautiously notes that 'a certain proportion' of reported stock levels are 'ghost stocks' resulting from fraudulent activity (USDA Foreign Agricultural Service, 2000, p.4).

One piece of evidence relating to the incidence of fraud in the southern Italian processing tomato industry emerged in a 1997 documentary film titled *Bitter Harvest*, produced for TV2 Denmark. This film brings together various threads of evidence allegedly demonstrating systemic corruption within this industry. The filmmakers point to two scales of illegality that run through the industry. First, they allege hand harvesting of processing tomatoes in southern Italy is connected to the recruitment and exploitation of illegal African and eastern European immigrants. They allege the existence of collusion between growers, labor hire operators (*caporales*) and individual piece-rate workers that serves to create an exploited and frightened workforce. Second, they allege that there is collusion on the part of growers and processors to falsify tomato purchase records with the aim of maximizing EU subsidies. According to EU regulations, at least 15 per cent of tomato factories should be audited for fraud annually. However, the filmmakers report a 1994 EU Court of Auditors finding that this requirement was continually breached, resulting in lax controls on the flow of subsidy payments. Fraud is allegedly perpetrated through over-invoicing of tonnages at the factory gate. Higher records of tonnages of tomatoes purchased at EU minimum prices flow through to higher EU quota payments. Thus, published production statistics contain an element of so-called 'paper' or 'ghost' tomatoes.

It is important to note that if correct, the incidence of fraud in the southern Italian processing tomato industry has been encouraged by structural characteristics of the EU's regulatory arrangements. Not to put too fine a point on it, the pre-2001 arrangements provided strong incentives for processors to over-state production as a way of protecting their quota levels. In an environment of 'no increase' in the EU's global allocation of quotas, the financial viability of processors rested on their ability at least to maintain their quota entitlements. The easiest way to achieve this outcome was through the creation of 'paper tomatoes' that existed only as book entries within processor company accounts.

Because they were in receipt of significant quota monies, processing companies had few incentives to criticize EU policies. Because of their dependence on quota-linked minimum prices, grower organizations were also silent over the need for change. The major voices advocating change originated in the EU bureaucracy itself, and with traders of EU processing tomato products, whose livelihoods were dependent on the cost competitiveness of product. From the latter group, Gandolfi, a leading Italian company, made an important intervention at the *Third World Congress on the Processing Tomato*, in 1998 in Pamplona, Spain, when it sharply criticized the inadequacies of EU processing tomato policies. Gandolfi identified the EU regulatory system as discouraging innovation and change within Europe. According to the company 'processors are forced to accept prices which have no realistic link to the true production costs' (Gandolfi, 1998, p.2). Consequently, the EU system

> protects obsolete processors who are too small or financially weak to compete. In the same time [sic] it inhibits those packers wanting to set their capacity at a level which would allow them to compete with the most efficient processors outside the EU (Gandolfi, 1998, p.2).

The company concluded:

> The true objective of the Regulation is to protect the price of [tomatoes] for growers … [but] the EU minimum price is set too high in comparison with either world market prices of EU production costs. The direct consequence of this situation is constant pressure for increasing production *and well-known fraudulent practice* (Gandolfi, 1998, p.3, italics added).

In summary, the 1978 regime enshrined rigidities and a lack of transparency. The allocation of quotas on the basis of past production levels encouraged processing company strategies to be built around the ownership and protection of quota allocations, rather than to seek cost efficiencies. Greenfield investment was effectively discouraged under the regulations. The 1978 regime was purpose-built for inefficiency, and tailor-made to encourage corruption and fraud. The fact it remained in place so long is indicative of the inertia of EU decision-making.

The 2001 EU processing tomato subsidy regime

In an acknowledgement of these inadequacies, the EU moved to a new regulatory regime in 2000. Through EU Council Regulation 2699/2000, adopted on 4 December 2000, a production threshold regime replaced the quota/minimum price arrangements first implemented in 1978.

The shift to a production threshold system reflects two conflicting themes within EU agricultural policy. On the one hand, it provides a stronger set of financial disincentives for over-production, and to this extent, the threshold regime responds to global trade liberalization agendas. Yet on the other hand, the threshold regime cements in place the new subsidy levels in EU Statutory Regulations without any provision for their reduction over time. Hence, while the threshold regime is more attuned to market signals than its 1978 predecessor, it does not pave the way for ongoing subsidy reductions.

The key change in the 2001 Regulation is a shift in the mechanism for payment so that growers, not processors, are the recipients of subsidies. Grower associations have become the conduits for payments, significantly strengthening their economic and political roles within the industry. According to the December 2000 regulation, the payment rate for processing tomatoes is fixed in perpetuity at Euro 34.5 per tonne. If growers exceed quota production limits, their payments per tonne are reduced proportionately in following years. Consequently over the period of a two-year cycle, any financial gains to growers in one year from increased production are offset by reduced payments in the following one. This also places a cap on the financial liability of the EU arising from subsidies.

In terms of economic efficiency, there are two distinct areas where the 2001 regime is superior to its predecessor. First, abolishing factory-level quota allocations eliminates incentives to fraudulently over-state tomato purchases. With quotas no longer attached to individual factories, processors' decisions on where and how much to produce are linked more closely to market mechanisms rather than they are to strategies to maximize the receipt of quota allocations. In a short space of time, this should result in a dramatic reorganization of production away from (generally smaller) factories and towards larger, more efficient, ones. Second, because producer aid payments are made directly to growers and statutory minimum prices are eliminated, flexibility is introduced into the economic relationship between growers and processors. Under the new arrangements, growers and processing firms sit down to negotiate prices for a forthcoming season, taking into account the fact growers will automatically receive a subsidy of Euro 34.5 per tonne. How these arrangements work in practice will differ over time and place, and will be dependent upon the types of institutions present in different environments. Outcomes and processes in France, where many growers sell within a cooperative and there is a strong inter-professional agency, will be different from the south of Italy, where production systems tend to be fragmented. At the time of writing, only one season had been completed under the new arrangements, and the system of price negotiations remains in flux. Aggravating the implementation of these arrangements was the fact that they were introduced at a time of high stocks and weak prices, creating considerable uncertainty for industry participants. During the first year of the new arrangements, farm-gate prices fell by about 10 per cent in many production regions.

Notwithstanding these improvements to the efficiency of domestic support arrangements for the EU processing tomato sector, Sumner et. al (2001) conclude that they will be more costly and trade distorting, compared to the previous quota system. This is because of the significant rise in the potential tonnage of processing tomatoes subject to subsidy. In the last year of the quota regime (2000), the EU paid producer aid on a maximum of 6.865 million tonnes. Under the new threshold regime, growers can be paid subsidies on 8.251 million tonnes. Accordingly, Sumner et al. (2001, pp.82–83) argue that compared with the previous quota regime, the 2001 arrangements will probably encourage a two per cent increase in

EU processing tomato output and will generate a net reduction in EU welfare of Euro 7.7 million annually.

As with any economic simulation, the veracity of the conclusions by Sumner et al. (2001) rests with the robustness of assumptions, comprehensiveness of cost and benefit measurements, and the ability to anticipate major changes in the future. As a partial equilibrium model concerned with analyzing price and quantity movements arising from various policy scenarios, this approach cannot encompass the entire range of implications of processing tomato sector restructuring. In particular, a comparison of policy-determined outcomes with free market alternatives, incorporates assumptions about what the EU processing tomato sector might look like in the absence of regulation. The concept of a free market alternative is necessarily presented as an abstract ideal. Hence, whereas the research by Sumner et al. (2001) is extremely valuable in charting the ways in which EU processing tomato regulations operate, it needs to be supplemented by analysis that problematizes the spatial and temporal specificities of regulatory change.

The remaining sections of this chapter focus on these issues. We undertake a grounded examination of the inter-relationships between regulation, industry structures, and the strategies of firms and growers. This approach considers the ways regulation and regulatory change shape industry outcomes. The EU's provision of domestic support and border protection has reconfigured the entire shape of the industry, establishing new loci of economic and social power and encouraging supply chains to take particular spatial forms. In assessing EU policies in this way, we take our cue from Hoggart and Paniagua (2001, p.64), who argue:

> EU regulations are not anonymous extra-local forces but are internalized and transformed so as to be manipulated and utilized in ways that correspond with dominant local social relationships.

The EU processing tomato trade-production complex

The EU processing tomato sector constitutes a regional trade-production complex involving significant intra-national and cross-border flows. This complex is linked inextricably to the development of the subsidy regime. Recent changes in EU processing tomato policies have contributed to intense restructuring at firm and regional levels that are challenging existing arrangements.

Figures 5.4 (tomato paste) and 5.5 (canned tomatoes) summarize major trade flows to and from EU production areas for 1998-99. Although these data are from official AMITOM publications, nevertheless their accuracy is open to question, because of the problem of verifying the reliability of official statistics in this industry.[3] For this reason, the Figures should be interpreted as illustrating only the general pattern of European processing tomato trade flows.

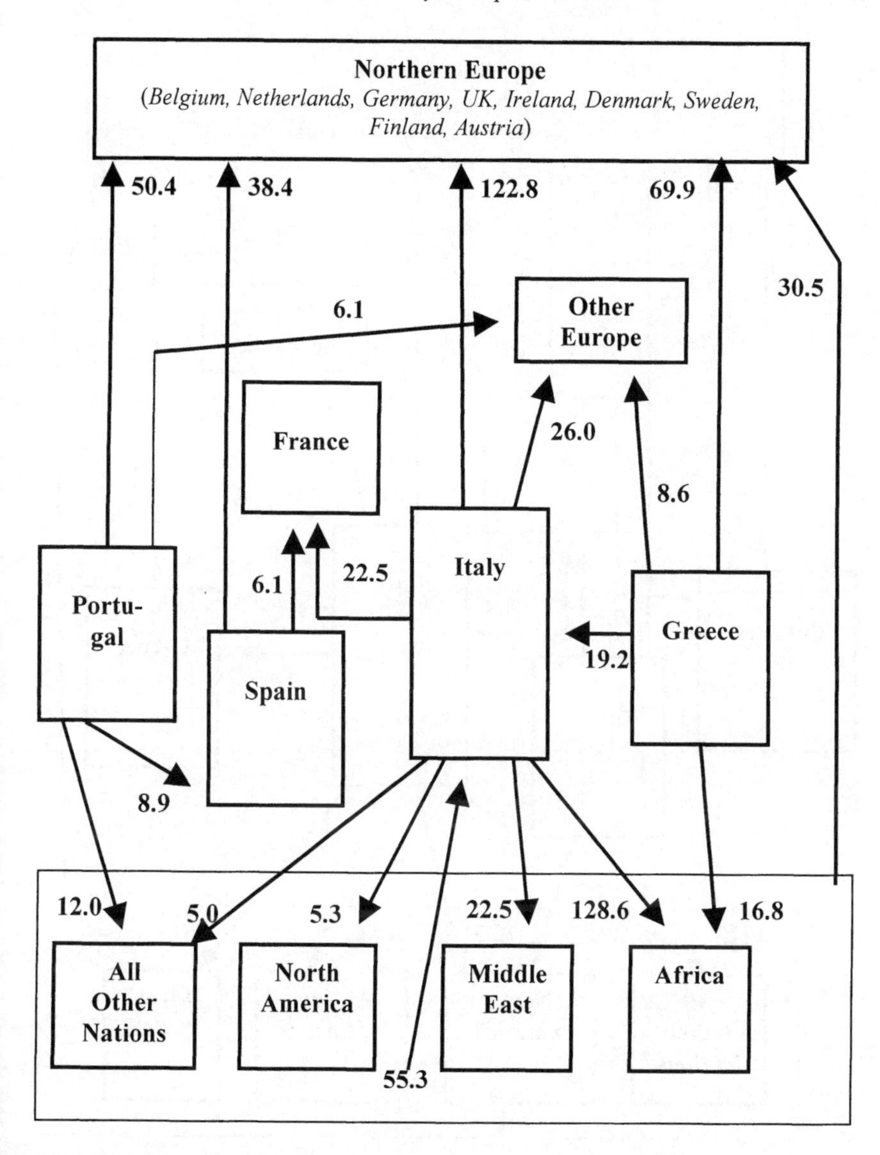

Figure 5.4 The European tomato paste trade-production complex, 1998-99 (thousand metric tonnes of finished product)

Note: Only flows of 5,000 tonnes and above are included.

Source: AMITOM (2000, various pages).

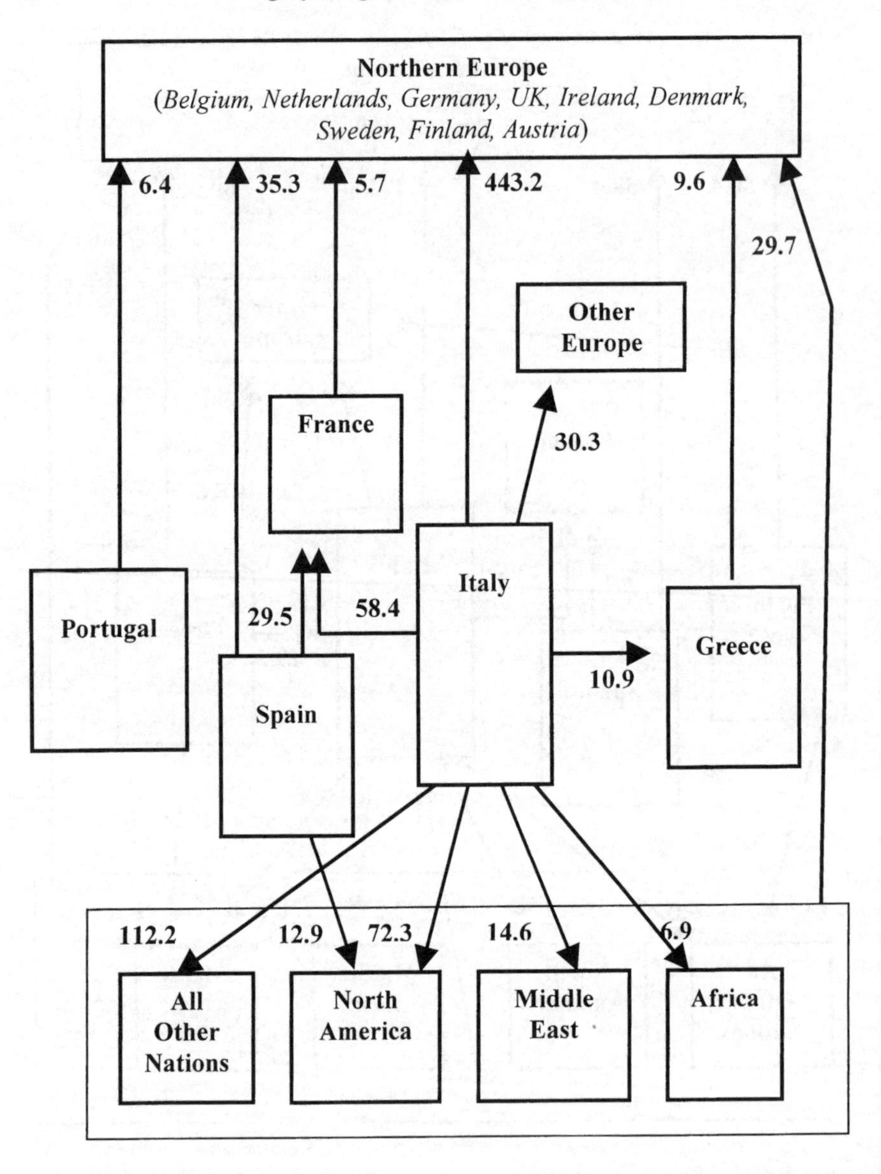

Figure 5.5 The European canned tomato trade-production complex, 1998-99 (thousand metric tonnes of finished product)

Note: Only flows of 5,000 tonnes and above are included.

Source: AMITOM (2000, various pages).

The most striking element of Figures 5.4 and 5.5 is the geographical complexity of flows, especially for tomato paste. There is considerable intra-industry trade, as represented by the fact that Italy, for example, is simultaneously an importer and exporter of paste. This demonstrates the fact that the market for tomato paste is not homogenous, but is distinguished by grades and quality standards. Tomato paste used for re-processing into consumer foods such as canned baked beans tends to be manufactured using the *cold break* process. Paste used for processing into ketchup tends to be manufactured using *hot break* technology (see Chapter Three). Within these categories, product is further fragmented by degrees brix. Largely for historical reasons, industrial food companies in Germany tend to demand cold break paste at 28-30 degrees brix, whereas in France and the UK cold break paste tends to be demanded at 36-38 degrees brix. (In contrast, most cold break paste from California is manufactured at 30-32 degrees brix, which tends to mitigate trans-Atlantic trade.) Some buyers also require paste to meet specific requirements. According to industry sources, Pizza Hut fulfils its European tomato paste requirements through the Portuguese supply of a hot break paste at 20-22 degrees brix, manufactured using a larger screen size to give the paste a rough texture.

Notwithstanding intra-industry complexity, Italy acts as the central node within this complex, being both a major importer and the largest exporter. Italy is the largest exporter of canned tomatoes from southern Europe. With respect to paste, Italy imports considerable quantities of paste from Greece and from non-EU sources, for re-processing and re-export. Markets in the Middle East and Africa provide important export destinations for Italian paste. Italy dominates the EU tomato paste export trade to these markets, although historically, Greece has also provided substantial exports to Africa. During the second half of the 1990s, Spanish and Portuguese tomato paste exports to the Middle East and Africa fell considerably, further cementing Italy's dominance in these markets.

The export of tomato paste and canned tomatoes to northern Europe represents the single most important set of trade flows for the region. According to AMITOM data, in 1998-99 approximately 32 per cent of all tomato paste produced in the EU was sold to northern Europe. In volume terms, this equates to about 250,000 tonnes of paste being transported northwards by truck and sea from production centres in southern Europe. The story is similar with respect to canned tomatoes, where in 1998-99 approximately 500,000 tonnes (equivalent to about 20 per cent of EU production) was sold to markets in northern Europe. Within the canned tomato trade, the UK is the single most important market. During the second half of the 1990s, UK buyers regularly purchased over 200,000 tonnes of canned tomatoes from Italian producers each year. The majority of this output is shipped from the ports of Salerno and Naples in southern Italy.

The size and importance of this trade invites a consideration of the ways in which tomato paste and canned tomato supply chains are organized. In 1998-99, according to AMITOM data, EU nations supplied 90.2 per cent of the tomato paste and 94.4 per cent of the canned tomatoes bought by northern Europe. No official data exist on the supply chain structures that underpin this trade, and because this issue is characterized by commercially sensitive information, industry participants tend to discuss the trade in general terms only. Notwithstanding these limitations, the broad supply chain parameters of Europe's south-north tomato trade are as follows. The key buyers in northern Europe are retail chains and industrial food companies. Canned tomatoes and consumer food products (such as tomato sauces) tend to be purchased through direct contract by distribution companies allied to retail chains. Of course, a considerable proportion of this trade involves generic and own-store labelled products, meaning that supply contracts are fixed prior to manufacture. According to industry sources, much of this trade is relatively stable, in that buyers tend to use the same suppliers year after year. As discussed below, in recent years key producers in southern Italy have attempted to further stabilize these channels via direct investment in northern European distribution companies. In contrast, industrial grade tomato paste tends to be purchased through a more diffuse set of arrangements. Some major purchasers (such as Heinz, Unilever and Kraft) utilize long-standing relationships with bulk suppliers to satisfy their needs. However, these purchasers may play one supplier off against another to achieve a lower price. During the late 1990s, Unilever developed preferred supplier arrangements with a small number of processing tomato firms for its productive activities in western Europe. These types of arrangements tend to be relatively permanent, with prices determined on the basis of agreed industry benchmarks (say, the published price of the *Camera di Commercia di Parma*). Independent traders also play a key role in this industry sub-sector. Traders act as intermediaries usually employed by buyers to secure supplies at the best possible price. They are individuals or family companies whose currency is information, and whose progenitors were the mercantile classes of pre-Renaissance Italy. Although traders deal in the entire range of tomato products (as well as other foods and food processing machinery), bulk tomato paste appears to be their mainstay. These operatives are critically important for the global processing tomato industry, because they are of central importance to the flexibility of this industry's supply chain. The northern Italian city of Parma is the main location of global tomato paste traders, with half-a-dozen traders' offices located within a square kilometre of the city center.

The extent to which these supply chain arrangements may be restructured is a keen issue of debate within the industry. The academic literature on supply chain structures gives little insight into the dynamics of the changes that are occurring. Gereffi (1996) typecasts supply chain structures as being either 'buyer-driven' (for example, the global clothing industry, where a few major companies outsource supply through relatively weak subcontracting relations) or 'producer-driven' (for example, the automobile sector, where production levels in particular regions shape global trade). Europe's processing tomato sector does not conform readily to either of these ideal-types. At the present time, changes are occurring at both ends of the supply chain. 'Near-consumer' agents are responsible for some of these shifts. Retail consolidation in northern Europe has narrowed supply channels and for suppliers, has heightened the importance to economies of scale and quality assurance. At the same time, industrial food companies, notably the large transnationals, also have restructured their operations over recent years. A key theme in recent restructuring has been the outsourcing of first-stage processing, and devoting capital resources to brand management instead. Yet changes are also occurring at the 'supply' end of the processing tomato chain. The reform of EU subsidy arrangements, in conjunction with the emergence onto the world market of relatively cheap Chinese tomato paste, is restructuring the European processing tomato sector. As these actions re-shape the industry, pan-European supply chains are reconfigured. The remaining sections of this chapter explore the current evidence regarding the nature of these changes.

Restructuring in northern Italy

As the central node within the European processing tomato production-trade complex, the restructuring of Italy's processing tomato sector has wide reverberations within this industry. As discussed above, Italy's processing tomato industry is sharply divided between paste production in the north, and canning and re-processing in the south. Coincident with the implementation of new EU subsidy arrangements, these regions have experienced major changes, however the contours of change differ widely between the north and the south.

Northern Italy is potentially vulnerable to the restructuring of pan-European supply chains because its key output, industrial grade tomato paste, is a standard product readily substitutable from a number of production areas. The efficient production of paste is heavily dependent on raw material costs and economies of scale in processing. Northern Italy is well placed with respect to the latter, as a number of relatively large factories dominate its production regime. There are however, potential problems with respect to the cost efficiency of its growing sector.

Because approximately six tonnes of raw tomatoes are required to make one tonne of paste (the exact ratio is dependent on brix and solids calculations), the competitiveness of tomato paste production depends vitally on raw product supply. Detailed and accurate statistical data on the costs and efficiency of the tomato growing community of northern Italy are not publicly available, so any discussion of this issue inevitably involves a degree of speculation.

An insight into the cost efficiencies of tomato growing in northern Italy is provided by examining the situation facing Copador (*Consorzio Padano Ortofruitticolo*), a large cooperative in the region. Copador was established in 1987 when a group of cooperatives in Parma and Piacenza acquired the productive assets of a large, privately-owned, processing tomato firm. Approximately 70 per cent of Copador's output is exported as industrial product for further processing in northern Europe. In the last season before the introduction of the 2001 EU regulations, Copador had a quota for 150,000 tonnes of raw tomatoes, but also processed a further 100,000 tonnes. The vast majority of its growers are located in a tight radius of 100 kilometres around Copador's factory near Parma, in the midst of the Po Valley. The geography of Copador's suppliers reflects the interplay of environmental and socio-economic factors. The alluvial soils and climate of the Po Valley provide good conditions for tomato growing; the high population density, pressures of urban expansion and high land costs in the Valley dictate that grower strategies are dominated by the attempt to extract a high rate of return per hectare, which in turn encourages on-farm diversity and multiple cropping. Consequently, tomato supply in the Po Valley is characterized by a relatively large number of relatively small producers (a situation not dissimilar to France; see Chapter Two). One hundred and sixty growers supply Copador, indicating an average tonnage per grower of 1,500 tonnes. This figure however, overstates the tonnages supplied by most growers, because there are a few larger growers who supply about 10,000 tonnes of raw product each. The high land costs in the valley coupled with the proliferation of relatively small tonnages suggest that northern Italian growers may have little scope for accepting reduced tomato prices, and in fact during the late 1990s, grower associations campaigned strongly against price reductions. Unofficial benchmark estimates by one key player in the processing tomato sector suggested that in 2001, the farm-gate raw product cost in northern Italy was about six percent higher than in the southern Italian province of Puglia, where larger farms produced Roma tomatoes for canning.

The cooperative structure of the processing industry also presents a set of structural issues for northern Italian tomato growing. At the onset of the new EU regulatory regime in 2001, four cooperatives traditionally linked to different political interests (Copador, ARP, Consorzio Casalasco del Pomodoro and Agridoro) controlled approximately half of northern Italy's 2.1 million tonne production complex. The organizational goals of these cooperatives, of ensuring the sale of members' raw product, encouraged these entities to increase their output over time. However, within any processing plant owned by a supplier cooperative, there are tensions between profit goals (implying that raw product prices should be kept as low as possible) and the goals of optimizing returns to grower-members (implying the maximization of raw product prices). It is difficult to ascertain how these tensions were worked through by northern Italian cooperatives. Because of minimum price regulations for in-quota tomato purchases, this issue only arose with respect to out-of-quota purchases. The price for out-of-quota purchases was determined on the basis of negotiations with grower associations. These negotiations produced a 'blend' price representing a proportion of the EU minimum plus 'market' prices. Growers, however, did not necessarily receive the 'blend' price for product, because cooperatives operated trucking and a range of agronomic services, the costs of which were deducted from grower payments. Hence, under the labyrinthine administrative arrangements of the pre-2000 EU subsidy regime, market processes were rendered ineffective, as measures of competitiveness and efficiency. The opaque nature of cooperatives' internal finances in the context of EU regulations, in combination with uncertainty about how these services were costed, means it is impossible to effectively interpret price trends over time.

Questions over the cost of raw materials and the operations of cooperatives created a situation whereby problems of excess supply and intense price-based competition during the 2000 and 2001 seasons caused great concern within northern Italian cooperatives and the grower community, although this should not be overstated. Northern Italy entered the new regulatory regime of 2001 in better shape than some other Mediterranean regions (see below). Nevertheless, with the introduction of the new regime it was apparent that northern Italy faced potentially serious competitive threats. The main source of these threats was the capacity of the region to dispose of its tomato paste stockpiles. Whereas northern European buyers generally remained loyal customers for the northern Italian cooperatives, the market was extremely weak for other sales. Traditionally, surplus northern Italian paste tended to be bought by southern Italian re-processors. In 2000-01 however, these flows were disrupted. Between January 1999 and January 2001, Italian paste prices fell by 34 per cent, from $830 (US) per tonne, to $551 (US) per tonne (Tomato News, 2001c, p.7). Contributing to these outcomes was the importation of the estimated 60,000 tonnes of Chinese tomato paste into southern Italy at the extremely low price of $350 (US) per tonne.

Although the impacts from these imports were felt most strongly in Greece and Turkey (see below), they also raised major concerns in northern Italy. As a consequence, in early 2001 three of the major cooperatives in northern Italy – ARP, Agridoro and Copador – merged their commercial activities. At the time of writing it is difficult to forecast the long-term implications of this step, however, it seems apparent that the newly formed regional institution, the *Consorzio Interregionale Ortofrutticolo*, may presage a new era for northern Italian cooperatives, as traditional economic/political rivalries begin to give way to market imperatives.

Neapolitan dilemmas: crisis and restructuring in southern Italy

The southern Italian tomato canning and re-processing industry will be transformed over the next decade. The industry consists of a concentration of canning and re-processing firms around the Bay of Naples (Figure 5.6). During the late 1990s, approximately 2.5 million tonnes of processing tomatoes were grown annually in southern Italy, mainly in the province of Puglia on the east coast of the Italian peninsula. Trucks take this output westwards to processing tomato firms south of Naples. In addition to canning tomatoes, many of these firms also re-process tomato-based products. Considerable quantities of bulk tomato paste are imported to the south of Italy to be used as an input to further-processed food commodities.

In 1999 there were 144 tomato canning and re-processing firms in the Bay of Naples area (Russo Group, 2000). Most of these firms were located in a tight, ten kilometre cluster, inland from Pompeii and Castellammare di Stabia, in and around the towns of Angri, Scafati and San Antonio Abate. In the main, these factories are very small, with many having annual raw product purchases of less than 4,000 tonnes. An estimated 40 per cent of these firms produce so-called 'bright cans', that is, cans without any ownership label. These non-branded products are sold either direct to traders for export (often as generic or store-brand commodities) or to large factories in the Bay of Naples area, where they are given an identity. This production is locked within an intense regime of price-based competition with minimal profit. For small processors, opportunities to expand out of these markets are circumscribed by retail structures. Some smaller factories have traditionally built sales relationships with small and medium-sized retailers but, as

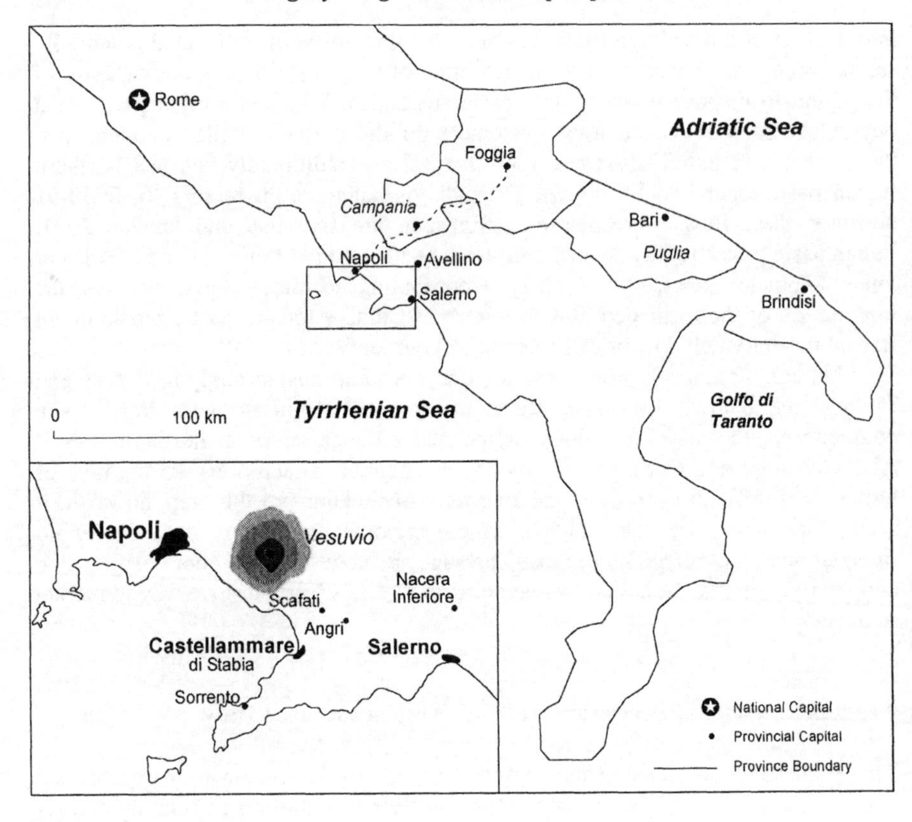

Figure 5.6 Southern Italy

European retail spending shifts towards supermarket chains, such opportunities become fewer and more difficult.

The evolution of this industrial complex raises major questions of economic geography. The tomato canning and re-processing sector was established in this area because of a confluence of three factors. First, the volcanic soils of Mount Vesuvius provided good growing conditions for tomatoes, especially the localized Roma and San Marzano varieties. Second, the port of Salerno was critically important for connecting local factories with import and export markets. Third, the proximity of Naples allowed for an urban-based manufacturing sector workforce with high levels of skills, and a competitive market for ancillary services such as machinery sales and repairs, transport and materials supply. Hence, tomato canning and re-processing in the Bay of Naples represents a flexible industrial district characterized by a considerable number of relatively small firms that engage in networks of association, sub-contracting and supply. This industrial architecture generates cost savings and, perhaps more importantly, facilitates flexibility. Embedded within the Bay of Naples is a strong entrepreneurial and trading culture based around sharp competition between rivals and strong collaboration among supply-chain actors.

At the present time, this structure is facing a great challenge. Urban expansion in the area of the Bay of Naples now means that few tomatoes are grown nearby. Urban development in the Bay area has also generated real estate inflation and urban congestion. Casual observation of this area reveals a manufacturing district consisting of narrow streets clogged by industrial traffic. Furthermore, the development of irrigation infrastructure in Puglia during the 1980s encouraged a shift of tomato growing 200 kilometres eastwards. Because of its recent evolution, the Pugliese tomato growing region has distinct differences from those of other European regions. Tomato growing in Puglia is not smallholder based, but is dominated by larger agriculturists. Nevertheless, poor road links between Puglia and Naples add significantly to the transport costs of the industry. Industry estimates suggest that raw product transport costs are over $20 (US) per tonne, compared to an average of $6 (US) per tonne in California. In short, the concentration of processing tomato firms in the Bay of Naples area faces considerable problems although, as a number of industry participants reminded us during interviews, 'never discount the entrepreneurial nous of the Neapolitans'.

At another level, EU subsidy restructuring will cut deeply into the existing industrial structure of the Bay of Naples processing tomato region. Under the pre-2001 tomato subsidy regime, small firms were effectively protected through their possession of quotas. The size of a firm's quota was the *de facto* arbiter of competitive advantage within the industry. The abolition of the quota system reshapes the competitive structure of this industry, giving greater weight to issues of efficiency. Under the new regime, advantages will flow to larger, more efficient firms, especially when it comes to the production of undifferentiated and generic products (for example, canned whole peeled tomatoes). These developments are encouraging rapid structural adjustment within the southern Italian industry. Production is coming to be concentrated in a few larger groups. The two pre-eminent players in this regard are *La Doria* and the *Russo Group* (also known as *Gruppo AR,* incorporated as *Industrie Alimentari S.p.A.)*. Analysis of the declared strategies of these firms highlights the inter-play between regulatory change, industrial restructuring and corporate behavior.

La Doria was established in 1953 as a small family-owned enterprise, which over subsequent decades, successfully marketed its *La Doria* brand in Italy and tapped into newly emerging export markets. Export growth and the rise of co-packing contract production arrangements proved the engine for the company's rapid growth. By the year 2000, La Doria's turnover had grown to approximately $300 million (US), of which 93.2 per cent was generated from the sale of products manufactured under contract for other parties (La Doria, 2000).

In 2001, La Doria operated three factories in the Bay of Naples region, at Angri, Fisciano and Sarno. The latter two were dedicated to tomatoes, whereas the factory at Angri also produced other fruit and vegetable lines. In 2001, the group purchased 280,000 tonnes of Italian raw tomatoes for processing which, when sold, represented about one-third of the group's revenue.

The group's expansion has been built on internal growth and a series of acquisitions. In 1995 it restructured its equity, floating 30 per cent of its stock on the Milan exchange, with the remaining 70 per cent of stock owned by the Ferraioli family. This injection of equity paved the way for the purchase in 1996 of a minority stake (25 per cent) in Delfino S.p.A., a specialist manufacturer of tomato sauces, and in 1998, the purchase of a smaller local tomato canning firm, Pomagra S.r.l.. The increased capacity gained through these acquisitions was used to rationalize production. In the same period, La Doria also established a distribution company in the UK, called LDH. This move strengthened the company's relations with key British retailers and food service companies. In the UK, LDH holds a 50 per cent market share for tomato and fruit-based products sold under private labels (La Doria, 2000, p.20). Then in 1999, Heinz bought 20 per cent of La Doria, forging a strategic alliance which was to be of great importance. By 2000, as a result of these strategies, La Doria could boast it was the preferred supplier to the major northern European retail chains including Tesco, Sainsbury, Safeway, Aldi and Danske Supermarket, among others (La Doria, 2000, p.17).

The Russo Group's recent evolution has strong similarities to La Doria's. The company has an annual turnover of approximately $200 million (US) (slightly smaller than La Doria), and is owned and managed by the Russo family. The Group is a diversified fruit and vegetable processor, although tomatoes are its main interest. In 2001, it operated four factories and three warehouses. Russo has expanded by acquisition and its four factories tend to be known by the names of their previous owners: Conserviera Sud (the head office, in San Antonio Abate), Ipa (also in San Antonio Abate), Pagani, and La Perla (in Scafati, although this factory does not process tomatoes). Of the Group's four warehouses, two were previously small factories that Russo transformed into storage capacity when the Group centralized production in its existing larger facilities. Over the period 2000-01, the company devoted considerable resources to coordinating production throughout its various factories, with the aim of generating cost efficiencies.

In 1999, the Russo Group purchased 428,122 tonnes of raw tomatoes, approximately the equivalent of 12 per cent of the entire Italian processing tomato crop. Because this volume significantly exceeded the amount of quota owned by the Group's factories (quotas were held for only 54 per cent of this amount: Table 5.8), it is reasonable to assume that Russo's profitability has hinged on its successful exploitation of economies of scale, and its astute buying and selling practices.

The Russo Group's key marketing and production strategy hinges on its ability to capture contract production for branded food companies and supermarket chains. An estimated 90 per cent of the Group's output is sold under other companies' brand names. According to a company brochure:

> While having its own trademarks, the Russo Group [has] managed its trade strategy to become a qualified and reliable partner, during many years, for leading market enterprises and for main distribution chains, both in Italy and abroad ... Requests arising from [the] catering industry, factories, purchasing groups, Italian and foreign

distribution chains are managed by means of an extreme versatility related to packaging, promotional policies and marketing requirements (Russo Group, not dated).

Table 5.8 Russo Group tomato purchases, 1999

		Thousand tonnes
Whole peeled canned tomatoes	In quota	95.01
	Out of quota	61.64
	Total	156.65
Tomato paste	In quota	97.44
	Out of quota	94.16
	Total	191.60
Other tomato products	In quota	38.30
	Out of quota	41.57
	Total	79.87

Source: Russo Group (2000).

Evidently, the Russo Group's success as a co-packer of branded products has depended not only on its capacity to deliver output at competitive prices, but the salesmanship of its export desks. To some extent, Russo's core business can be seen as its specialist trading expertise. Russo generally satisfies its customers' requirements through internal production, but on occasions outsources contracts to smaller companies. Contracts are negotiated directly with import companies or distributors.

With 70 per cent of output exported, access to the ports of Naples and Salerno plays a critical role in the Group's competitiveness. Although the Russo Group exports to a wide range of markets, the UK is by far the most significant. Approximately 40 per cent of Group exports are destined for the UK. This dependence led to a significant restructuring within the Group in June 2000, when it changed its incorporation to a shareholder company ('S.p.A.') and offered 10 per cent of its stock to Princes Food (UK), a major food importer and distributor owned by the Japanese conglomerate, Mitsubishi. Establishment of this strategic alliance was an important step in securing access to the valuable UK market, and under the terms of the investment, Princes Food agreed to purchase its Italian fruit and vegetable requirements through Russo. Processing tomato industry analysts perceived this strategic alliance as highly symbolic of the trajectory of change within the southern Italian processing tomato sector. Russo's corporate restructure and its embrace of external investment appeared to mark a shift away from traditional family ownership structures and commercial secrecy within the sector. From Mitsubishi's perspective, the investment seems to have reflected its belief the relationship would be conducted in an environment of transparency and trust.

The emergence of La Doria and the Russo Group also says something about the role of quality as an agent of competitive advantage in this sector. Both companies have devoted considerable resources to improving their quality credentials, and this provides a key point of difference between them and some of the smaller tomato re-processing firms in the region. Related to this, La Doria has established an organics product line for distribution to UK buyers.

The La Doria and Russo groups presage a new model for the processing tomato sector in southern Italy. They provide an example of a 'national champion' model of capitalist development, organized around regionally embedded production in family-owned enterprises, and strategic alliances with key external parties. On the one hand, the desire of many branded food companies to divest processing facilities and adopt co-packing arrangements fits neatly with the scale, strategy and efficiency of these companies. To date, buyers have remained relatively loyal to their contract processors, giving stability and certainty to La Doria and the Russo Group. Strategic alliances with larger transnational companies (Heinz and Mitsubishi, respectively) have played a critical role in cementing distribution channels. At another level, the abolition of individual factory quotas from the 2001 season shifts the basis of competitive advantage strongly in favor of larger processors. Based on industry experience elsewhere in the world, it is likely that factories will require a minimum of 50,000 tonnes of raw material input to survive in this new competitive environment. Retail restructuring in Europe, with its demands for larger volume, lower priced purchases, will reinforce these pressures.

Tomato sourcing: China meets Turkey

Southern Italy's restructuring has been accompanied by geographical shifts in the sourcing of tomato paste for re-processing. The central component of these changes has been the entry of low-cost Chinese tomato paste into the Mediterranean region. This event triggered increased volatility to production volumes and prices, and the long-term pattern of winners and losers arising from these developments is not yet clear.

Rapid increases in Chinese production levels (Table 1.1) occurred at precisely the time Californian and Mediterranean output was being cut back because of weaker demand and prices. In season 2000, when Californian and EU production volumes fell 16 per cent and 19 per cent respectively, Chinese output was estimated to have more than doubled, from approximately 0.8 million to 1.8 million tonnes. The magnitude and timing of this expansion reverberated through the entire global processing tomato industry, and particularly in the Mediterranean region. This area was especially vulnerable to Chinese competition because of the particular geographical dynamics of the Mediterranean processing tomato production-trade complex. Some countries in the region have been heavily dependent on tomato paste exports to southern Italy for re-processing. In this period the landed price of Chinese paste, according to industry sources, was approximately $350 (US) per tonne. By 2001, an estimated 84 per cent of Italian tomato paste imports came from China (United States Department of Agriculture Foreign Agricultural Service, 2002b, p.4). China's *de facto* dumping of tomato paste during this period seems to have been intended to shift massive stockpiles, in the context of 'compensation trade financing' arrangements that obliged some Chinese processors to sell paste in order to meet loan repayments on machinery purchases (see Chapter Six). Considerable industry debate surrounds the ability of China to sell product at low prices over an extended period. As indicated in Chapter Six, the financial contexts of key Chinese producers are very different from those facing western companies, creating problems of interpretation for outside analysts.

The entry of China into this market primarily affected Greece and Turkey. Excess supply of tomato paste and weak prices over the years 1999 to 2001 exposed the vulnerabilities of these countries' processing tomato sectors. Southern Italian re-processing firms traditionally provided an important market for Greek and Turkish paste. Turkey exports over 50 per cent of its tomato paste production. Until 2001, 30,000 tonnes of Turkish tomato paste exports were allowed to enter the EU before prohibitive tariffs were levied, although these arrangements were discontinued following a trade dispute over EU beef access to the Turkish market. Paste imported to Italy for re-processing and export to non-EU nations, however, does not attract a levy.

The impacts on Turkey from Chinese competition were especially profound. Between 1999 and 2000, the Italian market for Turkish tomato paste completely dried up. In 1999, Turkish producers exported 16,077 tonnes of tomato paste to Italy (roughly equivalent to 95,000 tonnes of raw product) (IGEME, 2000). The loss of this market contributed to a significant fall in export prices for Turkish paste. Between 1999 and 2000, the total tonnage of Turkish tomato paste exports remained almost static, but prices fell from $721 (US) per tonne in 1999, to $535 (US) per tonne in 2000. The following year, Turkey's problems were compounded by a collapse in the Turkish lira. Although ostensibly this currency collapse should have provided the Turkish processing tomato sector with an immediate and dramatic improvement in export competitiveness, in reality it provided a major shock to the domestic economy that damaged the industry. The fact that much debt and trading terms were denominated in foreign currencies mitigated the potentially positive effects of the depreciation.

The pain in Spain: rural restructuring and rural politics

Spain represents a further example of the different prospects facing producers within the Mediterranean. Although Spanish tomato production levels have increased over recent years, the country's industry faces considerable problems. The Spanish processing tomato sector is highly fragmented, with average production per factory the lowest in the EU (Table 5.6). Furthermore, the industry's institutional structures are extremely complex. In 2001, there were 70 producer and grower associations for an industry with 123 processing factories. Individual factories often have to deal with more than one producer/grower association. According to industry sources, the introduction of the 2001 EU regulations has probably complicated, rather than improved, administrative processes. Under the new arrangements, associations responsible for receiving and distributing EU production subsidies are in some cases not the same ones with whom processing firms contract, meaning that an individual grower must belong to two associations; one to receive the Euro 34.5 subsidy, and another through which processor payments are made. These complexities, and their inevitable inter-relationships with rural politics, give credence to Hoggart and Paniagua's (2001, p.63) argument that:

> Processes that analysts associate with rural restructuring are of little help in understanding the Spanish context. As regards capitalist markets, the Spanish countryside is not characterised by economic diversification, professionalism, environmentalism and consumerism on a scale that resembles anticipations derived from the restructuring literature. For state processes, lethargy is a more appropriate adjective than restructuring.

Yet despite Hoggart and Paniagua's gloomy outlook on the intransigence and complexity of rural institutions in Spain, some industry actors have demonstrated considerable scope for flexibility and innovation. In May 2000, for example, Heinz completed a restructuring of its Spanish tomato operations with two smaller factories being merged into a single unit with 100,0000 tonne raw product capacity. Moreover, because of cheaper land and labor in Spain compared to Italy, in 2002 the United States Department of Agriculture forecast that:

> Production of canned tomatoes under the new CAP will be located more and more in Extramadura [Spain] (USDA Foreign Agricultural Service, 2002a, p.4).

At another level, some Spanish producers have captured certain niche markets, including global dominance in tomato powder manufacturing. Although this is a small and specialist product, a number of dedicated Spanish companies have become world leaders in this sector. There is evidence too, of a few family-owned processing tomato firms that have experienced rapid growth over recent years by generating efficiency gains in paste and canning production. Spain's future position in the Mediterranean processing tomato complex will hinge on whether these changes generate wider momentum in the industry, or if they remain isolated pockets of change in an institutional environment beholden to agri-political interests.

Cross-border integration and supply chain restructuring within Europe

In the late 1990s, there was a surge in cross-border acquisitions by large European tomato firms. At the same time that a number of transnational branded food companies such as Heinz, Kraft, Unilever and Nestle, were concentrating on brand management and divesting some of their productive assets, a new generation of processing firms were developing international scale and scope.

This is a significant development in the evolution of the European sector, because traditionally, production has been demarcated strongly by national borders. A comparison of the four largest European markets for the processing tomato sector (the UK, Italy, France and Germany) indicates wide diversity in the product mix and leading companies of each (Tables 5.9 and 5.10).

Table 5.9 Key European markets for processing tomato products, 1999

Percentage of market	UK	Italy	France	Germany
Ketchup	20%	2%	18%	39%
Sauces	56%	17%	57%	38%
Passata	2%	33%	9%	8%
Chopped	12%	27%	5%	3%
Whole peeled	10%	21%	11%	12%
Total market (millions Euros)	605	578	329	302

Note: Product-mix excludes tomato paste sales.

Source: di Bella (2001, p.15), based on Neilson data.

Table 5.10 Leading processing tomato firms in key European markets, 1999

Italy	Cirio, De Rica [Cirio], Star, Valfrutta
UK	Dolmio [Mars], Ragu [Unilever], Heinz, Napolina
France	Panzani, De Rica [Cirio], Dolmio [Mars], Buitoni [Nestle]
Germany	Knorr, Thomy, Heinz, Kraft

Note: Product-mix excludes tomato paste sales.

Source: di Bella (2001, p.15), based on Neilson data.

Cross-border acquisitions have taken many forms, and have been based on a diversity of corporate strategies. Acquisitions have been made in Spain, France and Greece, but a notable target have been firms in Portugal. In the years leading to the 2001 subsidy restructuring, control of Portuguese production gained in strategic significance. Italian interests purchased four Portuguese factories: three to the Parmalat group, and one to Cirio. The high national quota possessed by Portugal was a major motivating factor behind these acquisitions. In 1999, Portuguese processing tomato output was (only) 17 per cent over quota, compared with 42 per cent in Italy. However, Portuguese facilities were also attractive because of their generally larger size. As revealed in Table 5.6, the Portuguese processing tomato sector has considerably greater economies of scale than others in Europe. With the prospect of more intense international competition and lower levels of processor aid, these factors gained in strategic importance for European processing tomato firms.

Recent events in France also highlight the ways in which supply chains and regional production complexes are currently being restructured. France is a relatively small processing tomato nation, traditionally reliant on imports from

Italy and Spain to meet domestic demand. French production is dominated by four producers; the Le Cabanon cooperative, which accounts for roughly 50 per cent of French output, and three smaller companies, Conserves France, the Louis Martin group, and Ciradour. Over time, this industrial structure has become increasingly meshed with Italian production. Italian interests own Conserves France, and Ciradour is partly owned by the Italian Cirio Group. Le Cabanon has also restructured its operations in line with new European market structures. During the second half of the 1990s the cooperative increased its contract production output, particularly of retail branded products such as ketchup. To this end, it has used considerable quantities of imported Italian paste in addition to its members' supplies.

Conclusion

In Europe, national and supra-national governments have played an extensive role in the processing tomato industry. Since 1978, the EU processing tomato sector has been the recipient of significant direct and indirect industry support, which was restructured in 2001. A complex array of rules and regulations has been developed to administer this support regime. Through the 1990s, the maintenance of this regime intersected increasingly with the multilateral politics of world trade reform.

European agricultural policies are at a critical historical juncture, with the politics of market liberalization being counter-balanced by the politics of rural development and food consumption. It is not evident, at the time of writing, how EU agri-bureaucrats, politicians and industry participants will reconcile these conflicting pressures. In the processing tomato industry, recent regulatory changes and intensified international competition appear to be leading to significant changes to industry structures and performance. The focus of these changes is Italy, where in both the south and north, industry participants are engaging in various restructuring endeavours.

The accelerated pace of change seen in the EU processing tomato industry over recent years will recast competitive structures and give impetus to new restructuring initiatives. As efficiencies are won, dependencies on the drip feed of subsidies will abate and new entrants are likely to be attracted to the industry. The emergent industrial structures of the EU processing tomato sector will not entirely resemble those elsewhere – the concept of a 'distinctly European' processing tomato industry will continue to hold sway – but at the same time, it also will hardly resemble the industrial structures which have historically dominated this sector.

Notes

1 However with EC Regulation 0161/99 the Commission amended this to take into account adverse weather conditions in Portugal.
2 Two per cent of national quotas were set aside annually to accommodate new industry entrants, but the system as a whole worked in favor of existing industry participants.
3 The AMITOM data contain many contradictions. Clearly, AMITOM is not the source of these problems, and the organization does the best it can with unreliable data from national statistical agencies and industry sources.

Plate 5 Hand harvesting of tomatoes, Xinjiang, China, 2001

Photo: Bill Pritchard.

Chapter 6

Cheap tomatoes from developing countries? The case of Thailand and China

The concept of agri-food globalization implies that developing countries are being gradually 'drawn in' to the logic of a global marketplace, particularly as sites for export. Relatively lower costs for land and labor, in addition to biophysical and environmental attributes, provide a strong competitive basis for agri-food industries in many developing countries. In the processing tomato industry, the standardization of tomato paste as a bulk industrial input, combined with the evolution of large, international buyers of processing tomato products such as retail chains and branded food corporations (see Chapter Three), have led some industry analysts to observe that competition from developing countries poses a major threat to tomato industries in developed nations.[1]

In interviews with industry participants during the period 2000-02, these issues were often raised in terms of a comparison with recent international trends in the garlic industry. It is unclear how the narrative of international restructuring in the garlic industry became such a popular point of reference within the processing tomato sector, but we were surprised at the number of occasions on which industry actors (in different countries) referred to this case. Clearly, through some process of discursive momentum, the recent history of the garlic industry was thought to provide a vision for what might happen to the processing tomato industry.

The popularity of this metaphor lies with the fact that international garlic production has undergone dramatic change recently. From 1992 to 2000, Chinese garlic exports rose from 128,239 tonnes to 383,860 tonnes, making China the world's largest garlic exporter (in 2000, the next largest exporter, Spain, exported only 65,070 tonnes) (FAO, 2002). Moreover, this growth occurred despite a decision by the US Government in 1994 to impose a 376 per cent anti-dumping duty on Chinese garlic, following an import surge the previous year (Iritani, 2001). The US Government decision stalled further import penetration by China in the US domestic market, but US garlic production during the 1990s continued to come under intensified import competition from other developing nations, particularly Mexico (Skorburg, 2002). US garlic imports rose from 17,320 tonnes in 1997, to 43,372 tonnes in 1999 (FAO, 2002). Also during this period, there was significant import growth from Thailand which, according to some US producers, may have represented Chinese exports being routed through Thailand with the aim of bypassing the prohibitive anti-dumping tariff against China (Skorburg, 2002).

There are several similarities between the garlic and processing tomato industries that may account for the way the garlic industry's experience resonated within the processing tomato industry. In both the processing tomato and garlic sectors, China became a key source of international competition during the middle years of the 1990s. The difficulties of the US garlic industry, based mostly in California, were seen by some observers as having parallels in the experiences of the Californian processing tomato sector over the period 2000-02. Generally weak market conditions in the processing tomato industry during these years, along with the high value of the US dollar, generated some negative assessment of the industry's future in California.

The key question, of course, is whether the experiences of the garlic industry provide any insights into the potential role of developing countries within the processing tomato industry. In other words, to what extent is the processing tomato industry engaged in a 'race to the bottom' in which the cheaper land and labor available in developing countries provides a key competitive dynamic for the industry as a whole? To answer this question, this chapter first considers the role played by developing countries in the processing tomato industry, and then provides a detailed analysis of the processing tomato industry in two developing countries, Thailand and China.

'New agricultural countries' and the processing tomato industry

Examples of rapid growth in agri-food exports from developing countries, such as the garlic industry, bring into focus recent scholarship on the role of these nations within the world's food systems. In the early 1990s, the term 'New Agricultural Countries', or NACs, was coined to describe developments that saw strong growth in agri-food exports from certain developing countries. This growth differed from the traditional colonial model of agri-food exporting (dominated by commodities such as coffee, tea or cocoa), because it was based on the production of a range of higher valued foods for western consumers, such as shrimp, counter-seasonal fruits and vegetables, and beef. In some instances, this growth was associated with policy shifts away from arrangements that accorded agriculture a privileged and protected status, and was often connected to the politics of international debt repayment and structural adjustment programs initiated by lenders and multilateral agencies (Friedmann and McMichael, 1989; Mingione and Pugliese, 1994, p.56). Such issues were most evident in the livestock sector in Brazil and Argentina, where the emergence of an export beef complex was underwritten by feed grain imports from the developed world (especially subsidized product from the US) (Friedmann, 1994, pp.270–71; Sanderson, 1986). In other instances, the impetus for growth came from dynamic national firms, which identified new export markets and the opportunity for new sources of profit. In Thailand for example, this process involved the export of products such as canned pineapple, poultry and shrimp to affluent markets such as Japan. Similarly, the expansion of export horticulture (especially fresh and counter-seasonal products for affluent western markets) has been a key sector in Chile (Friedland, 1994) and some parts of Africa (Friedberg, 2001). Export horticulture had become the third largest agri-food export from Sub-Saharan Africa by the middle 1990s, behind coffee and cocoa (Berry, 2001, p.137).

Transnational agribusiness firms from Europe, the US and Japan, have often been centrally involved in these transformations. However as Burch (1996, p.325) observes with respect to Thailand, the growth of export agriculture consistent with the NAC model does not necessarily involve foreign capital. In Chile, the burgeoning export fruit sector of the 1990s created 'a new agrarian bourgeoisie ... [including] urban industrialists and professionals ..., foreign investors including multinational corporations, and some of the previous landholding families' (Berry, 2001, p.139). Nevertheless, whether controlled by domestic or foreign interests, the export of agri-food products from developing countries tends to be broadly similar in the way it is positioned within international supply chains. Research on the São Francisco irrigated export agriculture complex of northern Brazil (the largest irrigated agricultural region of Latin America) reveals agri-food exporters as highly vulnerable when operating within tightly organized supply chains controlled by 'near-consumer agencies', such as supermarket chains based in developed nations (Marsden, 1997, pp.174–77). Friedberg's (2001) research comparing two African-European supply chains for green beans (the Zambia to Britain chain, and the Burkina Faso to France chain) emphasizes that, despite vast differences in the ways these chains operate, in both cases farm producers bear the major component of risk and remain in a subservient position with respect to buyers who are generally larger and more geographically mobile. Fold's (2002) research on the West African-European cocoa-chocolate supply chain documents how the rise of transnational branded chocolate companies in Europe, combined with the deregulation of state marketing boards in Africa as part of structural adjustment programs, has systemically weakened the bargaining powers of cocoa producers. These cases demonstrate that international agri-food supply chains operate in ways which endow some actors with greater abilities to add value and exercise control than others, and that these power relations generate particular environments for risk and profit. As the cross-national research cited above indicates, key contemporary trends in international agri-food restructuring have often tended to erode the economic position of farmers and their families in developing countries.

These developments are important to our understanding of the relationships between export agri-food production on the one hand, and poverty alleviation, development and inequality on the other. Many mainstream economists tend to argue that agri-food exporting helps to alleviate rural poverty in developing countries, because it raises wage-labor employment in places where employment and income levels may be considerably below urban levels (Berry, 2001, pp.126–28). In contrast, political economists 'are on average less optimistic' (Berry, 2001, p.129) because of the historical record which demonstrates that an increase in food exports is usually accompanied by a rise in food imports, with negative implications for rural food security (Barkin and DeWalt, 1988; Bessis, 1991, cited in Mingione and Pugliese, 1994, p.65). In this tradition, a recent study of agricultural performance in nine Latin American countries (Brazil, Mexico,

Argentina, Colombia, Peru, Chile, Costa Rica, Bolivia and Jamaica) since the 1970s concludes:

> Full market deregulation [encouraging export agriculture] did not work for the agricultural sector as expected in the sample of countries analyzed above. On the one hand, there was little, if any, noticeable change in the already positive trend of output growth that characterized earlier periods of little if no reform. On the other hand, developments during radical attempts of macro-reform did affect the agrarian sector in terms of increasing volatility and deteriorating distribution (Spoor, 2002, p.398).

Given these outcomes, the shifts in power relations associated with agri-food export strategies are of critical concern. For example, in some cases agri-food exporting tends to be predicated on land tenure reforms that facilitate the establishment of large corporate farming enterprises. In Brazil, for example, the promotion of soya cultivation has been central to strategies aimed at increasing agri-food exports, but because soya tends to be grown on large landholdings, these policies have been implicated in the displacement of smallholders (Coote and LeQuesne, 1996, p.195). By the end of the 1990s, less than one per cent of the Brazilian population owned 54 per cent of the fertile agricultural land (McMichael, 1998, p.107). In other cases, agri-food exporting is based on small-holder involvement in contract farming, in association with transnational agri-food corporations. In countries where this strategy is used extensively, for example Thailand, research has shown that contract production offers no significant benefits *vis-à-vis* alternative rural production arrangements, when it comes to issues of income generation or food security (Burch, 1996; Goss, 2002). Consequently, it has been suggested that 'the growth of agro-export "platforms" is an unstable strategy' (McMichael, 1998, p.105), because food security and rural incomes become dependent upon the monetary gains from export earnings, but control over these earnings is largely outside the domain of developing countries.

Although the expansion of agri-food exporting from developing countries over the past twenty years reflects an important shift within the global food system, the size and scope of this change needs to be kept in perspective. Export growth in 'new' commodity areas reflects only one element of recent agri-food restructuring in developing countries. Despite the significant export growth of certain agri-food products from a number of developing countries, when viewed in value terms, developing countries have actually *lost* global agricultural market share since the 1980s. The share of developing countries in total world agricultural exports fell from 46 per cent in 1986 to 42 per cent in 1997 (Private Sector Agricultural Trade Task Force, 2002, p.2). This decline results from two major developments. Firstly, falling commodity prices have reduced the monetary returns of agri-food exports from many developing countries, with significant implications for their balance of payments and their abilities to meet debt repayment schedules. Between 1980-2000, there were dramatic falls in prices for key developing country agricultural exports, including cocoa (71.2 per cent), coffee (64.5 per cent), palm oil (55.8 per cent) and rice (60.9 per cent) (Oxfam, 2002, p.151). Secondly, the global agricultural trading environment has remained dominated by the protectionist policies of the wealthier economies. Following the Uruguay Round Agreement on Agriculture (WTOAA, see Chapter Five), developing countries have been obliged to open their economies to market liberalization while at the same time the EU, US and Japan have been required to reduce protection by only modest amounts. These structural outcomes have had the two-pronged effect of limiting developing countries' export access to lucrative markets, and encouraging subsidized exports from developed to developing nations which destabilize local markets. They have also encouraged a situation whereby 'survival in agricultural markets depends less on comparative advantage than upon comparative access to subsidies' (Watkins, cited in McMichael, 1998, p.97).

These points emphasize the need for caution when interpreting the impacts of agri-food exporting from developing countries. The NAC model has relevance in the sense that it points to the possibility for export-oriented agri-food systems emerging in developing countries, at the instigation of either local or foreign investors. However, in the context of the global restructuring of agri-food supply chains, it is often the case that such industries are highly vulnerable because of their dependence on larger buyers, usually in developed nations, and the fact that agri-food industries in developing countries may not have access to the subsidies available to some of their developed country competitors.

These issues suggest two key questions with respect to the role of developing countries within the global processing tomato industry. First, to what extent is processing tomato production likely to shift to developing country locations, and second, is processing tomato production in developing countries vulnerable to the volatility of external markets? These questions are now considered, prior to detailed analysis of two case studies of the emergence of processing tomato exporting from Thailand and China.

The role of the developing countries in the world processing tomato industry

Developing countries account for slightly less than 30 per cent of world processing tomato production (Figure 6.1), and although this proportion has been relatively stable in aggregate terms over recent years, there have been considerable shifts in its composition (Table 6.1). Among the developing countries, the most notable change relates to an approximate doubling of Chinese production between 1993 and 2001. In addition to their roles as production sites, some developing countries are also significant consumers of processing tomato products, either from local or imported sources. Within the global processing tomato industry, important trade networks have been established, in which tomato paste manufactured in relatively lower cost regions (including, for example, Turkey and China) is exported to some relatively less wealthy markets (for example, the former Soviet Union). Additionally, as discussed in Chapter Five, Italy is an aggressive exporter of tomato paste to markets in the Middle East and Africa. Some Italian paste is manufactured with the benefit of European Union producer support payments, and some is re-processed using bulk imports from source nations such as Turkey and China. In recent years, developing country agricultural trade negotiators have cited the example of material damage to the Senegalese processing tomato industry, in their campaigns for the reform of European agricultural policies. In 1995, the World Bank forced Senegal to liberalize its processing tomato sector, allowing for significant import competition from subsidized Italian product, as part of a structural adjustment package (Tomato News, 1995c, p.30).

Table 6.1 Processing tomato output in major developing countries, 1993-2002 (tonnes)

	1993	1994	1995	1996	1997	1998	1999	2000	2001	2002
Mexico	350	360	275	140	250	290	380	216	136	100
Hungary	52	115	156	182	90	220	130	128	100	100
Bulgaria	200	130	150	150	70	156	150	90	30	130
China	450	600	550	610	480	780	800	1,800	1,000	2,500
Brazil	700	700	930	680	1,096	1,017	1,290	1,200	1,000	1,200
Chile	488	733	822	854	600	867	950	925	725	550
Argentina	113	240	190	284	270	224	330	260	255	210
India	150	180	80	85	100	45	120	140	120	120
Thailand	100	130	220	226	250	121	188	124	140	160

Note: Data for 2002 is preliminary.

Source: Tomato News (various issues).

In terms of developing country participation in processing tomato exporting, it is clear that the establishment of export-oriented processing tomato sectors is the exception rather than the rule. In most developing countries, production is geared either for domestic or near-neighbor markets. In the Middle East and North Africa, for example, Turkey is the only significant exporter (Table 6.2). The processing tomato industries of Bulgaria and Hungary are oriented to the markets of eastern Europe and Russia, largely as a result of their past involvement in the Soviet bloc. In Latin America, Brazil, Argentina and most other countries tend to produce mainly for domestic purposes. Mexico exports approximately 50 per cent of its annual processing tomato output, but the overwhelming majority of this is sold to the adjacent US market. Moreover, the Mexican industry is unique among the world's processing tomato sectors, in that it plays a subservient role to the fresh tomato sector; the volume of tomatoes that are sent for processing can depend on the price of fresh tomatoes, so that when fresh tomato prices are low, the *saladet* fresh tomato is diverted to processing (Tomato News, 1993c, p.17). Of the remaining major world producers, less than one per cent of Indian and South African processing tomato production is exported. Accordingly, there are only four significant examples in the developing world where the processing tomato industry has evolved to play a major role in export markets: Turkey, China, Thailand and Chile.

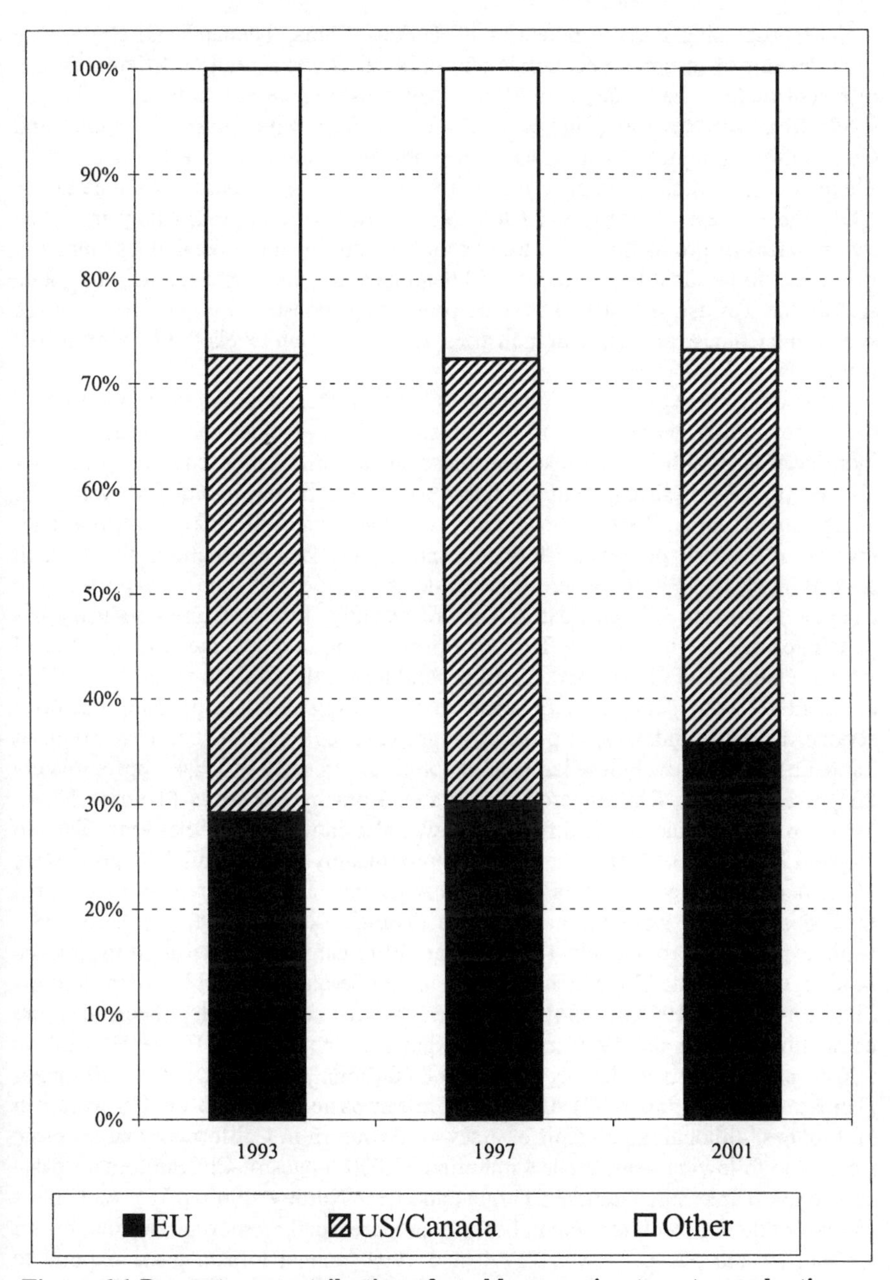

Figure 6.1 Percentage contribution of world processing tomato production, by major region, 1993, 1997 and 2001

Source: Tomato News (various issues).

The processing tomato industries of Turkey, China, Thailand and Chile have each developed in unique economic and political contexts, and at different points in recent history. Each has been affected by considerable volatility in production levels from year to year (Figures 6.2, 6.3, 6.4, 6.5), with Turkey, Thailand and Chile suffering significant downturns in production volumes since the late 1990s. There is considerable difference too, in the size of these industries. As recently as 1999, the Chinese, Turkish and Chilean industries were approximately the same size in terms of production volumes (Table 1.1), but by 2001, the Chinese industry was some 30 per cent larger than the Chilean industry, and approximately the same size as the Turkish industry. The Thai processing industry is about one-tenth the size of the Chinese and Turkish industries, and production levels fell by almost half from 1995 to 2000.

Moreover, there are further differences between these four countries in terms of the export characteristics of their industries. The Turkish processing tomato industry was established earlier than those in the other three countries, and has always possessed strong trading links with neighboring countries. With the exception of Japan (which historically has accounted for approximately one-quarter of Turkish processing tomato exports), the key destinations for Turkish exports have been Italy (where bulk paste is re-processed and re-exported: see Chapter Five), the Middle East and, more recently, Russia. Chile's industry has certain parallels with that of Turkey, insofar as its exports are also positioned within regional and global markets. In general terms, the Chilean industry is highly competitive, with economies of scale in processing (eight processing factories service the entire industry), a productive growing region (150 kilometres south of Santiago), and relatively low land and labor costs. As early as 1994, approximately 80 per cent of the Chilean crop was grown from hybrid seeds (Tomato News, 1994a, p.18), and mechanized harvesting was also introduced in that year (Tomato News, 1994c, p.18). During the 1990s, approximately half of Chile's exports were to the neighboring countries of Brazil and Argentina, and of the remaining exports, the US and Japan were the key markets (Tomato News 1994a, p.15). In 1989, Chile exported approximately 18,000 tonnes of tomato paste (measured in finished product terms) to the US, which was then about one-quarter of all US paste imports (Tomato News, 1993a, p.34). However, as the US industry became more competitive during the 1990s, Chile's exports to the US fell. By 1992, Chilean tomato paste exports to the US were only 8,000 tonnes, and in 1993, 5,000 tonnes (Tomato News, 1995a, p.33). By 1997, Chilean paste exports to the US were just 500 tonnes, although as a result of a severe downturn in Californian tomato paste stocks the following year, the US imported 31,000 tonnes of Chilean tomato paste as a 'one-off' measure (Rausser, Fargeix and Lear-Nordby, 2000, p.10). As Chile's exports to the US declined, Japan became an increasingly important destination for Chilean tomato paste. Between 1990 and 1994, annual tomato paste exports to Japan increased from 7,951 tonnes to 13,444 tonnes, and fluctuated between 10,000 to 15,000 tonnes for the rest of the decade (Tomato News, 1993b, p.38; Tomato News, 1995b, p.28; Tomato News, 2001d, p.9).

Table 6.2 Exports of tomato paste as a percentage of total production, selected countries, 1993, 1996 and 1999

	1993	1996	1999
China	63	53	108
Algeria	1	11	0
Israel	28	77	25
Morocco	21	10	15
Tunisia	8	8	28
Turkey	82	51	59
Chile	56	85	104
Brazil	24	20	9

Note: Data indicating tomato paste exports are the equivalent of over 100 per cent of production is explained because (i) in the relevant calendar year there was the sale of previous year stocks to foreign buyers, or (ii) there are statistical discrepancies between the two series (in some countries there are significant questions about the reliability of official production statistics).

Sources: Calculated using data from FAO (2002) and Tomato News (various issues).

Thailand differs from these two countries in that it has a greater reliance on the canned tomato sector. In 1994, canned tomatoes represented approximately 23 per cent of output, involving 30,000 tonnes of raw product out of total production of 130,000 tonnes (Tomato News, 1994c, p.21). Furthermore, as discussed later in this chapter, a considerable proportion of Thai tomato paste is not exported directly, but is used as an input for other processed food products (notably, canned fish), which are sold to export markets. For this reason, the export levels of the Thai processing tomato sector are difficult to estimate precisely, and hence Thailand is not listed in Table 6.2. Finally, China's experiences differ from all of the above, in its scale of development and explicit orientation to global markets. As discussed later in this chapter, the Chinese processing tomato industry was developed largely on the basis of external capital and technical advice (especially from Italy) and had as its aim the capture of distant markets in the Mediterranean and Middle East, in addition to markets in the Asia-Pacific.

An appreciation of these differences is vital to an understanding of the global dynamics of the processing tomato industry, and in particular, the role being played by developing countries. As noted at the outset of this chapter, the concept of agri-food globalization seems to imply progressively more enhanced participation by developing countries in world food complexes. Reductions in trade barriers, flexibility in the location of production, lower labor costs, lower shipping and transport costs, and greater international consistency in industry standards, point to increased opportunities for processing tomato exports from developing countries. However, the extent to which these developments are realized depends upon the particular ways that specific developing country processing tomato sectors are

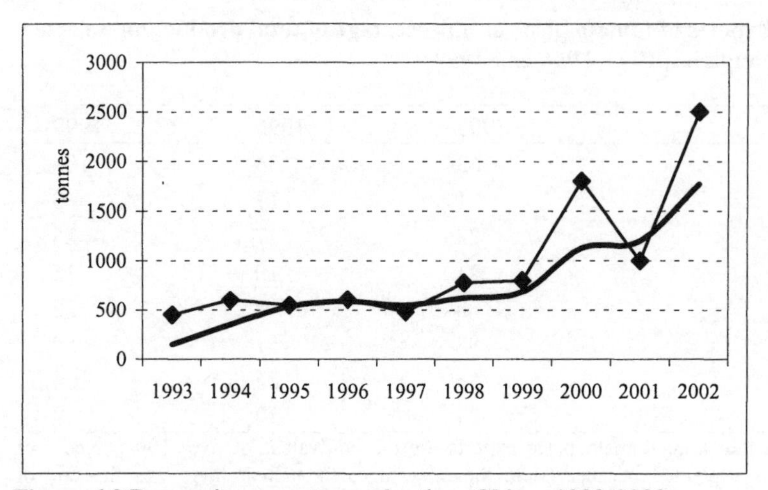

Figure 6.2 Processing tomato production, China, 1993-2002

Note: Smoothed line represents the three-year moving average.

Source: Tomato News (various years).

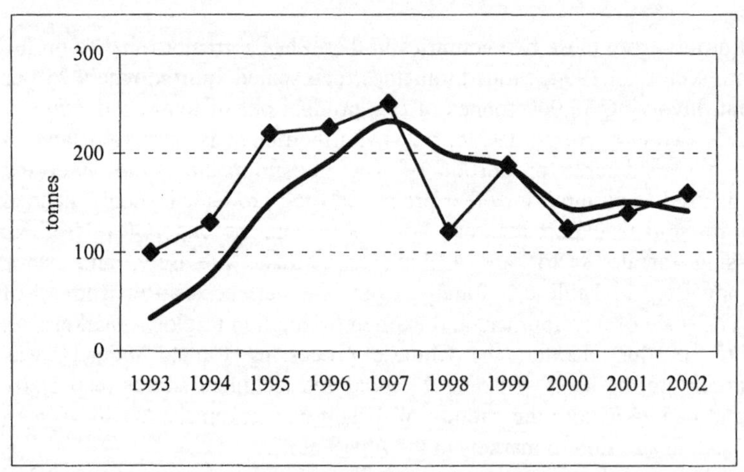

Figure 6.3 Processing tomato production, Thailand, 1993-2002

Note: Smoothed line represents the three-year moving average.

Source: Tomato News (various years).

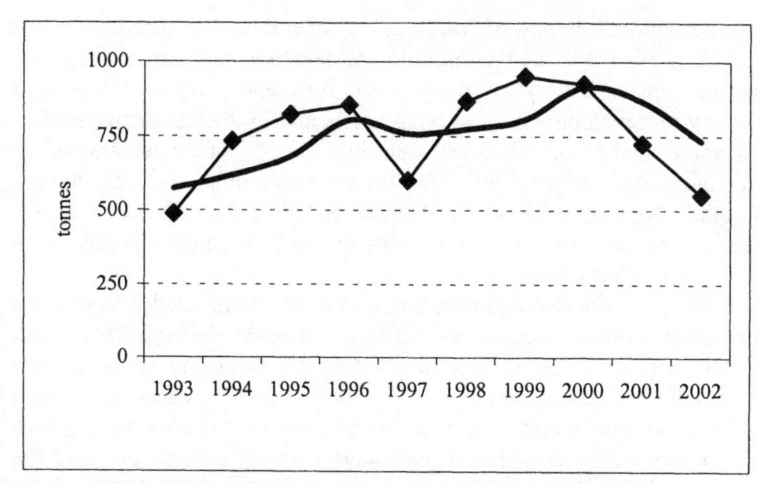

Figure 6.4 Processing tomato production, Chile, 1993-2002

Note: Smoothed line represents the three-year moving average.

Source: Tomato News (various years).

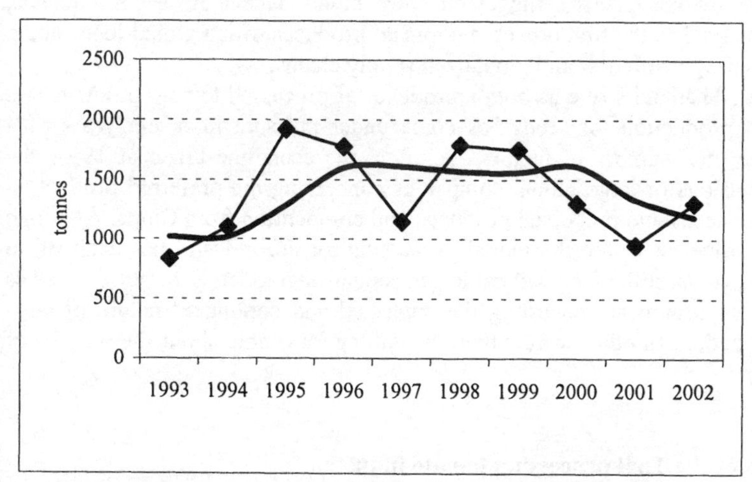

Figure 6.5 Processing tomato production, Turkey, 1993-2002

Note: Smoothed line represents the three-year moving average.

Source: Tomato News (various years).

connected to international arenas of trade and competition. Consistent with the general argument of this book, the incorporation of developing countries within the world processing tomato industry is a contested historical process, featuring a complex interplay of developments at varying scales and in different geographical territories. In short, global agri-food restructuring should not be understood in terms of an 'inevitable' process of intensified competition from developing countries (as might be assumed from the 'garlic model', discussed earlier in this chapter), but as a process by which countries are re-positioned globally, with varying (and unpredictable) consequences.

To support this position, detailed case study research is required to document the evolution of processing tomato industries in specific developing country contexts. To this end, attention is now directed to Thailand and China as two prominent examples of the evolution of export-oriented processing tomato industries in the developing world. The selection of these case studies corresponds to our desire to emphasize the diverse pathways of such production, and the varying impacts which may be experienced by other producing countries. Thailand is not a major producer in global terms, but has merit as a case study because it was one of the first developing countries to engage in the export of processed tomato products, and was particularly important in terms of the development of innovative sourcing strategies by a number of major corporate players, including major western supermarket chains selling their 'own brand' labels. As a consequence, Thailand emerged as the first processing tomato producer with a global focus and a capacity to engage with new and globalized supply chains.

However, Thailand's role as both a producer of processed tomato products and as a site of production for seeds has come under pressure in recent years as a consequence of a number of factors, including the economic crisis of 1997, the strategic decisions of large global companies concerning the preferred production sites for both seeds and processed products, and competition from China. Although Thailand's influence within the global processing tomato industry has declined in recent years, it is still of considerable conceptual significance because it gives support to our arguments regarding the contested and contingent nature of agri-food globalization. In other words, there is nothing inevitable about the process of globalization.

The origins of the Thai processing tomato industry

Tomatoes for the fresh food market have been grown in Thailand for many years, but until the early 1970s there were only a few growers and cultivation was carried out over limited extents of land in only two provinces (Kanchananaga, 1973, p.98). In the early 1980s however, fresh tomato production began to expand. Between 1981 and 1982, the production of fresh tomatoes increased dramatically, from 3,798 tonnes to 51,230 tonnes. Most of this increased production was sold on the domestic market, but some was exported to Malaysia, Singapore and Hong Kong (Government of Thailand, various years). The great bulk of this increased production was grown in the north and northeast of Thailand.

Given that fresh tomatoes were not a major part of traditional Thai diets and that export opportunities were relatively limited, the reasons for the rapid growth in tomato production during the early 1980s lie with changes in supply rather than increased demand. Of great importance in this regard were the activities undertaken by the King of Thailand, H.M. Bhumibol Adulyadej who, from 1969, established a number of major projects in the north and northeast of Thailand. These 'Royal Projects', as they were called, were originally designed as welfare and development activities which would assist the impoverished hill tribes people of northern Thailand. This was to be achieved firstly, by funding projects which would encourage the production of a range of crops which could serve as alternatives to poppy production for opium, and secondly by undertaking land conservation and management techniques in order to reduce the forest and watershed destruction which was threatening the livelihoods of disadvantaged minority groups (Judd, 1989). Under the program to develop alternative crops to poppies, tomatoes emerged as the most important commodity. Indeed, so successful were these operations that hill farmers began to produce a large surplus, which was difficult to market. As a result, the Royal Projects established a Food Processing Section in 1972, with the aim of creating a capacity to process the surplus of fresh produce in order to expand markets and maintain prices. The first of four processing factories was opened in the Fang District of Chiang Mai in northern Thailand in 1973 (IMI/KMIT, 1996). In subsequent years, this factory expanded its marketing operations, and not only purchased the surplus product of the hill farmers, but started to promote contract production. By 1985, the factory was engaged in contract vegetable production for a wide range of commodities (Table 6.3).

Table 6.3 Value of production of processed foodstuffs, Royal Project factories, 1985-93 (Baht, millions)

	1985	1987	1989	1991	1993
Baby Corn	5,208	14,874	25,894	42,537	55,257
Green Okra	—	—	—	1.973	12,977
Full Fat Soy Flour	2,937	2,973	4,064	4,185	9,500
Strawberry	—	—	—	19,106	23,381
Tomato	5,362	14,865	37,394	52,973	33,878
Longan	—	—	1,824	—	28,528
Lychee	—	—	2,168	1,1687	13,957
Passionfruit	—	—	1,909	1,724	11,547
Bamboo Shoot	—	—	1,188	2,984	9,493
Other products	1,363	4,241	1,459	6,528	43,842

Note: '—' indicates zero.

Source: IMT/KMIT (1996).

The second of the Royal Project factories was established at Mae Chan in Chieng Rai, soon after the first facility was built, and a third factory was established in 1980 in the province of Sakhon Nakorn in northeast Thailand. Significantly, the establishment of the third factory marked a more determined shift in Royal Project activities towards the use of contract farming in order to generate new sources of income. As well as establishing a childcare centre, infirmary, a temple and a well, the project also taught farmers to grow tomatoes as an off-season crop, which were processed into canned products. Subsequently, the project expanded further, taking in more farmers over a wider area and growing a wider range of crops for processing, including baby corn, bamboo shoots and papaya (IMT/KMIT, 1996, p.9). The fourth factory was established in the Laharnsai district of Burirum Province in northeast Thailand, in 1982. The main impetus here was not simply poverty alleviation and the delivery of welfare services to the rural poor, but also an attempt to counter the effects of the conflict between the Thai army on the one hand, and local Thai communist insurgents, assisted by the Khmer and Vietnamese communist troops who were active in the Thai-Cambodia border region, on the other. Five villages in the project area were selected for the establishment of a child nutrition centre, an infirmary and village services. At the same time, farmers in these villages were introduced to the cultivation of baby corn and tomatoes for processing. The processing facility was in operation by 1983 and, as in the case of other factories, contract farming operations were subsequently expanded to take in farmers in more far-flung districts, who were growing tomatoes, baby corn, passion fruit, papaya, bamboo shoots and more (IMT/KMIT, 1996, p.10). The Food Processing Section of the Royal Project remains one of the largest fruit and vegetable processing companies in Thailand, and tomatoes are a key component in the company's product range. Output from the Royal Project is mostly marketed under the proprietary *Doi Kham* label, although the company has also produced canned goods for supermarket labels overseas, and under the brand names of other companies. By 1993, the four food processing factories established by the Royal Projects had a turnover of nearly $10 million (US), and processed the output of some 12,000 farm families from 627 villages (IMT/KMIT, 1996).

At about the same time as the first of the Royal Project factories was established, the first private sector interests were attracted to the production and processing of tomatoes. In 1973 the Eisenberg Group of Companies, an Israeli conglomerate, established Mae-Jo Agricultural Industries Ltd, which set up a food processing company in Chiengmai Province, in northern Thailand (Laramee, 1975). Subsequently, this operation was incorporated into the Lanna Food Complex, a joint venture between the Eisenberg Group and a number of Thai businessmen. This company, with capital of $10 million (US) and based in Chiangmai, consisted of four distinct subsidiary companies: the Crown Frost Company (which operated a freezing plant); Thai Tomato Industries (which operated a canning plant); D.V Industries (involved in dehydration processes), and the Thai Farming Company, a farming operation which was to act as a nucleus estate, growing produce to supplement the wide range of vegetables being supplied under contract by small farmers in northern Thailand (Laramee, 1975). The company's plans involved the cultivation of sweet corn, baby corn, tomatoes, cabbage, onions, peppers and beans. Tomatoes were among the most important commodities for processing, with up to 350 hectares being required to meet the available processing capacity for canned tomatoes. However, production did not extend beyond one or two seasons, although in the absence of any documentary sources, the exact reasons for this are unclear.

However, notwithstanding the failure of the Lanna Food Complex and the welfare focus of the Royal Projects, both operations succeeded in laying the basis for the major investments in tomato processing which were to occur from the late 1980s. Both succeeded in developing new crops and new cultivation skills, and combined these with the routines of contract farming, which was reflected in the expansion of this sector from 1986 onwards.

Expansion in the late 1980s

The growth of the tomato processing industry in Thailand from the late 1980s was based on four major factors. The first was a growing demand for tomato paste from a number of domestic sources, most notably the seafood canning industry and the newly emerging fast food industry. In terms of the seafood canning industry, by 1984 some 6,300 tonnes of tomato paste was being used in the canning of sardines and tuna for both export and domestic markets, with imports accounting for about half of these volumes. According to other sources, this figure had risen to 20,000 tonnes per year by 1988 (Becker, 1989, p.12). In addition, there was the growth of demand for tomato paste that emerged with the establishment of western style fast food chains such as Pizza Hut. This particular chain established Thai operations in 1980 and by 1987 there were eight outlets at major centres such as Bangkok and Pattaya. By 1992 there were eighteen Pizza Hut outlets as well as numerous other fast-food franchise operations selling pizza and Italian-style food. Since that date, of course, the number of such food outlets has grown rapidly, with some 85 Pizza Hut restaurants and many hundreds of Italian-style and pizza outlets being established by 2000 (Burch and Goss, 1999, p.95).

Secondly, investors in Thailand recognized the potential of rapidly opening export markets, particularly for low-cost product lines such as canned tomatoes which were purchased by western supermarkets and sold as 'own label' alternatives to the proprietary brands. The importance of this factor has been discussed extensively in Chapters Three and Four.

Thirdly, the growth of the processing tomato industry in Thailand was connected with the incentive scheme provided by the Thai government through the Board of Investment (BOI). BOI incentives were designed to support a wide range of industrial projects, but were particularly beneficial to export-oriented agri-industrial activities involving the production, processing and marketing of quality agricultural produce. In an attempt to encourage companies to locate away from the south and central areas of the country, and Bangkok in particular, further generous incentives were made available. The BOI designated plants located in northeast Thailand as falling within 'Zone 3' of the BOI Investment Incentive Plan, and as such, were eligible to receive a range of subsidies:

(i) exemption from machinery tax;
(ii) six to eight years exemption from corporate income tax;
(iii) five years exemption from import duties and business tax on raw materials for export;
(iv) full exemption from import duties and business tax on goods for re-export;
(v) 90 per cent reduction on business tax for five years;
(vi) 50 per cent reduction of corporate income tax for five years after initial exemption period (of six to eight years);
(vii) ten years double deduction from taxable income of water, electricity and transport costs;
(viii) deduction from net profits of 25 per cent of installation/construction costs of infrastructure;
(ix) exemption from export duties and business tax on exports;
(x) five per cent of increment in export earnings over previous year deductible from taxable income;
(xi) provision of credits of up to 80 per cent of the value of an export order, made available through nominated banks at 10 per cent per annum interest, repaid on receipt of proceeds from overseas buyers (Canned Food Information Service, 1991, p.16).

Finally, the growth of the processing tomato sector was related to the prevailing economic and social conditions which facilitated the participation of local small-scale farming communities in the industry. For example, many farmers in the north and northeast of the country had had prior experience in contract farming, and particularly in the contract production of tomatoes for processing. At this time, contract farming was still very attractive to many smallholders in Thailand, because it held out the prospect of a secure return far in excess of what most growers could expect from other crops. In some cases, such expectations were borne out. Indeed, Sjerven (1991) has suggested that the contract farming project operated by NACO (see below) helped tomato growers in northeast Thailand to increase their annual earnings fivefold in some cases. However, whether these returns could be sustained was a critical issue. As discussed below, the experience of most contract growers was not particularly encouraging in this respect, and as Watts (1992) has noted, the introduction of contract farming in developing world contexts represents as much a social and political transformation of agriculture, as it does an economic one.

As a result of all these factors, the processing tomato sector expanded rapidly from the late 1980s, and by 1995-96 there were 16 processing tomato factories in Thailand; eight of these in northeast Thailand accounted for 70 per cent of Thai production, and six in the north accounted for 25 per cent of total output (two smaller factories were located in the Bangkok vicinity) (Tomato News, 1996, p.30). From 1993 to 1997, the Thai processing tomato industry increased its annual production from 100,000 tonnes to 250,000 tonnes (Figure 6.3).

As the 1990s progressed, four types of investors were attracted to the industry; (i) newly established specialist tomato processing companies; (ii) existing fruit and vegetable companies moving into the sector; (iii) foreign investors, combining with local interests to create new joint venture companies, and (iv) large and diverse Thai-owned conglomerates which had investments in a wide range of projects in a variety of sectors, and which moved into the processing tomato industry. The Thai processing tomato industry, therefore, was underwritten by capital from a range of sources, structured within a diversity of corporate entities. During the period of rapid industry growth of the late 1980s and early 1990s, until the sharp downturn in the industry following the 1997 Asian economic crisis, firms responded in various ways to the changing environments of opportunity and competition facing the industry. In the following section, a case study of one firm brings into focus the wider issues confronting the Thai processing tomato industry during this period of rapid growth.

The rise and fall of a specialist processor: the Northeast Agricultural Company Ltd

The Northeast Agricultural Company Ltd (NACO) was founded in 1986 in order to produce tomato paste mainly for the domestic canned seafood market. It established its tomato processing activities at Nong Khai, in northeast Thailand, close to the Mekong River and Lao border. Its founder was Mr. Chatchai Boonyarat of the Thai Fruit Canning Corporation, who was a leading figure in the emerging Thai agri-food sector, with interests in pineapples, dairying, palm oil and other, non-agricultural sectors. Chatchai's entry to the processing tomato industry, grew out of an encounter with a business colleague, Mr Paiboon Thienchaniya, a building contractor who also managed his family's 1,600 hectare tobacco-growing operation in Nong Khai province. Tobacco was generating low returns in the 1980s and Paiboon was seeking a joint venture to produce alternative export crops. According to Chatchai,

> tomatoes were the biggest food product in the world market place [sic] and the only exporter in Asia was Taiwan. And we could beat them: our labor costs were lower and we had a better year round climate (Business in Thailand, 1991).

Chatchai also sought equity from Mr Dumri Konantukiet, an exporter of canned tuna and sardines, and this group initiated a feasibility study (Becker, 1989). The initial proposal was for the production of 6,500 tonnes of tomato paste from 48,000 tonnes of raw product, with canned tomatoes and other related products to follow later. Total initial investment in the project was $8.81 million (US), with equity contributions from the Asian Development Bank (ADB) (15 per cent), the International Finance Corporation (IFC) (13.5 per cent) and Thai Military Bank (10 per cent), complementing the investments of Chatchai (31.5 per cent), Paiboon (20 per cent) and Dumri (10 per cent).

Equity participation by financial institutions proved critical for the project's establishment, with the ADB also extending an 11-year loan of $1.46 million (US) and the IFC providing a loan of $1.44 million (US). Expenditure on the project included $2.7 million for the purchase of the tobacco farm owned by Paiboon

Thienchaiya, which was to become the nucleus estate of the production facility, and $3.4 million (US) for state-of-the-art Manzini equipment purchased from Italy. It was proposed that one-third of the 48,000 metric tons of tomatoes produced each year would come from the nucleus estate, and the balance of production from some 3,000 contract farmers. Most of those contracted to produce tomatoes were farmers who grew rice and/or tobacco in the rainy season, but who left their land fallow in the dry season. Tomatoes planted and cultivated over the period December to April, provided a second crop outside of the rice growing period, and created an additional source of income (Becker, 1989; Sjerven, 1991).

The NACO factory began operations in January 1988. In the period January-April 1988, NACO processed 16,500 tonnes of tomatoes and produced 1,052 tonnes of paste. For the 1989-90 season, NACO processed 5,000 tonnes of tomato paste, generating export sales to Japan, Korea and the US, totalling $6 million (US). During this initial period, most of the tomatoes were grown on the company farm rather than by contract farmers. At the high point of the four-month season, up to 1,500 workers were employed on the nucleus estate and up to 75 in the factory (Becker, 1989; Sjeven, 1991).

The success of the first plant led to the establishment of a second processing facility about 60 kilometres away, in Nahkon Phanom. This completed its first season in April 1991, doubling NACO's tomato paste production and generating export sales of $10-12 million (US) (Rainat, 1991; Sjeven, 1991). In overall terms, by 1991 between 60 to 65 per cent of NACO's two factories were being supplied by approximately 6,000 contract farmers, who on average delivered 10 tonnes (Sjeven, 1991).

This contract farming regime was supported by company investments in a nursery and tomato-breeding program. The Bank for Agriculture and Agricultural Cooperatives (BAAC), a rural credit supplier operated by the Thai Government, advanced loans to farmers as working capital to pay for the tomato seeds and other inputs. Repayments to the bank were deducted from the crop payments. NACO also provided extension services to farmers under contract, through a network of 40 salaried field officers and agents who advised farmers in the production of processing tomatoes. Key local growers were also incorporated into the extension network. In small villages, a local farmer was chosen to operate a receiving station where farmers delivered their produce for weighing and grading. In terms of payments to workers on its estate, NACO paid unskilled planters and pickers $1.20 (US) per day (30 Thai Baht), which represented the minimum payable under law at that time. More skilled operatives were paid $1.60 to $2.40 (US) (40 to 60 Baht) per day (Sjerven, 1991).

The successes of the first few seasons, however, were reversed sharply in 1993 when tomato paste export sales collapsed in the face of intense competition from China and Turkey. In 1993, export sales represented just 20 per cent of turnover, compared with 80 per cent during the previous year (Burch and Pritchard, 2001). This led to a restructuring of equity (although the company remained Thai owned), and a subtle change in name, to the Northeast Agricultural Investment Company (NAICO). This change reflected the acquisition of a 10 per cent investment in the company by General Finance and Securities (GFS), a Thai investment company. In

1994, there was a further change in ownership structures, with the company coming under the control of a joint venture enterprise owned by GFS and Brierley Investments Limited (BIL), a New Zealand holding company headed by Sir Ron Brierley. This was, of course, the same Sir Ron Brierley, who also had a majority holding in Cedenco, the New Zealand-owned firm which operated Australia's largest processing tomato factory during the 1990s, until its sale to SK Foods of California in 2001 (see Chapter Four). The GFS/BIL joint venture (known as GF-Brierley) purchased 15 per cent of NAICO, which alongside the 10 per cent of the company owned previously by GFS, was sufficient to give the joint venturers managing control over the enterprise (Burch and Pritchard, 2001).

BIL's investment in NAICO was significant because it was clearly part of a larger investment strategy by the company to expand processing tomato operations globally. BIL's acquisition of a share of NAICO broadly coincided with its purchase of 25 per cent, later increased to 48.5 per cent, of Cedenco. While not having outright majority shareholdings in either Cedenco or NAICO, BIL was clearly in a strong position when its came to making decisions about the operations of both companies, and saw the significant gains which could be made from the integration of the activities of both companies. In the decision to invest in NAICO, for example, the executive director of BIL, Mr. Andrew Meehan, was reported as saying:

> Brierley ... has identified significant 'upside potential' for Northeast Agro-Industry by using New Zealand management practices with the help of a technical service agreement with Cedenco Foods Ltd., a New Zealand tomato paste producer (The Asian Wall Street Journal, 3 March, 1995).

According to John Rodwell, chief executive officer of the GF-Brierley Company:

> The management contract with Cedenco is to supply people and technology systems. Last year the project in the north imported 18 experts and this is now showing substantial improvement in production (Independent Business Weekly, 29 March, 1996).

One such expert spent 18 months at NAICO installing irrigation systems, after which he undertook an investigation into the feasibility of mechanization in order to 'replace the labor-intensive hand-picking, which in the peak season from January to April, can provide employment for up to 1,500 casual laborers' (Burch and Pritchard, 2001). This quote provides two important insights into BIL's investment in NAICO, and the competitive position of the Thai processing tomato industry more generally; firstly, that relatively low labor costs in Thailand were not a critical component to the industry's competitiveness, and secondly, that BIL's competitive vision for its investment was linked to an agenda involving significant transformation to the Thai industry – in particular, the expansion of large-scale tomato cultivation based on mechanized harvesting and hybrid seeds (see Chapter Two). This vision was consistent with Cedenco's corporate strategies at the time, which saw the company shift its processing tomato operations from New Zealand to Australia, with the goal of focusing on the bulk production of industrial grade tomato paste (see Chapter Four). As part of this international corporate strategy, Cedenco had also investigated the prospects for tomato paste production in Egypt, and in Xinjiang Province, China.

Whatever the role assigned to NAICO in BIL's overall strategy, it was undermined by the Asian economic crisis that began in Thailand in July 1997. BIL's joint venture partner, GFS, was one of the first casualties of the crisis, being shut down by the Bank of Thailand as part of a deregistration of 58 bankrupt finance companies (Burch and Pritchard, 2001). The joint venture company itself was also placed under administration and BIL undertook to restructure its investment, along with its other Asian investments which were also proving problematical. The joint venture in NAICO was concluded with considerable recrimination as BIL impounded containers of plant and machinery in Seattle belonging to other companies owned by the joint venturers, and took legal action in the US against GF-Brierley to recover a $400,000 (US) loan (Smellie, 1998). Amidst the weight of debt held by leading shareholders, the recriminations and court proceedings between the joint venturers, and the uncertainty created by the economic crisis, NAICO ceased operations.

There are three key lessons to be learned from the rise and fall of NACO/NAICO. The first relates to the role of investment and entrepreneurship. Local business elites, supported by financial institutions of various kinds, were the prime movers in Thailand's first significant, privately owned, processing tomato firm. Over time, this local control gave way to control by foreign and finance sector interests, but this pattern of control was unsustainable in the wake of the Asian economic crisis. Consequently, NACO/NAICO provides an exemplar for the wider arguments of this book, which suggests that global agri-food restructuring should be seen as a contested and volatile process of change, rather than as an inexorable process of incorporation by which industries come to be progressively dominated by large, usually foreign, interests (Fagan, 1997). Second, the NACO/NAICO example reiterates arguments about the vulnerability of export-oriented processing tomato production. The history of NACO/NAICO revolves around two economic shocks and their aftermaths (the collapse of export markets in 1993 and the Asian economic crisis of 1997). Ironically, the firm's response to the 1993 shock (attracting foreign and finance capital investment) led it down a path that ultimately created a new set of vulnerabilities that were exposed in the 1997 crisis. Third, the attempt by BIL/Cedenco to restructure NAICO's activities in line with its own aspirations for global profit accumulation highlight the ways that processes of global restructuring are manifested at national and regional levels, and change over time.

Thai processing tomato industry restructuring during the 1990s

The experiences of NACO/NAICO provide an insight into broader trends in the Thai processing tomato industry during its roller-coaster ride of growth and crisis over the 1990s. By the year 2000 there were 26 companies in Thailand involved in the processing tomato industry (Table 6.4). However, this gives a somewhat misleading indication of the political economy of ownership and control within the Thai industry. Most of the companies listed in Table 6.4 were not engaged in first stage processing, but instead were either re-manufacturers, using tomato paste as an industrial food input, or companies with a pre-eminent interest in marketing and trading, using a relatively few first-stage processors as contract manufacturers for their own line of branded products. Analysis of these processing companies reveals a narrative of volatility, vulnerability and crisis within the industry, including attempts to resolve these problems through the injection of non-Thai investment. Yet, such investment appears to have produced a hiatus in the industry's development, rather than generating a new round of growth. Declining production since the late 1990s (Figure 6.3) has been accompanied by shifts within market channels, with the Thai processing tomato industry largely retreating from export markets. In 2000, only 5.5 per cent of Thai tomato paste was exported (authors' calculations based on official data), and canned tomato exports were also at very low levels. Perhaps the most telling indicator of the Thai canned tomato industry's fortunes was the fact that in the late 1990s, when the Australian Customs Service was reconsidering the levying of anti-dumping duties on imported canned tomatoes, it decided to not impose duties on Thai product because of their insufficient volumes. In general, the Thai industry has largely returned to its roots, as a producer of paste for re-processing purposes in the southeast Asian market.

Consistent with the general pattern of industry development since its inception in the 1980s, the key firms in the Thai processing tomato sector are characterized by diversity in ownership structures and profit accumulation strategies (Table 6.5). A number of Thai-owned firms have pre-eminent positions in the industry, including Malee Sampran, Universal Food Company (UFC) Doi Kham, and Sun Tech. The first two of these are diversified fruit and vegetable companies, while Doi Kham is the food processing operation of the Royal Project (discussed above) and Sun Tech is an industrial firm that has moved into the processing tomato sector. Typically, these firms are deeply embedded within the rural economies in which they operate, depending on nucleus estates and contract farming arrangements to procure tomatoes. Tomato paste and canned products are manufactured either to be sold under their own labels, as contract products sold under other companies' labels, or as unbranded commodities for re-processing. Prior to the 1997 crisis, there were only three significant processing tomato companies in Thailand with external investment (excluding NAICO). Heinz Win Chance, a Taiwanese entity that was merged into a larger joint venture by Heinz in 1987, specialized in branded products using tomato paste as an industrial input. Heinz's investment in this operation was aimed at extending its market reach into the fast growing southeast Asian economy. Heinz invested $840,000 (US) of new registered capital to match the original investment of $800,000 (US) by Win Chance, and the joint venturers subsequently invested some $3 million (US) to expand its factory near Bangkok (Agribusiness Worldwide, 1987). Thai-Soon and Chico-Thai Plantations, the other examples of foreign capital investment in the Thai processing industry prior to 1997, were joint ventures between domestic investors and those from other Asian countries. These operations were established as joint ventures because Thai Government policy has long required foreign investors to operate in most cases with Thai partners.

Table 6.4 Tomato processing activities of Thai vegetable processing companies, 2000

Company	Canned	Paste	Sauces and ketchup	Other
Agr-on Co	X			X
Asian Union	X			
Daily Foods	X			
Doi Kham Foods	X	X		X
Food Blessing Co			X	
Globo Foods			X	
Grand Asia	X			X
Great Oriental Foods	X			X
Heinz Win-Chance			X	
KC Chiangmai Foods	X	X		
Malee Supply Co	X			
Northern Food Co	X			X
Peace Canning	X			
Premier Canning			X	
Royal Foods	X			
Royal Canning Co	X	X		
Roza Foods	X			
Siam French Co	X	X		X
Srichiengmai Industry			X	
Srithai Foods			X	
Sun-Tech Group	X	X		X
Thai QP Co			X	
Thai-Soon Food	X	X	X	X
Thai Therapos Food			X	
Tropical Premier Food	X	X		
Universal Food Co	X		X	X

Sources: Kompass (2000) augmented by field research.

The 1997 Asian economic crisis generated considerable challenges for all the companies in the Thai processing tomato industry. In 1997, the peak year of output, Thai farmers produced some 250,000 tonnes of processing tomatoes. In the following year, production fell by over 50 per cent to 121,000 tonnes, and the number of processing plants fell from 16 to 11. While the dramatic devaluation of the Thai Baht might have been expected to provide the Thai industry with a competitive edge, any positive effects were over-shadowed by the collapse of financial liquidity within the Thai economy, the closure of financial institutions, and sharp reductions in domestic consumer spending. This latter effect was especially relevant for those companies supplying the emerging fast food and pizza industries of southeast Asia. A case in point was Chico-Thai Plantations, which ceased trading in the wake of the 1997 crisis. A key customer for the company was the Pizza Hut chain in Thailand and elsewhere in Southeast Asia (Kompass, 1997). The immediate effect of the 1997 crisis was to reduce Pizza Hut's growth rate in Thailand from 30 per cent per annum to 10 per cent per annum. The chain experienced not only a decline in patronage, but in addition, the average value per transaction at Pizza Hut restaurants to fall from 260 Baht to 240 Baht (Kwanchai, 1998). Clearly, reduced margins at Pizza Hut had a flow-on effect to suppliers.

Notwithstanding the severity of the 1997 crisis, and the fact that some companies were coming under intense competitive pressures in the period beforehand, many processing tomato companies managed to survive. Malee Sampran, for example, had begun to experience some difficulties in 1995, largely as a consequence of the decision of the United States Government to impose anti-dumping duties on imports of Thai canned pineapple supplied by Malee and three other companies (Polkwamdee 1998, p.17). In 1997-98, the company came close to being delisted by the Stock Exchange of Thailand, but survived following an injection of funds by overseas investors, and the transformation of Malee Sampran from a wholly owned Thai company to a joint venture involving substantial control by overseas interests. In the case of Malee Sampran, Credit Suisse First Boston Private Equity acquired nearly 29 per cent of the company, and emerged as the largest single investor.

Malee Sampran was not the only wholly owned Thai company to seek an injection of foreign capital as a way of overcoming the crisis of 1997. The UFC also ran into trouble as a consequence of the losses registered by subsidiaries. In 1999, in order to repay outstanding debt and to generate working capital, UFC raised a $6.2 million (US) loan from the Commonwealth Development Corporation (CDC), which carried with it the option of converting the debt into equity, which would give the CDC a one-third share of the company (Somluck, 1999). In short, as with Malee Sampran, the economic crisis of 1997 served to transfer some of the ownership and control of Thai capital assets to overseas investors.

The Sun Tech Group also came under pressure during the 1997 crisis. As noted in Table 6.5, the Sun Tech processing tomato operations are part of a large industrial conglomerate, with a diverse range of interests. The problems of the Sun Tech Group go back to the collapse of its export sales in 1992-93. The Group responded to this slump in exports of tomato products by diversifying into new areas, including the processing of steel scrap. However, as is often the case with large diversified conglomerates, there appeared to be little logic in this pattern of diversification, since there were few linkages or economies of scale to be exploited as a consequence of investments in both processing tomatoes and steel scrap. However, there were benefits that could be realized by coordinating the steel activities that were at the core of the various activities within the large parent company, the NTS Group. In concentrating on such activities, the agri-food businesses were accorded a low priority. In 1995, the NTS Group boosted its exposure to the steel industry, which further marginalized the tomato activities within the overall corporate entity. The processing tomato arm of the Sun Tech Group experienced consistent losses and, consequently, was not well placed to

Table 6.5 Summary of leading processing tomato companies in Thailand

Company	Key characteristics
Malee Sampran	Wholly Thai-owned until 1997. The largest fruit and vegetable processing company in Thailand. Established in 1964 at Nakhon Pathom, in the northeast. As well as producing pineapples and a range of other fruit and vegetable lines, it has been involved in the processing tomato sector since 1992, through its wholly owned subsidiary KC Chiang Mai Food Processing.
Universal Food Company	Wholly Thai-owned until 1999. Thailand's second largest fruit and vegetable processor. Established in 1969 and built its first processing plant at Lampang, in the north, in 1971. From its very inception, the company focussed on export markets, as a contract canner for companies such as Libby's, Del Monte, and for UK supermarket chains. From 1993, UFC began to export canned goods to global markets under its own brand name. The company was a major producer of canned tomatoes, tomato juice and tomato ketchup throughout the 1990s.
Doi Kham	A processing operation that grew out of the Royal Projects (discussed above).
Sun Tech Group	One of 12 subsidiaries controlled by the NTS Group, a large Thai-owned conglomerate, with interests in food, scrap metal, steel production, video rental, entertainment, real estate and retailing. Established in Nong Khai Province in 1990 as a fruit and vegetable processing company. It was the largest Thai producer of canned tomatoes in the mid-1990s.
Chico-Thai Plantations	Ceased trading in 1998. It was 60 per cent Thai-owned, with 40 per cent of equity shared equally between investors from Malaysia and China. Manufactured pineapple juice concentrate, tomato paste and pizza sauces at Nong Khai and Kalasin in northeast Thailand.
Thai-Soon	Thailand's largest processing tomato company, with a factory at Srichiengmai, near Nong Khai in the northeast. Established in 1989 with shared ownership between Thai (52 per cent), Taiwanese (44 per cent) and Japanese (four per cent) interests. Tomato products accounted for 60 per cent of total output in 2001, of which tomato paste is the most significant component (3,000 tonnes), followed by canned tomatoes, juice, ketchup and pizza sauces. A supplier for a number of leading fast food chains in Thailand, as well as exports. Japan is its leading market, and it is the only Thai firm exporting processed tomato products to Japan.
Heinz Win Chance	Originally a Taiwanese company, which in 1985 established a Thai subsidiary, in order to manufacture and market cereals and flavored milk. In 1987, Heinz purchased 51 per cent of the company, and added ketchup and chili sauce to existing product lines, all bearing the Heinz brand name, and aimed at the local market as well as other countries in the region.

Sources: Field research; Board of Investment (1990; 1996); Yoa (2001).

survive the economic crisis which occurred in mid-1997. The Chief Executive and Owner of the NTS Group, Mr. Sawasdi Horrungruang, emerged from the crisis as the second largest individual debtor in Thailand, with debts totalling more than $2.1 billion (US) (Sasithorn, 2000). The Stock Exchange of Thailand ceased to trade in Sun Tech group shares in May 1999, after the company failed to submit quarterly results. When these were eventually submitted, they revealed losses of $13 million (US) (The Nation, 1999). To avert de-listing, the Group was required to submit a restructuring plan, and in April 2001, the Group's creditors approved a corporate restructure (Bangkok Post, 2001). Although technically bankrupt, at the time of writing the Sun Tech Group continues to operate, and still produces canned tomatoes for sale by western supermarkets.

The view from below: contract farmers and industry transformation

The discussion so far has focused on the rise and fall of the Thai processing tomato sector, in the context of corporate responses to changing conditions at the national and global levels. In all of this, it is easy to overlook the impacts at the farm level, and to analyze what happened to smallholders who committed themselves and their families to contract production. Did the growth of the industry generate any significant long-term gains to poorer rural communities in the north and northeast of Thailand, or did this industry offer only marginal returns which left people largely unchanged?

In order to try to answer these questions, a survey of farmers in the Nong Khai district in northeast Thailand (Table 6.6) was undertaken in 1994 (Burch, forthcoming). Although only 37 respondents in three villages were involved in this survey, the interview schedules were comprehensive, and provide important insights into the village-level dynamics of contract tomato production. The survey reveals key attributes of contract tomato growing in northeast Thailand, as well as differences at the village level. In general, contract tomato growers also grew rice for subsistence, and various other cash crops, including tobacco, baby corn, chilli, melons, maize, cucumber, long beans and soya beans. Hence, growers had varying degrees of dependence upon processing tomatoes for their cash incomes, and importantly, had experience with contract farming arrangements beyond the tomato industry. At the time the survey was undertaken, farmers in the three villages surveyed had been engaged in contract tomato growing for between seven and 12 years.

According to the survey results, contract growers produced about 10 tonnes of tomatoes each, from lots that were approximately one-third of a hectare in size. Reported yields in villages 'A' and 'B' were broadly comparable to what might be expected in a situation of low input/ low output tomato cultivation (see Chapter Two). The reported yield in village 'C' was roughly half that of the other two villages.

The prices reportedly received by growers varied considerably between the three villages, and deserve closer inspection. In village 'A', all but one surveyed grower was contracted to the Srichiengmai Industry Company, a producer of sauces and ketchup. According to these interviews, Srichiengmai operated a grading system that separated produce into three categories: Grade 1 (Big Pink), Grade 2 (Small Pink) and Grade 3 (Red). However, survey respondents appeared to have only limited awareness of these arrangements, with most receiving a base price of 1.5 Baht per kilogram, roughly equivalent to $60 (US) per tonne. In Village 'B', 14 respondents grew tomatoes for the Chico-Thai Plantation Company, a producer of tomato paste and pizza sauces for Pizza Hut and other fast food chains. Chico-Thai procured its contract-grown tomatoes via 'traders', who acted as buying agents for the processing firm. According to survey results, these traders offered a fixed price, which was significantly below that offered in village 'A'. In general, the range of prices in villages 'A' and 'B' is broadly comparable with that in Australia and California during the same period, suggesting that, on the basis of these data, the Thai processing tomato industry did not demonstrate significant cost competitiveness at the farm-gate.

Table 6.6 Survey of contract tomato growers, northeast Thailand, 1994

	Number of growers surveyed	Average length of time contract growing tomatoes	Average size of individual contract lots, and yields	Prices received $ (US)[1]
Village A	17 (11 male; 6 female)	7 years	Size: 0.31 ha Yield: 31.5 tonnes/ha	$60/ tonne
Village B	14 (5 male; 9 female)	7.3 years	Size: 0.29 ha Yield: 34.6 tonnes/ha	$48/ tonne
Village C	16 (6 male; 10 female)	12 years	Size: 0.4 ha Yield: 17.6 tonnes/ha	$60/ tonne

Notes: 1. original price data reported in Thai Baht, converted to US dollars using prevailing exchange rate of the early 1990s (25 Baht).

Source: Burch (forthcoming).

Unlike the previous two villages, where most farmers grew for one processing firm, the growers in village 'C' supplied wide variety of outlets. Five respondents produced partly for sale to traders and partly for sale to specific companies, including Chico-Thai, Srichiengmai and Thai Soon; three growers produced exclusively for sale to specific companies, and eight sold their produce exclusively to traders. This suggests a more competitive market for processing tomatoes and, as a result, an average price which like Village A, was towards the top of the range of variability.

Processing companies provided varying levels of support to contract growers. Most respondents indicated that the processing company provided credit in the form of seeds, pesticides and fertilizer, and that field agents were on hand to offer advice. Payment for inputs supplied on credit was deducted from growers' returns. Some growers had sought additional credit from the BAAC or other commercial banks. The survey data suggest that in approximate terms, the *gross* tomato income per grower in these villages was of the order of $600 (US) and that, growers on average received *net* incomes of approximately $400 (US) from contract tomato growing. These data appear to be broadly consistent with other studies (Benziger, 1996; Dolinsky, 1991).

Of course, the most important factor determining the returns to growers is the overall level of demand for their product, and in the early 1990s, increased demand for Thai tomatoes on the world market led the numerous processing tomato companies in the north and northeast of Thailand to compete with each other for the growers' product. Accordingly, many farmers sold their production 'off-the-contract' at high prices. In 1991, this led to the situation whereby some processing firms required growers to pay a deposit, in order to be offered a contract. If the agreed volume of tomatoes was not delivered, the grower would lose the deposit (Sjerven, 1991, p.21). There is little doubt that when demand for tomatoes declined in the late 1990s, the returns to small scale producers in Thailand fell as quickly as they had earlier increased. Subsequently, as the impacts of the 1997 economic crisis were registered in the decline in the number and size of processing tomato companies, farmers would have seen their net returns decline even further. To most rural producers, then, the processing tomato sector might have provided short-term gains, but was not a viable source of income in the long term.

Lessons from Thailand

Thailand is a relatively small player in the world's processing tomato industry, but its recent history provides important lessons relating to the global restructuring of the industry as a whole. The dominant theme that emerges from the Thai industry's rise and fall relates to the volatility and vulnerability of export-oriented production.

Thailand is a story of transition. The first significant private sector investors in the industry, NACO, interpreted the country's potential by reference to competition from Taiwan. In the mid-1980s, exchange rate appreciation and generally higher land and labor costs, impaired the competitiveness of Taiwan as a production site within the industry. Observing these trends, investors eyed the flat, alluvial soils of north and northeast Thailand, adjacent to the Mekong River, as the basis for an industry. Opportunities were opened by growth in demand for tomato paste as an input to the Thai seafood canning industry, and by the emerging market for private label canned tomatoes from Western supermarkets. Small scale farmers, familiar with contract production because of their earlier experiences with tobacco, tomatoes and other vegetables, were able to be mobilized to grow the raw materials.

Yet, within a short period, the industry faced crisis. A temporary loss of export markets in 1992-93 was compounded by the much larger problems associated with the Asian economic crisis of 1997. The collapse of some processing companies and a generally unstable environment for others, has resulted in a period of immobilization, with minimal new investment in processing facilities over recent years. In the 2001-02 season, only 160,000 tonnes of tomatoes were processed in Thailand, and only a relatively small proportion of this found its way onto world markets. In the global processing tomato industry, Thailand has been dismissed as a competitive threat, with attention instead turned to China.

In a broader sense, the Thai processing tomato sector affirms some of the key insights from literature on New Agricultural Countries (NACs), discussed earlier in this chapter. Thailand's processing tomato sector attempted to compete on the world market through price, in an environment of intense volatility in the trading conditions for this section of the market. Many of the key customers for Thai products, notably the private label businesses of Western supermarket chains, rely on highly flexible, global sourcing strategies. Supermarket 'own brand' products typically sell on price, and supermarkets are continually seeking cheaper sources of supply in order to maintain a competitive position. In the case of Thailand, this placed the industry in a vulnerable position, and in combination with widespread debt financing by some of the major firms, has raised questions about the future viability of some leading companies. It is significant, for example, that whereas some of the 'high flying' Thai processing companies were casualties of the Asian economic crisis, companies such as Doi Kham, with strong links to rural economies and a combined commercial-social charter, have been less affected.

However, there is another side to the history of the Thai processing tomato industry. As discussed in Chapter Two, processing tomato cultivation tends to be best suited to production sites around 40° latitude North and South, in generally drier regions either on continental west coasts or inland locations. These locations tend to generate climates with more predictable characteristics, and during the summer months, there are longer sunlight hours, which are critical for tomato ripening. Whereas the Thai experience certainly suggests that commercial cultivation can occur in tropical areas, the prevailing wisdom of many in the industry is that such areas do not possess the environmental attributes that allow them to compete globally, over a sustained period. Thus, whereas the history of the Thai industry exemplifies the capacity for globalizing industries to shift among production sites with increasing speed and volatility as relative cost structures change (in this case, from Taiwan, to Thailand, to China), it also seems to emphasize a working out of the industry's geography, as some relatively less productive sites are replaced by others.

The processing tomato industry comes to China

The processing tomato industry was established in China in the mid-1980s, at roughly the same time that the industry was established in Thailand. The subsequent history of the industry in China, however, bears virtually no other resemblance to its Thai counterpart. In almost all respects, the experience of the Chinese processing tomato sector is entirely different from that of Thailand. These differences include farming systems, investment patterns, domestic competition, export marketing, and most importantly, the industry's growth trajectory. From a near zero production level in the mid-1980s, the Thai industry grew to 250,000 tonnes output by 1997, before falling to less than 150,000 tonnes by 2000. In contrast, China's processing tomato industry grew from a near zero production level in the mid-1980s, to 500,000 tonnes by 1993 and 1.8 million tonnes by 2000. The growth rate of the Chinese industry is unprecedented in the world's contemporary processing tomato sector, and at international tomato industry meetings and conferences, the talk tends to be dominated by discussions about the future prospects for the Chinese industry.

Of course, there have been many examples of industries experiencing rapid growth in China over the past two decades. Since Deng Xiaoping's famous edict of the early 1980s to put development ahead of ideology ('it doesn't matter if a cat is black or white, so long as it catches mice'), China's economy has experienced rapid and sustained economic growth. However, the rapid growth of China's processing tomato industry has demonstrated distinct differences from the pattern of growth in other sectors. Since the mid-1980s, China's growth has been characterized by the strong geographical pull of urban areas and coastal provinces. An 'east-west' divide in China has caused concern for national policy-makers aiming for equitable distribution of economic progress. This has been most pronounced with the extraordinary development of Guangdong Province, adjacent to the Hong Kong Special Autonomous Region, which in the space of a decade has been transformed into one of the world's largest industrial regions, generating massive amounts of wealth for investors and some residents.

The Chinese processing tomato industry however, does not fit within the standard geographical narrative of Chinese development. It is located in the extreme far west of the country, remote from the key growth poles of China's economy. The Chinese processing tomato industry is located in Xinjiang Province, close to the Mongolian and Kazakh borders (Figure 6.6). From a Western perspective, there are probably few places on earth more remote and difficult to get to than Xinjiang. In winter, the Province is blanketed with deep snow (the average temperature range in January for Urumqi, the provincial capital, is –20 to –10 degrees Celsius), but during the brief summer months, temperatures rise to 30 degrees Celsius on a consistent basis, there is very little rain, and the days are long; in short, such conditions provide an almost perfect climate for tomato cultivation. How the world processing tomato industry reached this far-flung corner of the globe, and how Chinese investors, industrialists and government officials have helped build the industry to the point at which it accounts for almost 10 per cent of world production within the space of a decade, elaborates the key themes of this chapter and this book. At one level, it highlights the volatilities of export agri-food production. In contrast to the situation in Thailand, which was poorly equipped in terms of institutional arrangements to respond to such challenges, the social and political context of the Xinjiang processing tomato industry provided substantial capacity for industry actors to sustain the industry in rapidly changing commercial environments. At another level, the Chinese example reiterates the general theme of this book, which relates to the uneven character of global agri-food restructuring. As discussed below, the key to understanding how and why Xinjiang developed its processing tomato industry so rapidly in the 1990s, relates to the ways the industry was embedded in specific temporal and socio-spatial arenas.

Xinjiang: the cross-roads of Central Asia

Xinjiang Province, officially known as the Xinjiang Uyghur Autonomous Region, has always held a pivotal geo-strategic position at the crossroads of invasion, trade and migration (Barnett, 1993; Becquelin, 2000). Xinjiang, also known as 'Chinese Turkestan' or 'East Turkestan', is often translated into English to mean 'New Dominion' (Barnett, 1993). The region occupies a sixth of China's land mass and has 5,500 kilometres of international borders with eight nations (Russia, Kazakhstan, Kyrgyzstan, Tajikistan, Mongolia, India, Pakistan and Afghanistan). This vast array of borderlands signifies Xinjiang's location in a network of social, cultural and economic partnerships and homelands across territorial borders (Ferdinand, 1994, p.272). Twelve non-Han ethnic groups live within Xinjiang: Uyghurs, Kazakhs, Hui, Mongols, Daurs, Khalkas, Manchus, Xibos, Tajiks, Uzbeks, Russians and Tartars. In 1997, these minorities constituted 63 per cent of the Province's population of 17 million. Of the minorities, the Uyghur people are the most numerous, constituting 47 per cent of the Province's total population (Becquelin, 2000, p.66). The potential volatility of this population mix – especially in the eyes of China's central government – has resulted in Xinjiang attracting sustained attention in terms of regional front-line defence and geo-political stability.

Chinese influence in Xinjiang goes back many hundreds of years, however formal Chinese political control of the Province was established only in 1954. During the preceding few decades, nominal Chinese sovereignty over the region was not backed up by effective control. In the 1940s, Soviet-backed insurgents gained control over much of the territory of Xinjiang and proclaimed the independent East Turkestan Republic (McMillen, 1981, p.67). This was an anti-Han Chinese movement led by ethnic nationalists. At the time of the 1949 Revolution, Chinese Communists had not developed a significant presence in Xinjiang and, in fact, only 10 per cent of the region's population of 4.2 million were Han Chinese (McMillen, 1981, p.66).

The assertion of Chinese political control over the Province was, therefore, a contested process that reverberates to the present day. Calls for the independence of 'East Turkestan' continue to be made by ethnic Xinjiang minorities. An international diaspora of ethnic Xinjiang minorities monitors and supports East Turkestan independence, and within the Province there has been episodic violence by ethnic Xinjiang minorities against Han Chinese. In the wake of the 11 September 2001 attacks on the World Trade Center in New York, Chinese authorities clamped down on Xinjiang separatists, arresting 3,000 people (Dwyer, 2002, p.41).

Fears of ethnic separatism have been central to Chinese administration of Xinjiang since the 1949 Revolution. On 25 September 1949, the Guomintang

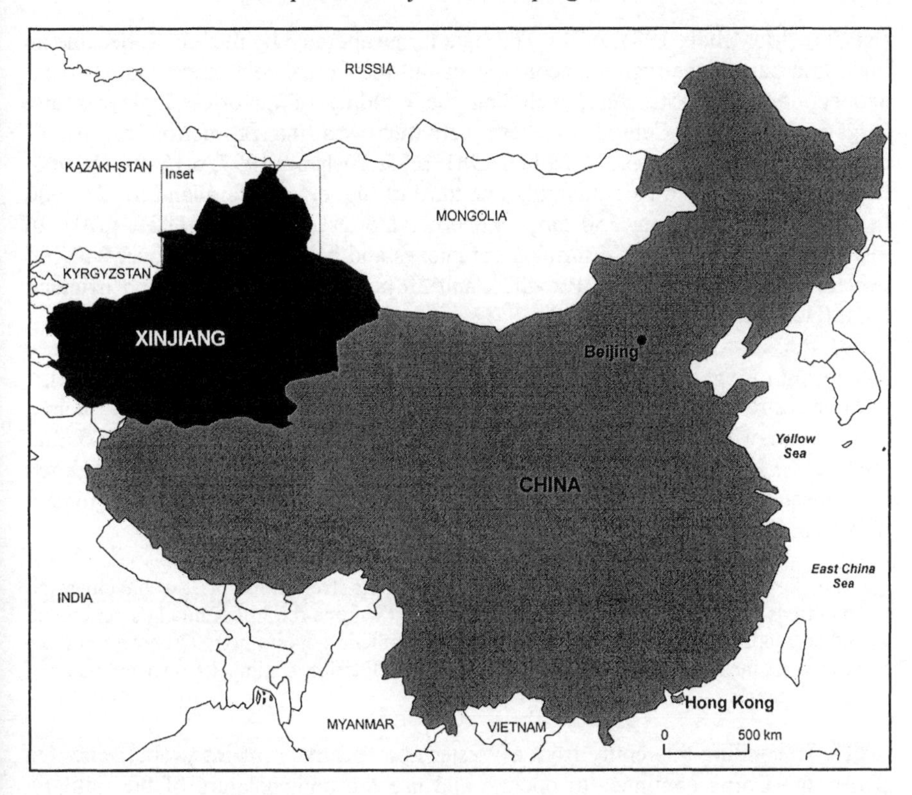

Figure 6.6 Xinjiang, China

Note: 'Inset' refers to the area described in Figure 6.7.

(Nationalist) forces in Xinjiang surrendered to the Peoples' Liberation Army (PLA), leading to the peaceful annexation of the Province into Communist China. Nationalist forces subsequently were re-educated and incorporated into PLA units. As the PLA advanced through Xinjiang it established a series of garrison camps, implementing effective political control over the entire Province. On 5 December 1949, Mao Zedong issued a directive to demobilize the PLA troops, 'turning the army into a working force' (McMillen, 1981, p.69). This created a collective labor force for land reclamation, water conservancy, agricultural production and infrastructure development, strategically located along major borders and transport routes to guard against internal or external threats.

The transformation of PLA army corps into a collectivized working force shaped Xinjiang's subsequent political economy. In August 1954, the army units were merged under the single organization *Xinjiang Shengchan Jianshe Bingtuan*, which is usually translated as Xinjiang Production and Construction Corps (XPCC), or more recently, the Xinjiang Production and Construction Group

(XPCG) (McMillen, 1981, p.71). The state farms operated by the XPCC became an important part of the region's economy, as did the Corps' role in expanding social and economic infrastructure, including the building of factories, highways and irrigation projects. Central to these activities was the promotion of inward migration by Han peoples (McMillen 1981, p.75; Fusheng, 1997, p.1539). In 1965, the Corps was said to be cultivating one third of the region's farmland (or 730,000 hectares), and operating 550 large modern factories (McMillen, 1981, p.74). In 1999, the Corps owned 174 farming complexes and held jurisdiction over 740,000 hectares (48 per cent of the Province, and 30 per cent of all arable and irrigated land) (Becquelin, 2000, p.78).

At the core of XPCC's identity is an integration of politics and economics. According to one analyst quoted in *The New York Times*: 'over time, "defending" became more a matter of occupying real estate than of bearing arms ... it is a corporation whose purpose is to colonize' (James Seymour, cited in Eckholm, 1999, p.1). Each of the Corps' farms is a Han stronghold (farms can be inhabited by thousands of persons), and 80 per cent of the Province's industrial assets remain in state hands. According to Eckholm (1999, p.1):

> The Corps, as it is known, is practically a world unto itself. It has built whole cities and hundreds of farms and factories producing more than one-third of Xinjiang's cotton and other goods. With workers and their families numbering 2.4 million, 97 percent of them ethnic Chinese, it runs its own university and television station, its own militias and prison camps.

Understanding this entity from a Western perspective remains problematic. On paper, the Corps continues to operate and use the nomenclature of the military. Factories and farms tend to be known in terms of the historical regiments that initially established them (for example, the tomato paste factory at Tiandan is known as the '22 Regiment' factory). According to one Western scholar researching the institution:

> The Corps was described to me as an organization that 'integrated agriculture, industry and commerce' and 'combined Party, government, army, and economic functions'. It ran its own police force and courts as well as its own educational institutions and hospitals (Barnett, 1993, p.400).

Reflecting these combined economic, political and military functions, the Corps is administered directly by the State Council of China's National Cabinet in Beijing. It has an equivalent status 'to a Province or municipality, which means that its commander-in-chief enjoys the same rank as a provincial governor' (Kwang, 2000, p.52).

Writing in the early 1990s, Barnett (1993, p.398) contended that Xinjiang was a 'company region', in contrast to the western term 'company town', used to describe cities where one corporation or industry plays a dominant role. It was also suggested that the XPCC seemed to operate as a 'corporate government' or a 'parliamentary corporation' that maintained a stranglehold on its immediate provincial and regional economy, with repercussions for China's border control and the whole nature of economic development in the Province. Since that time, Xinjiang has undergone considerable change. The XPCC has been forced to adapt but nevertheless has remained the central player in Xinjiang's political economy.

These issues are entwined within contemporary central Asian geo-politics. The demise of the Soviet Union and political instability in many bordering Central Asian Republics have changed the way in which Xinjiang is seen by Beijing. On the one hand, these developments reinforce the strategic importance of the XPCC as a political-economic agent of control in the Province. On the other hand however, economic growth in Xinjiang has become a national priority, as a means of strengthening Chinese authority in Central Asia. China has sought to 'seize an historic opportunity' to 'open up Xinjiang to the world' by setting up economic zones, expanding border trade and foreign investment into infrastructure and capital construction in the Province (Becquelin, 2000, p.67). The central government legislates for large amounts of 'special' foreign and domestic capital expenditure in Xinjiang due to its unique geo-political position, and to help boost economic development outside of the booming coastal provinces (Ke, 1996, pp. 5–6). In 1999, the freight capacity of the railway leading into Xinjiang was doubled, and the Province's central highway was extended by 500 kilometres between Turpan, Urumqi and Kuitan; this was a $300 million (US) project supported by $150 million (US) in World Bank loans (Becquelin, 2000, p.74).

'Pushing back the desert': The political economy of agriculture in Xinjiang

The centrality of the XPCC to Xinjiang's modern economy immediately raises questions about the role being played by agri-industrial development in the Province. The important point to consider is that the shift of agri-industry to Xinjiang (including tomato production) must be understood in terms of a broader geo-strategic imperative to bring development and security to this pivotal region of Central Asia.

Focusing on the XPCC's agricultural activities, it is apparent that over recent years, the Corps has been confronted by significant change in the operation of markets and production arenas. During the 1990s, there was a series of reforms aimed at encouraging a transition from communal farms to family-farms organized within rural cooperatives, although these have remained under the umbrella of state-control (Ling and Selden, 1993). These policy shifts however, should not be understood solely within the framework of 'liberalization'. Chinese agricultural policy necessarily involves the management of complex economic, social and political tensions. Much of the rural population in China depends upon the base level subsistence livelihood which agriculture provides (Ling and Selden, 1993, p.5). Consequently, Chinese agricultural policy has been caught between competing demands; on the one hand, to promote commercial agriculture as a strategy to lift farm productivity, and on the other, to retain rural structures as a means of preserving basic livelihoods and contain potentially massive rural-urban migration.

Under Deng Xiaoping's reforms of the 1980s, the state farms of the XPCC were seen to provide strong support for the central government through land reclamation and frontier defence, while running a 'reasonable' loss which underpinned the socio-economic livelihood of the rural population (Woodward 1982, p.232). In the 1990s, Beijing's directives to restructure agriculture were translated in Xinjiang in ways that emphasized the need to boost farm productivity, but without the XPCC losing control. The 'household responsibility contract system' (*chengbao zhidu*) reforms, introduced in the 1980s, underline the influence exerted by the quasi-liberal market over China's recent policy initiatives. The critical point here is that while 'responsibility' policy appears to give farmers autonomy over their labor power, their mobility and the marketing of their products, the indirect influence of state control remains dominant in the production and distribution of goods and services (Ling and Selden, 1993; Barnett, 1994). The XPCC implemented the above market-led reforms by dividing much of its land into 17,000 household farms, although these operated 'within the framework of the state farms system' (Barnett, 1994, p.400). In other words, the XPCC nominally *de-collectivized* its state-farm system while at the same time maintaining overarching political and economic control over the structures of Xinjiang agriculture. Hence, notwithstanding Beijing's policy of introducing structural change within Chinese agriculture, the XPCC has remained central to the Xinjiang agricultural economy. Although liberalization has opened Xinjiang's agricultural sector to new players (such as the Tunhe Corporation, discussed below), it has not resulted in a significant dilution of XPCC's influence within the Province. This factor is critically important for an understanding of the Xinjiang processing tomato sector. While over the past twenty years China has constructed an increasingly liberal and plural economy, Xinjiang has remained a bastion of paramilitary state-control.

Spaghetti westerns: the new 'Marco Polos' of western China

The aim of the Chinese Central Government to promote development in Xinjiang, coupled with the existence of a strong central institution (XPCC) with the ability to mobilize resources and labor power, facilitated considerable experimentation with potential agri-industrial developments during the 1980s. The first processing tomato factory was constructed in 1983 at Shihezi, Xinjiang Province, using imported Italian tomato seeds (variety IT875), which were tested for local conditions by the Shihezi Vegetable Institute, an agronomic research center linked to the XPCC. During this early period, relatively small-scale processing tomato operations were established over a wide region of western China, including Inner Mongolia and Gansu Provinces, as well as Xinjiang.

The early participation of Italian interests in western China's embryonic processing tomato industry provides some interesting insights into the international dynamics of this industry. For example, although the US is the single largest producer of processing tomatoes in the world (in 2001, it contributed approximately 35 per cent of world output), no American processing tomato industry representatives seem to have visited the region until the late 1990s, by which time Chinese tomato paste was already considered a credible import threat by US producers. In contrast, from the mid-1980s to the turn of the century, several hundred Italians were involved in various capacities in the Xinjiang processing tomato industry. Notably, these have included traders and employees of Parma-based food machinery companies.

Within the global processing tomato industry, the Italians in western China have been labeled the new 'Marco Polos'. Clearly this is meant as a comical reference, but ironically, it also captures a central element of the motivations that created this relationship. In the second half of the thirteenth century (scholars dispute the dates), Marco Polo and his brothers supposedly traveled from Venice to the court of Khubilai Khan, in Beijing, and crossed the area that is now Xinjiang (Larner, 1999). In contrast to some later European explorers, who were motivated by a colonial agenda, Marco Polo and his brothers were merchants in search of trading opportunities. According to most mainstream accounts, the Italian merchant venturers did not travel with the aim of extending Venetian geo-political influence, but with the aim of outflanking their commercial rivals in lucrative trans-Asian trading routes. The contemporary Italian influence in the Xinjiang processing tomato sector is based on a comparable strategy. The Italians in Xinjiang are traders, *not* investors. Their role has been to aid the Xinjiang tomato companies to compete on the international market, and then to position themselves as entities able to control that trade.

The linchpin for this role has been the relationship between Parma-based food machinery companies, and Parma-based tomato paste traders. As discussed in Chapter Five, the history of the northern Italian processing tomato industry has been enmeshed with the evolution of Parma as a central node in the world food machinery sector. Northern Italian companies including Manzini, Rossi & Catelli, Ing. A. Rossi and Sasib are pivotal players in the global development of the processing tomato industry. The American food machinery company FMC FranRica (created in 1995 following the FMC Corporation's acquisition of California-based FranRica) also has a significant operational base in Parma. These companies emerged in the post-1945 years, in the context of investment by Heinz and others, which was designed to modernize Italian tomato production. The growth of the processing tomato industry in new international locations provided major opportunities for these food machinery companies. As the processing tomato industry expanded into new production sites around the world, it was usually the case that sales representatives of companies such as Manzini, Rossi & Catelli, Ing. A. Rossi and Sasib were centrally involved in the negotiations with governments and local investors. Obviously, trading opportunities followed the expansion of the processing tomato industry to new production sites. Thus, the global expansion of the processing tomato industry tended to take place via engagements between investors, food machinery companies, and tomato paste traders.

This was certainly the case in Xinjiang. During the second half of the 1990s, the expansion of this industry proved to be a boon for food machinery companies. By the year 2000 it was estimated that there were more than 40 processing tomato production lines installed in western China. A significant number of these had been purchased second-hand and transported to China, but at the same time, all the major processors were intent on acquiring some state-of-the-art equipment, as this was seen to be vital to the task of improving product quality and securing export orders. Throughout the 1990s, the major northern Italian food machinery companies had a strong presence in Urumqi, as each vied for orders from Chinese processing tomato firms.

The question of how Chinese processing firms financed the Italian-made equipment installed in China is still a vexed issue within the industry. Of course, the terms of these transactions are commercial-in-confidence to the parties involved. However, given the global significance of Chinese tomato paste export capacity, speculation on these issues is rife amongst industry participants. In the course of researching this book, many informants offered the view that the Italian food machinery companies sold equipment on generous credit terms, with the primary goal of establishing long-term relationships with leading Chinese processors. If this was the case, it would certainly have provided an important element of competitive advantage for the Chinese industry.

Whatever the financial arrangements for equipment purchasing, it seems clear that very little hard currency changed hands between equipment suppliers and Chinese processors. Following a visit to Xinjiang in 1996, the industry analyst Barry Horn reported:

> Much of the processing tomato investment in China has been made possible through 'compensation trade' financing, available mainly from Switzerland. Effectively, loans are repaid in tomato paste. This system is workable because Chinese banks (effectively, the Chinese Government) have been prepared to guarantee the loans (Horn, 1997, p.13).

Interviews with industry informants in western China did not add significantly to the conclusions arrived at by Horn in 1997. Indeed, since Horn published his observations on this topic, the only public acknowledgement of such 'compensation trade' financing is one sentence in the 2002 annual review of the Italian tomato industry by the United States Department of Agriculture (USDA), which states:

> Italian companies have secured export sales of technology and processing machinery to China in exchange for commitments to contract for Chinese tomato production (United States Department of Agriculture Foreign Agricultural Service 2002b, p.4).

Regardless of the exact nature of contract terms, it is clearly the case that export marketing in the Xinjiang processing tomato industry has been closely related to the commercial aspirations of Italian food machinery companies and traders, with major implications for the sale of Chinese paste into Mediterranean markets. As indicated in Table 6.7, Italy has accounted for approximately one-fifth of Chinese tomato paste exports, and the volume of Chinese paste exported to Italy increased by approximately 100,000 tonnes (in raw product equivalent terms) between 1992 and 1999. By 2001, an estimated 84 per cent of Italian tomato paste imports were Chinese in origin (United States Department of Agriculture Foreign Agricultural Service, 2002b, p.4). In the main, these were used for re-processing and export to countries outside the European Union. Italian personnel in food machinery and trading companies provided the conduits to facilitate this trade. Moreover, because of compensation trade financing (requiring Chinese firms to meet interest repayments through the sale of paste), it is likely that a number of Chinese processing firms were obliged to offload tomato paste in adverse market conditions, giving rise to the situation described in Chapter Five where, in 2001, Chinese paste was allegedly being sold at extremely low prices. Hence, these arrangements probably generated considerable market instability in the global processing tomato industry.

Table 6.7 Chinese tomato paste exports, 1992 and 1999

	1992 Tonnes (raw product)	Per cent	1999 Tonnes (raw product)	Per cent
Yemen	35,319	15.21	31,958	3.62
Saudi Arabia	n/p	—	31,958	3.62
Libya	58,078	25.02	n/p	—
Total North Africa	n/p	—	23,315	3.58
Rest of Africa	n/p	—	14,664	2.25
Japan	49,465	21.31	151,811	23.31
Italy	41,743	17.98	138,293	21.33
Rest of Europe	n/p	—	42,096	6.46
Hong Kong	9,986	4.30	n/p	—
Rep. of Korea	6,905	2.97	45,854	7.04
United States	n/p	—	47,226	7.25
Others	30,665	13.21	124,085	21.54
Total	232,161	100.00	651,260	100.00

Note: n/p = data not published separately.

Source: Tomato News (1994b, p.22); Tomato News (2000d, p.41).

The evolution of the Chinese processing tomato industry

Since its beginnings in the 1980s, three distinct periods can be identified in the evolution of the Chinese processing tomato industry. In the initial years of experimentation, a diverse group of investors constructed processing tomato factories across a range of locations in western China. Taiwanese investment played an important role during this period. With the flight of processing tomato firms out of Taiwan in the late 1980s and early 1990s, some companies physically transported processing lines from Taiwan to western China in order to establish a productive base. By 1992, processing tomato production in Taiwan was less than 40,000 tonnes, but Taiwanese firms based in China and southeast Asia undertook production of a further 100,000 tonnes (both figures are raw product equivalents) (Tomato News, 1992, p.14). The first Taiwanese investors in China tended to be small firms using second-hand equipment, but in the early 1990s there was a second wave of investment which established larger scale operations and purchased the latest processing equipment. The most notable of these investors was President Foods, one of Taiwan's largest agri-food conglomerates, which in 1992 established a joint venture factory in the Province as part of a wider investment drive into mainland China (Lee, 1992, p.1).

In the mid-1990s, this initial period of experimentation gave way to second phase, characterized by a geographical concentration of production in two regions of central Xinjiang Province; a 300 kilometre stretch of arable land, west of Urumqi, fed by snowmelt from the Tianshan Mountains, and a smaller area adjacent to Bosten Lake, 400 kilometres south of Urumqi (Figure 6.7). This period saw a proliferation of new processing tomato factories, with many operations running just one processing line. In 1996, in the first published analysis of the Chinese industry, Barry Horn identified over 30 tomato factories in China. At least 17 of these were in the area west of Urumqi, nine were in the southern area near Bosten Lake, four were in Inner Mongolia, and there was one each in Gansu and Hubei Provinces (Horn, 1997, p.11). However, this was an unsettled period for the Chinese industry, with equipment apparently being moved between factories on a regular basis. Reflecting the general upheavals of this period, the industry journal Tomato News reported in 1995:

> Factories can sell their products themselves on the domestic market, however they are not currently allowed to export themselves [sic]. International sales are made by large Chinese trading companies, like Xinjiang Import-Export. However, the situation is not that simple since some of the trading companies also own factories. It is generally difficult for foreign buyers to know the exact origin of the merchandise they purchase (Tomato News, 1995d, p.18).

At the time of Horn's visit, most factories had a production capacity of approximately 30,000 tonnes of raw product, with the largest factory (at Manas, west of Urumqi) having a capacity of approximately 65,000 tonnes (Horn, 1997, p.11).

Consistent with the patterns of control in Xinjiang's agricultural political economy, described above, the XPCC played a central role in the growth of the processing tomato industry during this period. Two factors are vital for understanding how and why the XPCC moved into the processing tomato industry. First, liberalization processes in the Chinese economy generated qualitative changes in the operational activities of the XPCC during the 1990s. The most important of these were steps to 'civilianize' the Corps, involving on the one hand the entry of non-military personnel into some executive positions, and on the other, a reorganization designed to bring internal structures into line with corporate rather than military models. World Bank funding played a key role in these initiatives (Laogoi Research Foundation, 1996). Accordingly, the XPCC established the Xinjiang Production and Construction *Group* (XPCG) as its commercial arm.

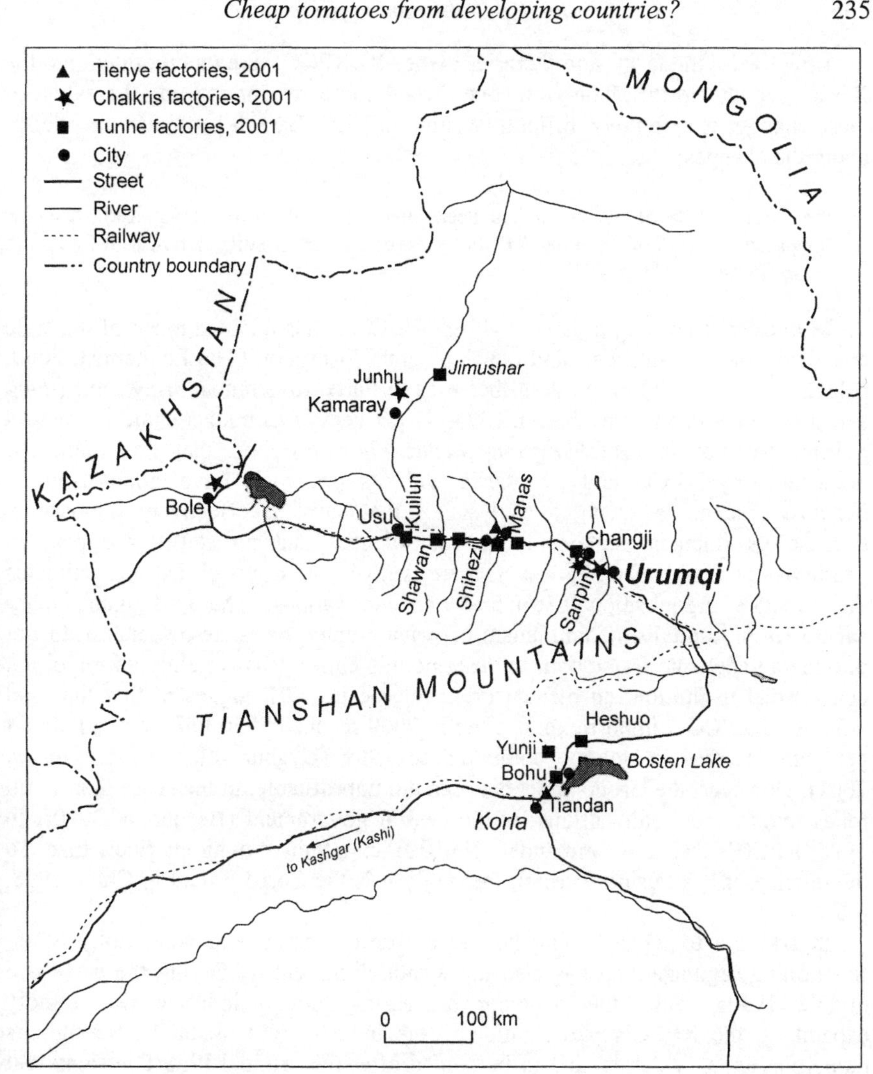

Figure 6.7 Processing tomato factories in Xinjiang Province, 2001

Effectively, the farms and factories owned by XPCC were incorporated into the XPCG, which had the status of a state-owned company. Interpreting the extent of these changes is extremely difficult. Writing in 1993, Barnett expressed skepticism about the changes:

> the formal name of the corps had been changed ... the title of its head was now 'president' instead of 'commander', but everyone I met ... still referred to it by its old name (Barnett, 1993, p. 401).

Moreover, according to later analysts, the Corps has retained much of the style and flavor of a paramilitary and developmental institution (The Economist, 2000, p.1). Becquelin (2000, p.78) describes it as a 'Party-government-army' unit (*dang zheng jun zuzhi*) and notes that, in 1994, 81 per cent of its budget came via central government subventions (*zhongyang caizheng bokuan*). Since that time, published documents by XPCG and interviews with its principals have emphasized its business credentials. Detailed analysis of the Corps is effectively impossible, because its financial statements are confidential and no statistics appear in yearbooks or newspapers. However, there is no Western model that approximates the XPCG's organizational structure and motivations. The traditional binary categories of capitalism/communism or private enterprise/state-ownership do not provide an adequate framework to describe this entity. It is certainly an important commercial institution; an official press release in 2001 suggested that the total volume of XPCG's import-export trade in 2000 reached $750 million (US), an 84 per cent increase in volume compared to 1999 (Uyghur Information Agency, 2001). However, the Group appears to remain unprofitable, in the sense that it still relies upon annual subventions by the central government (Becquelin, 2000). In 1998, XPCG's Deputy Commander, Zhu Jianfan, admitted in an interview that 'To the outside, it is a business group: internally it is the Corps' (cited in Chan, 1998, p.6).

In addition to XPCG being neither a wholly commercial nor wholly State-functioning organization, it is also not a monolithic entity. During the early and middle 1990s, when the Xinjiang processing tomato industry was rapidly expanding, the XPCG's participation in the industry was organized through its numerous units, which tended to be named after the original PLA Divisions that wrested control of the Province for China in the late 1940s. The XPCG is organized according to 13 Divisions, each with individual Regiments. Altogether, these 13 Divisions have 2.5 million 'members', meaning individuals whose livelihoods are entwined within the fortunes of the Group (The Economist, 2000, p.1). The XPCG's participation in the processing tomato industry seems to be best described in terms of the emerging, but largely uncoordinated efforts, of competitive units within the XPCG, rather than a single strategy emanating from XPCG's central decision-making institutions. In 1996, Horn identified nine factories owned by seven different XPCG units (Table 6.8). Altogether, these operations accounted for approximately one-half of Chinese production at the time. Local governmental entities, Taiwanese interests, or (relatively small-scale) Chinese investors owned the remaining factories.

The emergence of Tunhe

A third period in the recent history of the Xinjiang processing tomato industry dates from the late 1990s, and is characterized by dramatic shifts in ownership and control within the industry. The driving force for change in this period was the emergence of a new entrant to the Xinjiang processing tomato sector, the Xinjiang Tunhe Investment Co. Ltd ('Tunhe'). The arrival of Tunhe encouraged greater concentration in the industry's production arrangements, with the relatively diffuse production geography of the industry (a large number of facilities with relatively small production capacities) being transformed into a more limited number of factories that made use of newly installed Italian equipment. Furthermore, Tunhe threw down a major challenge to XPCG, which has responded in quite radical ways.

The rise of Tunhe occurred with startling speed. Over the period of a few years towards the end of the 1990s, Tunhe emerged to become Xinjiang's largest processing tomato firm. By 2001, it operated 11 factories and accounted for 80 per cent of Xinjiang's tomato paste production (Xinjiang Tunhe Co. Ltd, 2001, p.2). This transformation in the ownership structure of the industry underlines the transitional character of the contemporary Chinese economy; unlike XPCG, with its deep political roots in China's state system, Tunhe represents the face of contemporary Chinese capitalism.

The origins of this company date to 1983, with the construction by the local municipal government of the Changjizhou Toutunhe cement factory at Changji, west of Urumqi. With the enormous pace of building and construction activities in Xinjiang over the 1980s and 1990s, this factory was immensely profitable for its owners. In 1993, it was re-incorporated as the Xinjiang Tunhe Investment Co. Ltd, and then, in 1996, the company was fundamentally restructured. Tunhe Investment Co. Ltd was floated on the Shanghai Stock Exchange, with a controlling interest being held by the newly formed Xinjiang Tunhe Group Co. Ltd, which, in turn, was majority owned through the D'Long group of companies. As illustrated in Figure 6.8, this has resulted in Tunhe's processing tomato factories being incorporated within a huge and diverse Chinese conglomerate company. At the head of this conglomerate are Tang Wanli and Tang Wanxin, two brothers who established this commercial empire from a film processing and computer shop in Urumqi, in 1986. By the year 2000, the group of companies owned by the Tang brothers generated annual revenues of $2.4 billion (US), and, according to Forbes magazine, the brothers were the 37th richest business family in China, with a personal fortune of $145 million (US) (Forbes, 2001, p.42). The evolution of the D'Long conglomerate underlines the massive and sustained pace of economic growth in China since the 1980s, and the opportunities for wealth creation it has generated. Like a number of other entrepreneurial Chinese conglomerates in this period, the D'Long group evolved and expanded through its ability to 'pick winners' across an array of industry sectors, in the country's tumultuous investment climate. Possibly the most-widely cited example of the Group's investment skills was its 1998 purchase of a former Soviet aircraft carrier ('the Minsk'), which at the time was destined to become scrap metal. D'Long towed the carrier to the southern Chinese city of Shenzhen, and made it the center-piece of a large-scale military theme park, 'Shenzhen Minsk Aircraft Carrier World'.

Table 6.8 Processing tomato factories associated with the Xinjiang Production and Construction Corps, 1996

Military unit	Factory name/location	Date established	Capacity (raw product tonnes)	Main product
124 Division	Green Leaf/Go Chang	1994	45,000	Paste
124 Division	Green Leaf/Go Chang	1995	45,000	Paste
102 Division	Waffer Food Co./Wu Jia Chi	—	36,000	Paste
145 Division	Shihezi General Field/ Shihezi	1994	12,000	Paste/ dice
145 Division	Shihezi Food Factory/ Shihezi/ Y1 (joint venture with local government)	1988	51,000	Paste
122 Division	Xia Ye Di/Shihezi	1983	6,000	Paste
7 Battalion	Kuiton Food Factory/ Kuiton	1990	10,500	Paste/ dice
131 Division	Factory 131/Kuiton	1990	18,000	Paste
21 Regiment	21 Regiment Factory/ Tiandan	1989	36,000	Paste

Note: '—' indicates data not known

Source: Horn (1997, p.11).

According to reports in the financial media, Tunhe's tomato operations generated $170 million (US) (Yuan 1.36 billion) in turnover and $22 million (US) (Yuan 180 million) in profits during the year 2000 (Quanxing and Feiyu, 2000). The speed of Tunhe's growth is illustrated by the number of processing lines installed during the late 1990s and early years of the twenty-first century. In 1997 and 1998, Tunhe installed only one new line each year. In 1999, the company installed four new lines; and in 2000, a further five lines (Xinjiang Tunhe Co. Ltd, 2001, p.4) (Table 6.9). The installation of this capacity is linked directly to the growth of Chinese paste exports to Italy during 2000 and 2001. Many of the newly installed lines were purchased from Rossi & Catelli and, according to industry sources, may have been financed through compensation trade arrangements, creating a situation whereby Italian trading companies received, and were involved in, offloading significant quantities of (relatively low-priced paste) in Mediterranean markets.

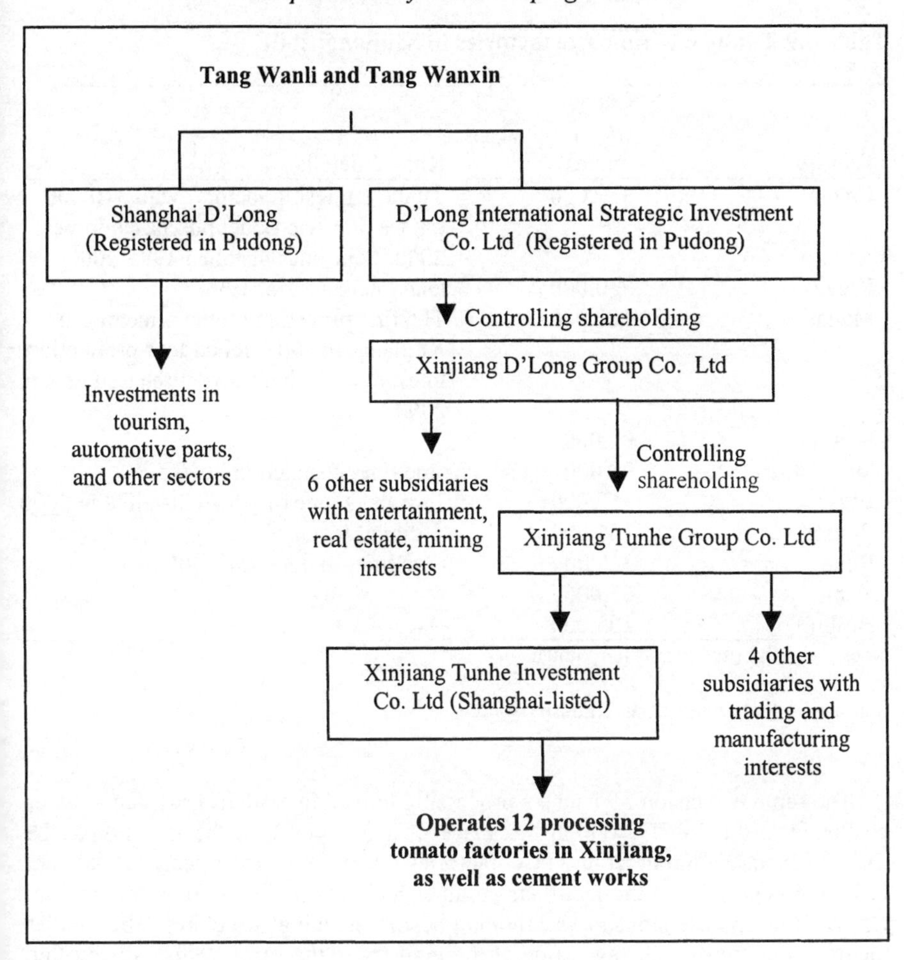

Figure 6.8 The D'Long group of companies, 2001

Source: Fieldwork data, based on official company documents.

Table 6.9 Tunhe tomato paste factories in Xinjiang, 2001

Factory	Capacity (raw product tonnes)	Known details
Changton	350,000	Tunhe's flagship factory, with 310,000 tonnes of raw product purchased in year 2000. Four lines installed 1995-2000.
Kleen	220,000	One line installed, 1999.
Manas	180,000	The first processing tomato factory in Xinjiang. In 2001, it had four production lines, one of which was purchased new in 2000.
Huaxin	90,000	—
Shawan	90,000	One line, installed in 2000.
Wusu	160,000	Two lines; one of which installed in 2000.
Beiting	165,000	A dicing factory.
Bohu	210,000	Five lines in operation, 2001.
Yanji	67,000	—
Heshuo	105,000	—

Note: '—' signifies details not known.

Source: Field research and industry sources.

The rapid expansion of Tunhe's processing tomato operations provided a major challenge to the XPCG. Within the space of a few years, in the late 1990s, the share of industry output from XPCG factories fell from approximately 50 per cent to 20 per cent, despite the fact their volumes had grown slightly over this period. Clearly, the Chinese industry was moving beyond its initial structures. The smaller factories that typified the evolution of the industry in the early 1990s, which often utilized second-hand machinery inside factories converted from other uses, had become increasingly ill-suited to the needs of Chinese processors as they sought to meet international demands. Although there remained many potential customers for cheaper, lower-quality Chinese paste, the future of the industry ultimately depended on its ability to improve quality and capture new (higher-priced) export markets. The path towards this goal rested with the installation of new state-of-the-art Italian machinery, in purpose-built factories. Tunhe had demonstrated an ability to rapidly expand on the basis of extensive investment in new plant and equipment. The XPCG now also followed this strategy.

The XPCG's response was to restructure its processing tomato interests into two corporate groupings, both of which raised equity on the Shanghai Stock Exchange. In 1997, the XPCG had taken its first significant steps towards attracting outside equity, with the formation and flotation of Xinjiang Tienye Co. Ltd ('Tienye'). This was the first public float involving XPCG, and marked an important evolution in the Group from its military/regional development roots.

Tienye operated as a conglomerate holding company for nine branch enterprises, across a wide range of business sectors including plastics, chemicals, and the manufacture of irrigation pipes. In 1999, XPCG interests used Tienye as the corporate vehicle to recapture production share in the processing tomato industry. Tienye built a state-of-the-art processing tomato factory at Shihezi, with two FMC lines and an annual capacity of approximately 180,000 tonnes of raw product, and took control of the '21 Regiment' factory in the Bosten Lake region. During this same year, XPCG also established and floated the Xinjiang Chalkris Tomato Products Co. Ltd ('Chalkris'), for the specific purpose of operating the newly constructed processing tomato factory at Bole (the most westerly tomato factory in China, less than 50 kilometres from the Kazakhstan border). In 2000, Chalkris added two more factories to its capacity (at Sampin and Junhu; see Figure 6.7), and was listed on the Shenzhen Stock Exchange. Hence, within the space of two years, XPCG interests had built two significant processing tomato groups with access to external financing and with control over five relatively large and modern factories.

The construction of new facilities by both the XPCG companies and Tunhe has improved quality in the Xinjiang industry. Most of the new facilities have received ISO9002 accreditation, which facilitates their ability to secure export contracts. Additionally, all exports are required to be accredited by CIQ, the Chinese Import/Export Inspection Agency. Some individual facilities also have organic, kosher and halal certification, and key buyers (especially from Japan) have undertaken extensive testing to ensure genetically modified crops are not used. However, product quality can be impaired by the tortuous logistics that still dominate the transportation of tomatoes from field to factory. During harvest, it is not uncommon for several hundred small trucks to be waiting for many hours outside factory gates, which leads to spoilage of product. Unlike California, Canada or Australia, where sophisticated scheduling practices attempt to optimize the flow of product, China's performance in this respect is still rudimentary.

The expansion in XPCG processing capacity generated a larger demand for tomatoes cultivated on State farms. During this period, Tienye controlled two State farms with 30,000 hectares of arable land (Xinjiang Tienye Co. Ltd, 2001), while Chalkris controlled three farms with a total area of 20,000 hectares and which employed approximately 4,800 persons (Peng, 2001). These landholdings have played a critically important role in the evolution of the Xinjiang processing tomato industry. Because of their landholdings, the XPCG companies have considerable ability to control the terms on which they are provided with raw product. In general, State farms are given instructions to grow sufficient tomatoes to ensure factory capacity is met. In 1996, farm workers in these enterprises were reported to be earning wages of approximately $1 (US) per day (Horn, 1997, p.12). During field research in Xinjiang undertaken in 2001, the authors were told that a 'good picker' could earn $4 (US) day during harvest. It is unclear how the XPCG companies account (in a financial sense) for the transfer of raw product from farm to factory under these arrangements. In Western commercial discourse, this would be termed a non-arms length, related party, transaction. In the 'half way house' of Chinese capitalism and state-ownership, such arrangements are not measured against a fully operating market system, and hence bear different consideration.

The key point, nevertheless, is that the XPCG companies have had access to large tracts of land, effectively at zero cost, as well as a large agricultural workforce that could be deployed at will. These conditions suggest an important element of competitive advantage for these entities.

As a new entrant to the industry, without extensive landholdings, Tunhe has been required to procure its raw product through contracting with peasant farmers. According to industry sources, these farmers till approximately 1.5 hectares each, and in 2001, were paid approximately $30 (US) per tonne (Yuan 240). In Xinjiang, the contract price for tomatoes tends to vary in line with the cotton price, since the latter serves as a substitute crop. Because cotton does not perish but tomatoes do, farmers tend to demand a relatively higher return on tomato cultivation, *vis-à-vis* cotton growing. Moreover, in situations of excess demand for tomatoes, growers can and do extract a premium price from processors, beyond what was agreed to in the contract.

Yet, in a reflection of the way the organizational structure of processing factories changed rapidly in the late 1990s, tomato cultivation arrangements may also be restructured dramatically in the next few years. As discussed in Chapter Two, there are powerful links between the introduction of seedling transplanting, the use of hybrid seeds, and mechanized harvesting. To date, the Chinese industry has relied heavily on open-pollinated seed varieties and hand harvesting. Demands by some key customers for improved consistency of product however, are encouraging shifts within these arrangements. The use of imported hybrid seed grew in the seasons following 2000, despite its high cost. In the view of some industry participants in Xinjiang, hybrid seeds deliver product characteristics that make up for its higher cost. Consistent with these changes, there is a general increase in the use of transplanting, which in 2001 was estimated to account for 25 per cent of Xinjiang tomato cultivation. Tienye imported China's first tomato harvester in 1999 (an FMC model), and in 2000, the company imported a further seven. During the 2001 season, the harvesters were trialled at Shihezi. At a superficial level, it may appear counter-intuitive that this industry is considering the introduction of mechanized harvesting, when hand harvesting labor costs are the lowest of any major tomato region in the world. However, mechanized harvesting needs to be seen as part of a general shift in the industry towards greater consistency and quality of product, which is facilitated by the use of hybrid seeds with uniform ripening characteristics. The introduction of mechanized harvesting is easiest to implement in large State farms (as opposed to relatively small, peasant holdings), and it is perhaps not surprising that an XPCG company is at the forefront of this innovation. Clearly, should mechanized harvesting prove successful, it may have dramatic social consequences in terms of significantly reducing demand for rural labor during harvest periods.[2]

The recurring debate: how to interpret Xinjiang's processing tomato industry?

Ultimately, the question remains as to how the dramatic development of the Xinjiang processing tomato industry may be interpreted. This question is at once geographical and temporal; in terms of the geography and economics of trade and production, does the global processing tomato industry face the prospect of being swamped by massive amounts of relatively cheap Chinese tomato paste? In terms of temporal considerations, is the production of relatively cheap Chinese tomato paste sustainable over an extended period?

Clearly, the emergence of the Xinjiang processing tomato sector has reverberated within the global industry. During the years 2000 to 2002, Xinjiang's tomato paste exports were at least double those of either Chile or Turkey, and roughly twenty-times the size of those from Thailand. In addition, more than any other developing country exporter of processing tomato products, the Xinjiang industry sells to an *international* marketplace, as opposed to being reliant on near-neighbor countries (c.f. Chile and Turkey).

Yet some industry participants point to the particular circumstances that have created Xinjiang's assault on world tomato markets. As discussed above, the large volumes of Chinese tomato paste exported to Italy in recent years are apparently to be explained (at least partially) by the requirements of Chinese processors to meet the terms of their loan finance. The sale of this paste may have generated the hard currency required at a particular point in time, but in the opinion of many industry observers, it is not sustainable at such low prices. Consequently, some analysts perceive Xinjiang's primary long-term future as a supplier of bulk industrial paste to regional markets in China, the Asia-Pacific and Central Asia.

There is, of course, already a strong network of tomato paste trade connections between Xinjiang and Japan, mediated by Japanese general trading companies (*sogo shosha*). Over the 1990s, China emerged to become Japan's most important supplier of tomato paste. Xinjiang tomato paste exports have also found important markets in Korea, the Philippines, and Taiwan. H.J. Heinz Co.'s restructuring of production capacity in Australia and New Zealand, discussed in Chapter Three and Chapter Four, also seems premised on a long-term expansion of Chinese tomato paste exports. Added to these markets, a considerable volume of Xinjiang's tomato paste exports are transported by road westwards, through Kazakhstan and into Russia. In recent years, Urumqi has (re)-emerged as a vital trading post within cross-Asian trading routes, with ethnic Kazakhs reportedly playing a central role in the organization of this trade.

Growth of the domestic Chinese market for tomato paste, however, provides a potentially massive market for the Xinjiang industry. Processing tomato products are not a traditional component of Chinese culinary cultures, and at present, only five per cent of Xinjiang's output is destined for the domestic market (Xian, 2002). However, many industry analysts believe that there may be significant dietary changes occurring in China over coming years associated with urbanized, western influences. Western fast food chains have made major in-roads into China during the past decade. There are over 500 Kentucky Fried Chicken (KFC) restaurants in

China, earning the company sales of approximately $250 million (US) per year (Xinhua News Agency, 2001). Moreover, of more direct relevance to the processing tomato industry, the first Pizza Hut restaurant in China opened in 1990 (San Francisco Chronicle, 1990, p.C7), and there are now over 800 such restaurants in the country. Nevertheless, although impressive, the per capita consumption levels of processing tomato products in China remain very low. For its part, Tunhe has attempted to raise domestic consumption through the development of tomato juice and sauce products, for the retail market. These products have been accompanied by a sales slogan (*Biological revolution, Red Tunhe, Red products*) that draws an association between the company's 'revolutionary' growth and the red coloring of tomatoes.

Conclusion

The concept of agri-food globalization seems to suggest an expanding role for developing countries, especially as sites for production and export. This chapter has evaluated these arguments with respect to the processing tomato industry. Firstly, it is apparent that there are only a handful of export-oriented processing tomato producers in the developing world and, moreover, in aggregate terms they did not capture a significantly larger share of the global market for processing tomato products during the 1990s. Case studies of Thailand and China reveal the complex circumstances from which these outcomes have emerged.

In the case of Thailand, the processing tomato industry evolved in response to domestic demand from seafood canneries and the international product sourcing strategies of some Western supermarkets, in a context in which village farmers in north and northeast Thailand had been exposed to contract growing systems, and government subsidies encouraged the establishment of factories in these regions. Moreover, the industry was assisted by flight capital from Taiwan, as processing firms sought lower cost production sites. The Thai processing tomato industry however, has gone through a series of peaks and troughs. Because processors sought export markets for relatively cheaper products in price sensitive environments, the industry was at the mercy of international volatility in demand and supply. The liquidity collapse that accompanied the Asian economic crisis of 1997-98 undermined the industry's attempts to generate sustained export growth. This example gives credence to general arguments in the research literature on the market vulnerability of export agriculture.

China, however, presents a different picture. Located in Xinjiang Province, one of the most remote and isolated regions of the world, the Chinese processing tomato industry has exhibited strong growth since the late 1980s. Of course, there are many obvious differences between this case and Thailand. In particular, Xinjiang's climate and environment appear better suited to processing tomato cultivation. However, beyond these factors, Xinjiang's industry has been propelled by significant investment in state-of-the-art equipment, financed in ways that forced the export sale of Xinjiang tomato paste through particular market channels. Extensive state-ownership of land and economic resources in the Province, via the

XPCC/XPCG, has proved a stabilizing influence for commercial activities. Unlike *laissez faire* market economies, a series of cross-subsidies underpin the activities of these entities, with the effect that financial losses can be cushioned in any given year. Added to this environment, the aggressive entry of Tunhe to the Xinjiang processing tomato industry in the late 1990s launched the sector into a new phase, based centrally on the large-scale production of bulk tomato paste in newly-constructed facilities. The entry of Tunhe, in turn, encouraged a restructuring of XPCG's processing tomato activities, leading to the formation of the Tienye and Chalkris groups.

The conclusion from these analyses is that so far, there has not been an overall shift in the processing tomato industry towards developing countries as cheap production sites. The dynamic growth of Xinjiang illustrates, quite clearly, that the driving forces of industry growth have been specific state-centered political factors, and the attempts by food manufacturing firms to make equipment sales. These insights support the fundamental argument of this book, about the contested and contingent character of agri-food globalization. However, because of the inherent dynamism of international competition and trade, it cannot be assumed that these outcomes will necessarily be the case in the future. Contemporary processes of globalization are in their earlier stages in this industry, and the possibilities remain open.

Notes

1 Debate on this issue was a strong undercurrent of industry reports at the *IV World Congress on the Processing Tomato* at Sacramento, in 2000, and the *V World Congress* at Istanbul, in 2002.

2 Although, because many of Xinjiang's agricultural crops are harvested at the same time, there is often a temporary labor shortage, filled by seasonal migrants from neighboring provinces.

Chapter 7

Unpacking the tomato: interpreting agri-food globalization

The world processing tomato industry consists of hundreds of thousands of farm and factory workers, tens of thousands of tomato farms, thousands of processing tomato factories, hundreds of specialist processing tomato companies, a dozen key transnational corporations, tens of thousands of individual products, brand names, trademarks and patents, and millions of consumers. As previous chapters have documented, these actors relate to one another in complex and ever-changing ways. The key question, therefore, is what general conclusions can be made about these interactions: in short, what is revealed when we 'unpack' the international processing tomato system?

Some accounts of global agri-food restructuring, for example within the agricultural economics literature, interpret these issues in ways that encourage a relatively narrow appraisal of these processes. These accounts tend to define globalization as encompassing activities that facilitate the operation of market processes internationally, which are assumed to generate efficiencies in resource allocation by way of the rational 'invisible' hand of the market. Using the deductive logic of positivism, such accounts are driven primarily by a concern to assess resource optimization, and point to the economic benefits arising from freer trade and resource flows within globalizing economic structures.

Yet therein lie both the strengths and weaknesses of such an approach. The literature on agricultural economics rightfully emphasizes the roles of the market as a powerful institution for resource allocation, and this approach has obvious merit for particular technical purposes, such as forecasting changes to incomes arising from shifts in trading or production systems. However, a singular focus on issues of market rationality may preclude wider consideration of how markets are embedded in political, cultural and social contexts, and ignore the question of how and why such developments relate to society. In the forthright language of Friedland (2001, p.82) the problem with much of the agricultural economics literature is that it is 'usually devoid of human beings'. In other words, it emphasizes the operation of markets at the expense of socio-historical context.

This book has adopted a quite different approach to questions concerning global agri-food restructuring. While we do not jettison the contributions of agricultural economic literatures, we do argue they need augmentation. From our backgrounds in critical agrarian political economy (see Chapter One), we identify global agri-food restructuring as a complex and contested set of processes, occurring in different ways at multiple sites of production, consumption and decision-making.

Such an approach does not tend towards a simple, persuasive answer to the problem of understanding global agri-food restructuring. The framework of critical agrarian political economy documents agri-food restructuring for what it is, rather than present it as a stylized model of what it might be. As argued in the Preface to this book, we contest simplistic accounts of 'a globalized food system' that conceptualize it as a seamless and vertically coordinated structure dominated by profit maximizing and rationally behaving transnational corporations. The material in the preceeding chapters assert that what passes for 'the global processing tomato system', actually consists of a set of heterogenous and fragmented processes, bounded in multiple ways by the separations of geography, culture, capital and knowledge. At times these separations are compressed, via corporate takeovers, global branding or the emergence of dramatic new trade relations linking erstwhile diffuse arenas of production and consumption. But these compressions are neither as smooth, seamless or permanent as some analysts suggest. The processing tomato sector, as with the food industry as a whole, is characterized by low returns to capital, significant exposures to risk, the specificities of culturally and spatially embedded consumption, and sharp fluctuations in the ability to capture competitive advantage. Local and regional factors, as much as global ones, shape the sector's organization and restructuring.

This final chapter reviews and refines these ideas, in order to provide insights which will be of interest to both industry practitioners, and analysts of agri-food restructuring. In the following section, we question the extent to which the processing tomato industry can be construed as 'globalized', before turning to a broader discussion of our findings, in the final section of the book.

How 'globalized' is the processing tomato industry?

Our first task is to assess the extent and character of globalization in the world processing tomato industry, with particular attention being paid to changes since about 1990. The notion of 'globalization' is poorly conceptualized in much published research, and is often used as a catch-all phrase to describe an ill-defined set of processes linked to internationalized flows of finance, products, knowledge and people. The concept can mean many things to many people, depending on the author's intentions. Clearly, debate on this issue would be enhanced if there was a commonly accepted analytical framework. While this may be too much to ask, nevertheless it seems to us that the approach adopted by Allen and Thompson (1997) has considerable merit. These researchers argue that the extent of globalization should be assessed according to a threefold schema, which focuses on the *scope* of investment flows (that is, the international geographical spread of large corporate investments), the *depth* of investment flows (that is, the ways these investments encourage the coordination of markets and greater price equalization in the markets for capital and goods), and the *pace* of investment flows (the effects of investment flows on the qualitative character of institutions and 'ways of thinking'). In similar terms, the United Nations Conference on Trade and Development (UNCTAD, 1997) draws a distinction between so-called 'shallow integration', that is, economic linkages via product flows in the form of exports and imports, and 'deep integration', that is, cross-border investments that encourage the coordination of production arrangements internationally.

These perspectives inform our assessment of globalization in the processing tomato industry. We propose that the extent of globalization in the processing tomato sector can be assessed with regard to four issues. First, in common with Allen and Thompson's notion of the 'depth of globalization', the concept is assessed with regard to convergence in international product prices. If processes of globalization are to be equated with the emergence of international markets, it is to be assumed that product prices will converge towards a 'global market price' over time. Second, we assess to which globalization with regard to internationalized product flows; in other words, the extent processing tomato products are traded in a global market. Third, globalization is assessed with regard to international investment linkages; to what extent do globalizing actors (such as transnational corporations) coordinate production internationally. Finally, globalization is assessed with regard to the international transfer of ideas and knowledge, which in the processing tomato industry have been a critically important element of change since the early 1990s.

Analyzing globalization through these categories recalls the discussion of agri-food globalization in Chapter One (pp.9–15), where it was argued that globalization does not simply 'exist' in the abstract, but is created within and through particular places and institutions. In turn, this implies that the analysis of globalization should be connected to its historical and socio-political contexts, in other words, explicit attention should be given to the political and commercial frameworks that underpin decisions by economic agents (such as firms or traders) to create particular international linkages.

Globalization as the construction of a world market price

Among other things, the concept of economic globalization implies the progressive replacement of separate national markets in goods and services, with the construction of international markets for these same goods and services. In general, the clearest evidence supporting such developments is the convergence of product prices towards a global market norm, so that industry participants in different parts of the world are obliged increasingly to compete in terms of a world market price, subject to transport and related costs.

In the processing tomato sector, the past decade has witnessed increased levels of industry awareness of the concept of the 'global market price' as a competitive yardstick. Not surprisingly, these developments have been most prominent with regard to bulk, aseptic-packaged tomato paste, the most tradeable component of the industry. Yet, notwithstanding the considerable discourse around this subject, the concept of the global market price for tomato paste is not as straightforward as it seems. Compared to some other markets (say, capital markets or those for some of the most widely traded agricultural commodities, such as wheat, coffee, frozen concentrated orange juice, or pork bellies), the buying and selling of bulk tomato paste market does not occur in accordance with the strict regimentation of a global market price.

Although there have been considerable changes in the industry over recent years towards product uniformity and standardization, the global market for this product is still fractured. As discussed in Chapter Three, paste production is fragmented along lines of production technique (hot break or cold break) and degrees brix. Paste is manufactured to varying brix levels dependent upon end-users requirements, or according to industry norms. For historical reasons, the US traditionally has produced paste at 30-32 degrees brix, whereas much paste in southern Europe is produced at either 28-30 degrees brix or 36-38 degrees brix. These differences may mean that some end-users in Europe (such as food processing firms using paste as an industrial input) may not regard Californian paste as suitable to their needs, and vice-versa. In these circumstances, European and Californian paste may not be in direct competition with one another. Furthermore, as discussed with reference to China in Chapter Six, end-users also have different expectations with respect to quality standards, measured by such factors as mould counts and Bostwick (viscosity levels), with the implication that paste from different source countries (or individual factories) may vary considerably in terms of price. Added to these considerations is the fact that a significant proportion of global tomato paste production is not traded in open markets, but is exchanged between supply chain partners in accordance with durable commercial arrangements (for example, long-term supply contracts). Pricing conditions in these market channels may depart from industry norms, because of the roles of trust, reliability and sales security. Therefore, it needs to be recognized that the concept of a 'global market price' for bulk tomato paste largely serves as a simplifying metaphor that does not give full recognition to the multiplicities of market channels involved in the buying and selling of this product.

However, despite these major qualifications about what constitutes the market for bulk tomato paste, since the mid-1990s there appears to have been some global convergence in published estimates of the product's prices in different national markets. This is most clearly apparent when comparing the free-on-board (FOB) prices for tomato paste in California and Italy, the world's two most important production regions (Figure 7.1). Price levels and trends in these two markets have converged markedly in recent years, suggesting that these two individual paste markets have been constructed with increasing reference to one another. This is all the more relevant because there is virtually no significant bulk tomato paste trade between the key markets for Californian and Italian paste, respectively (see the discussion on this point in the following section). What this suggests, therefore, is that discourses of global competition have facilitated the construction of a global market price for bulk industrial tomato paste, regardless of whether this price becomes the basis for determining the direction of trade flows.

These arguments reinforce the general conclusions reached in much of the economic geography and agrarian political economy literature about the role of competitive threats as an agent of industrial restructuring. According to Webber et al. (1991), global competition acts as a restructuring agent because of its *understood possibilities*, as much as its reality. As Koc (1994) argues, a major component of 'globalization' relates to its properties as a discourse about the political and economic organization of the world. In the case at hand, this is manifested in the ways firms and growers across many countries have increasingly perceived their positions in the industry through studies of international benchmarking and competitiveness. Knowledge about price trends and competitive cost structures has circulated with increasing currency, shaping participants' decision-making processes and leading towards outcomes that hinge on the ability to match international prices. For example, over the past decade the annual price offers for raw product made by Australian processing firms to their growers have been developed with explicit attention to conditions in the Californian market (Australian processors make their offers to growers in the period from May to August, during the same time in which the northern hemisphere season takes place). Often, processors have pointed to excess supply conditions in the Californian market as justification to make relatively low initial prices to Australian growers. However, as growers tend to retort, processing companies appear less keen to make Californian comparisons in conditions of excess demand.

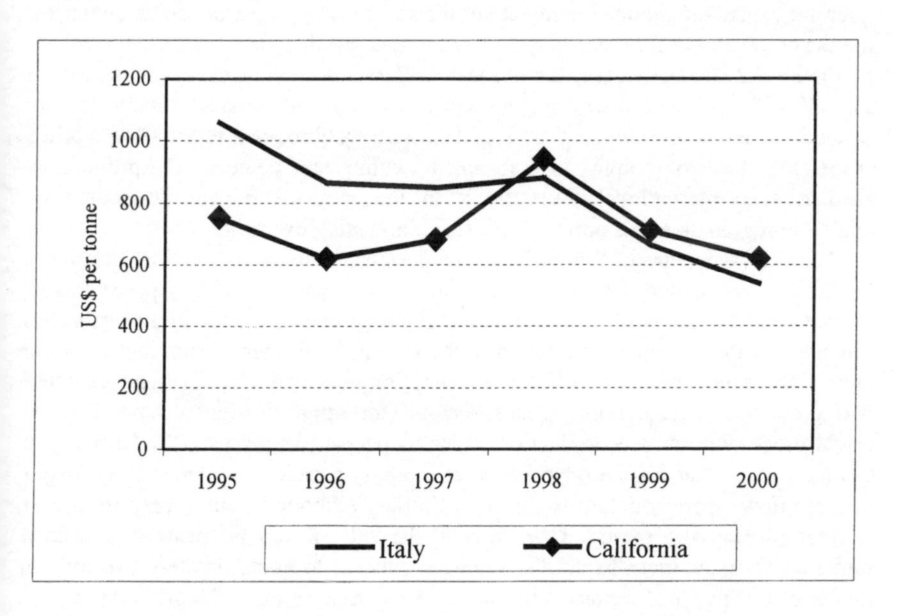

Figure 7.1 Price of Californian and Italian tomato paste, 1995-2000

Note: Free-on-board price for tomato paste at equivalent 28-30 degrees brix.

Source: Tomato News (various years).

Globalization as the flow of products across national borders

Second, globalization in the processing tomato industry would seem to imply an increasing tendency for products to be traded internationally, as cost/quality efficient producers capture markets from their less competitive rivals. Reduced international transport costs over recent years, in combination with the lowering of tariffs and other forms of border protection by many countries, have created a more 'open' environment in which international trade can take place.

Analysis of production and trade flows in the global processing tomato industry, however, reveals that international exchange within the sector is relatively limited. Figure 7.2 presents a general model of the global processing tomato production-trade complex in 'first-tier' tomato products (canned tomatoes and paste), during the period 1999-2001. To compensate for the year-on-year volatility in production and trade that is a feature of this industry, the percentage figures in this diagram represent approximate values only; the primary purpose of Figure 7.2 is to portray the broad patterning of global production and trade flows, rather than to attempt to provide a model indicating absolute statistical precision (in any case, as discussed earlier in this book, such an exercise would be futile given the considerable uncertainty about the accuracy of data from some countries, and the complexity of trade in re-processed tomato products).

Figure 7.2 suggests that if the EU is considered as a single economic entity (so that intra-EU product flows, for instance, between Italy and Germany are not defined as being *international* trade), then at least three-quarters of the world's processing tomato output is consumed within its country of production. Furthermore, international trade flows in the industry remain dominated by relatively regionalized, as opposed to trans-continental, exchange.

The central feature of Figure 7.1 is the dominance of the US and EU as arenas for the production and consumption of processing tomato products. Approximately 39 per cent of the world's processing tomato production over these years was produced in the US for consumption in the US, and a further 27 per cent of world output was produced in the EU for consumption in the EU. Taken together, these two production-consumption arenas account for approximately 66 per cent of world processing tomato production over this period. As discussed extensively in Chapter Five, the EU production-consumption complex is marked by major product flows from southern to northern Europe. Although figures vary from year to year, during the period under review the sale of tomato paste and canned tomatoes from southern to northern Europe equated to approximately two million tonnes of raw product, representing roughly nine per cent of total world output.

The remaining one-third of world production tends also to be dominated by regionalized production-trade networks. Most production in the Middle East, Africa and Latin America tends to be consumed within these regional markets. As discussed in Chapter Six, significant export volumes from Chile and Turkey are sold mainly to neighboring countries in, respectively, Latin America and the Middle East. Furthermore, over three-quarters of EU exports are sold to the proximate markets of eastern Europe, the Middle East and Africa. Even in the case of the US, approximately one-third of exports tend to be sold to the neighboring country of Canada, which, since the passage of the NAFTA, is effectively within the same market.

Trans-continental trade in processing tomato products comprises only a relatively small proportion of the global industry. During the period 1999-2001, the proportion of processing tomato products shipped over large, trans-continental distances (defined as exports from Europe to North America or Asia; from China

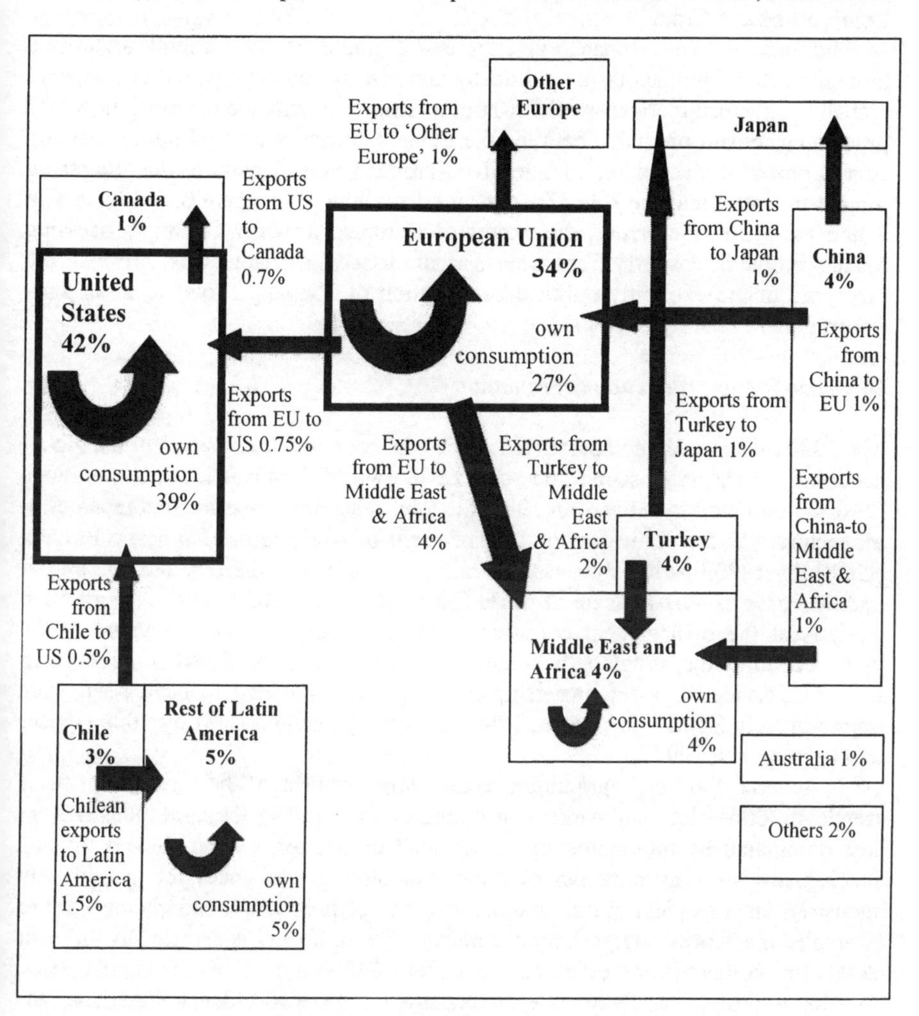

Figure 7.2 Stylized model of major production sites and trade flows in the processing tomato industry

Source: Own work, based on published industry data.

Note: Percentage data refers to raw product equivalents, in approximate terms only, for the years 1999-2001.

to North America, the Middle East, Europe or Africa; from Turkey to Japan, or, from Chile to Japan) probably would have accounted for no more than six per cent of world output. Across the world, Japan represents the only example of a major importing country heavily dependent on trans-continental processing tomato trade. Although China has captured a significant share of Japanese tomato paste import requirements over recent years, the country still imports large amounts of paste from Turkey and Chile.

The quantitative estimates in Figure 7.2 highlight the general arguments throughout this book about the politically and historically constructed character of global restructuring. The world's largest single international trade flow for processing tomato products, between the EU and countries of the Middle East and Africa, occurs in the context of extensive European subsidization of the processing tomato industry, and the role of Italian food machinery companies and traders in acting as agents facilitating the export of European (mainly Italian) processing tomato products. Exports from China, as discussed in Chapter Six, are also the outcomes of the explicitly political construction of Xinjiang Province as an agri-export zone.

Globalization as production coordination

The 1990s witnessed an unparalleled process of economic integration in the global economy, as the size, scope and scale of the world's transnational corporations (TNCs) continued to expand. In 1982, the total sales of transnational companies in all sectors was the equivalent of 23.2 per cent of world Gross Domestic Product (GDP); by 1990 this had increased to 25.5 per cent of world GDP, and by 2000, it had increased to 45.1 per cent of world GDP (UNCTAD, 2000, p.4)[1]. According to analysts at the management consulting firm McKinsey & Co., the value of the world economy that is 'globally contestable', that is, open to global competitors in product, service, or asset ownership markets, was estimated to have risen from approximately $4 trillion (US) in 1995, to over $21 trillion (US) by 2000 (Fraser and Oppenheim, 1997).

In general, food and agriculture sectors have *not* been at the forefront of these trends. In terms of global market concentration, food and agricultural industries are less dominated by internationally coordinated production systems linked through mechanisms such as common ownership or strategic alliances, compared with industries such as electronics, automobiles, energy production and chemicals. For example, the world's largest food company, by turnover, is Nestle. In the year 2000 this company possessed approximately $48 billion (US) in annual sales, making it (only) the 59th largest company in the world. By comparison, the world's largest corporation, energy company Exxon Mobil, had annual sales of $210 billion (US).

The reasons for the apparently less advanced state of globalization in the food industry would appear to relate to factors such as the propensity for differences in national and regional food consumption patterns, as well as the fact that this industry has been subject historically to extensive regulation that has often had the effect of reducing foreign ownership. However, during the late 1990s there was evidence that these patterns were changing. Between 1995 and 2000, the total value of international food industry mergers was $50.1 billion (US), but with an accelerating pace of international mergers as major food companies sought strategic acquisitions, in the single year 2000-01, the total value of such mergers was $69.2 billion (US) (Rogers, 2001).

The processing tomato industry, as a sub-set of the broader food sector, reflects these developments. In general, 'national firms' (defined as those operating solely within one country) play a significant role in the global industry. In California, national firms control approximately two-thirds of all first-tier processing capacity, including the entire production system for bulk industrial tomato paste. National firms similarly dominate the European processing tomato industry, however as discussed in Chapter Five, these have become more 'pan-European' in their operational scope over recent years. It is also the case that national firms are either the sole or dominant type of corporate enterprise in the processing tomato industries of China, Turkey, Chile, and countries in the Middle East and Africa.

Transnational corporations, in contrast, tend to play a relatively minor role in the first tier processing of tomatoes. The two companies with the greatest size and scope in this regard, are H.J. Heinz Co. and Unilever. Both of these companies operate first-tier processing tomato factories in North America, Australia and Latin America, whereas Heinz also has major facilities in Europe (Spain), and Unilever operates processing tomato facilities in India. Unilever asserts that it purchases five per cent of the world's processing tomatoes (Unilever, 2001) and industry sources suggest that Heinz's purchases would be somewhat larger than this amount. This implies that the two largest transnational firms in the sector together purchase only slightly more than ten per cent of global product, a relatively modest contribution when compared to patterns of industry control in other sectors.

However, the situation with regard to the first-tier processing of tomatoes does not fully describe patterns of control and coordination in this industry as a whole. Since the early 1990s there has been considerable merger and acquisition activity in the consumer foods sector, leading to increased market concentration in many product areas relevant to the processing tomato industry. In general, these trends emerged during two key phases of industry rationalization. The first involved a series of strategic acquisitions and divestments as major companies re-oriented their structures and strategies during the 1980s, when extensive merger and acquisition activity reshaped Anglo-American food industries (Pritchard, 1993). In many cases, this wave of mergers created unwieldy corporate structures built on unsustainable debt levels. Stock market plunges in 1987 and 1989 coupled with the recession of the early 1990s, exposed the weaknesses of financially-driven strategies and encouraged firms to identify and focus on their 'core businesses'. As these business models gained influence through the 1990s, divisions emerged within the world food industry between companies concentrating on brand management and consumer food businesses, and those intent on generating profit through the internationally cost efficient production of food and agri-inputs. Heinz's 'Project Millennia' strategy, discussed in Chapter Three, is an exemplar of both these processes.

More recently, a second phase of industry mergers and acquisitions has emerged as key corporations in major markets have sought even greater size and scope. Unilever's acquisition of Best Foods, in 2001, is the most notable example of such a process. Changes in the selling markets for food products have been a major driver of these developments. During the 1990s there was steady concentration in retail sectors and growth in food service industries, especially the franchised take-out component, based around 'fast-foods' such as pizzas and hamburgers. This growth and concentration has forced changes in the selling strategies of food manufacturers. To take one example, Del Monte, one the largest branded food companies in the US, now sells its tomato products into a fiercely competitive market where supermarket private labels (or 'store brands') comprise 32.3 per cent of market share. In 2001, Del Monte's fifteen largest customers accounted for 61 per cent of the company's sales, with one customer, Sam's/Wal-Mart, handling 15 per cent of sales (Del Monte, 2001, p.10). In an attempt to ensure its marketing practices were increasingly attuned to retailers' requirements, in 2001 Del Monte appointed a single national broker to represent it to the entire retail trade, across the US. This initiative was unprecedented for Del Monte and is rare within the food industry generally (Del Monte, 2001, p.35).

Concomitant with these developments was the rise of discount and warehouse retailers which, as Buttel (1996, p.31) observes, represents the antithesis of value-adding: 'Warehouse discounters confine their merchandise lines to those in which manufacturers are willing, or can be coerced, to supply product at a significant discount'. Over the period 1994-98, retailers' private label sales grew six per cent per annum, and by 2000 were a $40 billion (US) business within the US (Colusa Canning Company, 2002). The rise of private labels has created new market channels in the industry. Specialist players, such as the US food corporation Ralcorp, formerly known as Ralston Purina, service this market across a range of product niches. The Ralcorp story presents a fascinating glimpse into these developments, with direct implications for the processing tomato sector. In the late 1990s, Ralcorp undertook a spate of acquisitions and divestments, in order to develop four key business units with national scope in the private label trade: Ralston Foods (cereals); Bremer (cookies and crackers); Nutcracker/Flavor House (snack nuts); and Carriage House, formerly Red Wing/Martin Gillet (sauces and dressings). The fourth of these areas uses substantial quantities of tomato paste as a bulk industrial input. To this end, Ralcorp operates a facility in Williams, north of Sacramento, to produce industrial tomato paste for its private label products. Excess paste is sold to the Campbell Soup Company under a preferred supplier agreement (Colusa Canning Company, 2002).

These processes bring into focus two important conclusions with respect to international production coordination in the processing tomato industry. First, control of first-tier processing remains relatively diffused, and is largely still in the hands of national firms, although many of these are currently engaged in restructuring and expansion strategies. Second, major consolidation has occurred in the consumer food products and retail/distribution sectors, with profound implications for the processing tomato industry. However, at this stage the major effects of this rationalization seem to have been to narrow market channels *within* key markets (for example, within North America and within Europe) rather than to create conditions for extensive *international* production coordination.

Of course, industry restructuring in the food sector is continuing, and these patterns may change significantly in future years. In particular, many industry analysts have identified the rapid international concentration of supermarket retailing as a key new development in the global food economy, with potentially major implications for downstream actors in food processing (Burch and Goss, 1999; Wrigley and Lowe, 1996). At present however, there is scant evidence of comprehensive, international production coordination within the industry as a whole.

Globalization as international flows of information, knowledge and people

The previous three sub-sections have essentially analyzed globalization with respect to material economic processes; the construction of global market prices, the international flow of products, and the coordination of production through ownership and control. However, globalization processes also correspond to international flows of information, knowledge and people. On the one hand, these flows establish the pre-conditions for globalized flows of product and investment; on the other, they are a form of globalization in their own right, as they forge linkages between parts of the industry that would otherwise remain separate.

The emergence of these international 'soft' flows (as opposed to the 'hard' flows of investment capital and product) has been a relatively recent industry development. Until the 1990s, international knowledge about the industry was very sparse, and held by relatively few individuals. During the course of conducting the research for this book (over the period 1999-2002), it has seemed inconceivable that many of the industry's apparently indispensable structures and institutions did not exist ten years earlier. Until the early 1990s, statistical data on production trends was very rudimentary, there were no formal international meetings or conferences within the industry, and there was very little exchange of personnel between major production regions. Communication within the industry was relatively infrequent; fax machines did not gain widespread usage until the mid-1980s, and the Internet did not effectively exist until the mid-1990s. From the perspective of 2002, the relatively few published industry reports published prior to the 1990s tend to read like descriptive travel journals; their main objective seemed to be to record basic production information. For example, as late as 1987 a major research project at the University of California at Berkeley was funded with the objective of collecting primary data on production trends:

> To enable US decision-makers — farmers, farm group leaders, and processors — to be more knowledgeable about current trends in the world's processing tomato industry, we have made field investigations ... in many countries where tomato processing is important (Garoyan and Moulton, 1987, p.1)

The justification for such a project would seem inconceivable in 2002, when such statistical information is readily available on the Internet.

The evolution of three inter-related international industry institutions has been critical to the development of improved flows of information and knowledge. The first of these is the official international organization representing the industry, the World Processing Tomato Council (WPTC). This organization is funded largely from annual membership dues paid by processing tomato industry associations within individual producer nations, and has a general oversight role for the industry as a whole. Within the WPTC are three Commissions, reflecting the core areas nominated for international cooperation in the industry: (i) the Exchange of Information Commission, which facilitates statistical reporting in the industry; (ii) the Tomato and Health Commission, which has responsibilities with regard to the promotion of research on the health properties of tomatoes, and (iii) the Codex Commission, which acts as a clearinghouse for industry representations to their national governments with respect to the *Codex Alimentarius*, the United Nations-affiliated body charged with making recommendations on international food standards.

The second key institution involved in information flows is Tomato News, the industry's bilingual (English/French) journal, and its most important source of data. Tomato News has changed considerably since its first incarnation as a mimeographed and stapled newsletter in 1988. By the early 1990s, production values of the publication had become more professional, and the journal contained a more comprehensive set of regular industry data. In 2001, the print version of the journal was complemented by a subscription-only electronic version, and a business-to-business Internet trading and information exchange site, Tomatoland.com. Tomato News has a quasi-official status within the industry. It is published by the World Information Center for the Processing Tomato, in Avignon, France, and its operational manager is also the Secretary-General of both AMITOM and the WPTC.

The third key institution to facilitate international exchange of knowledge and information is the biennial meeting of the *World Processing Tomato Congress*, an event which aims to provide a 'discussion space' for participants to review developments in the industry. In 2002 the *V World Processing Tomato Congress* was held in Istanbul, in conjunction with a technical symposium on tomatoes convened by the International Society for Horticultural Science. This model of jointly convening technical and industry conferences followed the successful development of this structure at the *IV World Processing Tomato Congress* at Sacramento, California, in 2000. Previous Congresses, at Pamplona, Spain (1998), Recife, Brazil (1996), and Avignon, France (1994) had more limited scope and structures. World Congress meetings often provide an opportunity for sectional groups within the industry, such as international grower organizations (see Chapter Four), to meet.

Alongside improved flows of knowledge and information, there have been flows of personnel. During the 1990s, industry study tours became commonplace for extension officers, research scientists, growers and processing firm representatives. Additionally, a growth in informal strategic alliances between processor firms has led to an increased exchange of personnel between firms on a regular basis. Indeed, the seasonal nature of the industry lends towards such developments, with personnel from European or US companies frequently being employed in Australia or Chile during the northern hemisphere off-season. These personnel exchanges mainly involve technical staff, such as mechanical engineers or logistics experts. However, one of the most surprising examples of international personnel exchange occurred during the 2000 northern hemisphere season, when a temporary shortage of truck drivers in California led to a number of Australian drivers experienced in the processing tomato industry to be employed for the season.

In addition to these seasonal-based personnel movements, many engineers and technicians with the major Parma-based food machinery firms are employed as international 'trouble-shooters', being shunted between various production sites to solve equipment or installation problems on a short-term basis. Other participants in the industry, notably those employed in trading firms and those involved with seed trials, are also extremely internationally mobile. This mobility of significant numbers of persons creates and perpetuates networks in the industry, for the production and exchange of information and knowledge.

Conclusions: the extent of globalization in the processing tomato industry

In Chapter One we noted that the tomato has been an internationalized commodity for at least four centuries, since Spanish conquistadors brought the fruit to Europe. Tomato cultivation spread rapidly across Europe and the Middle East, leading to it being re-introduced to the Americas as a decorative garden plant whose fruit was regarded as poisonous. Increased tomato consumption in the nineteenth century, coupled with the development of embryonic canning and bottling technologies, saw the emergence of a *processing* tomato industry, based on specialist production of tomatoes for factory conversion into non-perishable food products such as ketchup. In the 1950s, mechanized harvesting technologies further revolutionized the industry. Hybridized tomato varieties developed specifically for processing encouraged an expansion in average farm sizes. These changes initially took root in California, before being disseminated internationally. The development of aseptic storage technologies for tomato paste in the early 1970s, enabled this product to become an industrial input for further processing, and ultimately led to the rise of massive specialist paste marketeers, such as the Morning Star Packing Co., Ingomar, and SK Foods. These industrial and technological developments dovetailed with shifts in consumption arenas for processing tomato products, especially the rise of pizza and convenience foods. Steady global growth in demand for processing tomatoes encouraged new entrants to this industry, most notably, China. By the year 2001, some new and apparently radical trade-production complexes had emerged. In one instance, a southern Italian canning company fulfilled a contract for private label canned baked beans for a British supermarket, using Mexican-sourced beans and Chinese tomato paste. This brought into effect a highly flexible and footloose set of production arrangements: the British supermarket's decision to source its canned baked beans from the southern Italian canner was made on the basis of short-term price-based contract, and the southern Italian canner in turn sourced its inputs on the basis of short-term price-based contracts.

In Chapter One we asked whether such arrangements represent paradigmatic shifts within the processing tomato industry. Are processes of 'globalization' fundamentally reworking industry structures towards an environment in which supply arrangements and regional futures are increasingly dictated by short-term shifts in competitiveness, thereby detaching supply chains from their territorial roots in a volatile price-based bidding war staged at a global scale, and resulting in a 'race-to-the-bottom'? Subsequent chapters of this book used detailed empirical material to address this issue. The overarching conclusion from these chapters is to temper 'radical globalization' readings of industry change. Whereas the dimensions and importance of global restructuring in this industry are profound, changes are occurring within the contexts of production-consumption complexes that remain organized at national or regional scales.

Chapters Two and Three set the scene for these arguments by analysing the distinctive characteristics of the processing tomato sector. Chapter Two focused on the cultivation of processing tomatoes, whereas in Chapter Three, attention was turned to spheres of processing and consumption. In turn, Chapters Four, Five and Six each emphasized separate, enduring elements of socio-spatial differentiation in this industry. Chapter Four critiqued the suggestion that intensified global competition leads inevitably to an erosion of the roles of national and sub-national spaces as arenas for industry regulation, especially with respect to grower-processor relations. A comparative analysis of Australia and Canada was used to support our argument that debates on industry regulation need to take account of how the norms, practices and behavior of participants are embedded within unique socio-spatial contexts. Although these production sites have certain similarities (that is, they are both English-speaking, with high input-high output farming styles), comparative research reveals major differences in the ways that the industries of both countries responded to the challenges of global competition.

Chapter Five critically evaluated those arguments which suggested that the world's agri-food industry is inevitably heading towards a situation in which global markets will replace national and regional ones. In particular, we focused on changes in Europe in the context of restructured agricultural producer subsidies. In the wake of the WTO Agreement on Agriculture, the EU formulated its *Agenda 2000* program to restructure farm subsidy payments in line with concepts of the 'multifunctionality of agriculture' and the 'post-productivist transition'. These initiatives were aimed at shifting the emphasis in farm support from quantitative measures to more sophisticated strategies entailing the explicit valuation and preservation of the social and natural elements of the rural environment. In the processing tomato industry, these policy currents led to the complete re-drafting of the tomato support payment regime in December 2000. Coinciding with these developments there was (and continues to be) extensive restructuring within the European tomato industry. However, we argued that this does not seem to be leading towards an erosion in what we label 'the distinctly European tomato', namely an industry based around production and consumption practices still firmly embedded within various national and regional arenas.

In Chapter Six, we critiqued the claim that contemporary global agri-food restructuring is witnessing 'across-the-board' growth in export competition from developing countries. Since the late 1980s, considerable attention has been focused on the shift of agri-food production to low-cost production sites in Asia, Latin America and Africa. The extent to which these processes have materialized in the processing tomato industry was explored through detailed case studies of Thailand and China. In both these cases, we suggested that the impetus to shift production to these sites was identified as being contingent upon particular historical opportunities when economics and politics conspired to make the development of this industry feasible. Yet Thailand and China have had widely divergent experiences. In Thailand, tomato production has experienced significant difficulties in recent years, while in the western Chinese province of Xinjiang, production volumes grew rapidly in the late 1990s, leading to significant restructuring of the industry at an international scale.

The overarching theme of Chapters Four, Five and Six is their interrogation of debates that are often expressed in terms of abstract 'market processes'. They underline the political and historical construction of markets, as well as the persistence of major contemporary differences in the world's processing tomato industry. In short, apparently *globalized* processes need to be understood as being rooted in particular socio-historical contingencies. Thus, the evolution of global restructuring in the processing tomato industry reads as a narrative of particular episodes of change, related to one another in varying and complex ways, rather than an uninterrupted process involving the 'working through' of market rationality.

With these points made, the current chapter asks questions about the extent to which the processing tomato industry may be considered 'globalized'. The fourfold framework used to address this question emphasized both the unevenness and limitations of globalization, as a means of describing the industry's contemporary state-of-play. Product flows and the level of international production coordination in the industry remain dominated by national-scale processes. However, critically, there have been important shifts towards global price convergence in product prices, and considerable globalization in flows of knowledge, ideas and people. What this suggests, therefore, is that the experience of globalization, in the form of perceptions and possibilities, is more advanced than its reality, in the form of material economic flows of product and capital. The globalization of the processing tomato industry, therefore, is not *fait accompli*, but will be the product of ongoing social processes and political decisions.

Relevance for agri-food theory: there is no 'end game' in agri-food globalization

These conclusions on globalization in the processing tomato industry accord with the detailed analysis of global restructuring and economic change carried out in other sectors. In his assessment of industrial restructuring in northern England, Hudson (1988, p.485) comments:

> [T]here can be no general theory of the geography of production or spatial divisions of labour. Attempting to construct such a theory will inevitably involve a flawed attempt to over-generalize from an historically and geographically specific set of circumstances. Rather, new and old forms of organizing production profitably over space have combined to produce changing spatial divisions of labour and have become intertwined in different ways in different times and places.

Further, Sadler (1992, p.15) has suggested:

> [C]hanges taking place in the world economy should be interpreted not as an inexorable process, but as a series of tentative corporate experiments with contingent local expressions.

In these contexts, the challenge facing researchers is to construct accounts of global agri-food restructuring which recognize the fact that these processes are 'messy', indeterminate and vary across space, but which, at the same time, need to be defined and analysed in order to find broader meaning in these developments.

Pre-eminently, the sectoral analysis presented in this book speaks to a particular historical experience of global, agri-industrial restructuring. Increased knowledge about the global conditions of the industry, aided and abetted by the role of internationally mobile finance capital, provide the conditions whereby economic actors become exposed to potentially alternative profit structures, and correspondingly are able to adopt a variety of strategies to extract value and profit from different geographical sites. Consequently, there is 'unity in diversity' (McMichael, 1994) in that the diverse manifestations of global restructuring are underpinned by a common process which allows owners of capital to identify, separate and exploit profit opportunities more intensively. These developments have been positioned within, but also facilitate, an ongoing restructuring of agri-food supply chains.

Research within the tradition of agrarian political economy provides the framework for connecting the disparate restructuring episodes in this industry. Certain activities within processing tomato chains are being devalued (i.e., some farming and first-stage processing) while others are being re-valued (in general, those activities or processes with significant intellectual property component). Because of the ensuing uncertainty of global restructuring, the pressures to restructure this industry are affecting some activities and places more than others. Some value chain components (e.g., the bulk production of industrial grade tomato paste) are especially vulnerable to change, as competitiveness washes from one site to another and geographically flexible operators emerge to take advantage of these shifts. Other components of these chains, particularly those rich in intellectual property, (e.g., hybrid seed development) are exposed to a different set of restructuring forces. This helps explain why global restructuring of processing tomato chains is simultaneously producing vulnerability and opportunity.

As such, this research brings into focus what we consider to be critical shortcomings in two of the important sub-fields of inquiry within critical agrarian political economy over recent years. On the one hand, our research questions the efficacy of attempts to classify and categorize contemporary agri-food restructuring within the framework of a 'third food regime' (c.f., Le Heron, 1993; Le Heron and Roche, 1995), that is, a global agri-food system characterized by a relatively stable set of relationships between the structures of political regulation, and those of profit accumulation. Our analysis does not lead us to suggest the imminent development of a 'stable order' within the processing tomato industry. In terms of food regimes terminology, the current period seems more appropriately classified as an interregnum, with an uncertain outcome. Secondly, the research also brings into focus the limitations of some elements of the 'food networks' approach. This approach uses post-structuralist theory in ways which suggest that the focus of agri-food research should be tailored to the task of documenting flows and networks (including the 'spaces' they inhabit), as a means of bringing into focus the heterogeneity of social processes. Consequently, what emerges from the 'food networks' literature is a series of unrelated studies focusing on the socio-spatial particularities in the ways food production and consumption is embedded. Such analyses generate worthy empirical detail but, as Araghi and McMichael (2000) argue, their particularist focus is somewhat short-sighted when it comes to analyzing the broader, world-historical contexts in which individual industries are situated.

The current study has sought to develop a narrative of global agri-industrial restructuring that transcends the macro-theoretical generalizations of food regimes theory, and the particularism of food networks approaches. Our applications of critical agrarian political economy have attempted to attach a wider political-historical meaning, coherence and relevance to what is otherwise a highly complex, confusing and at times apparently contradictory set of restructuring processes.

Final words: the future of the processing tomato industry

The future trajectory of the world's processing tomato industry lies in the hands of industry participants and other stakeholders. However, on the basis of the research of this book, we identify four key areas of change that will impact upon on the industry during the early twenty-first century.

First, the competitive requirements of end-users (for example, pizza chains, processed food manufacturers and own-brand retailers) will increasingly depend upon a valorization of 'quality', which in this context, means the reliability and consistency in the biophysical attributes of processing tomato products. For supply chain actors to extract value and profit, an ever-narrower range of specifications needs to be met. Satisfying the 'quality imperative' requires increased attention to linkages between supply chain actors, which blurs the boundaries between firms and industry sectors. Examples of this include the ways in which downstream users of processing tomatoes actively determine farming activities (through specifying harvesting practices and agri-chemical usage), and the increasingly tight integration of harvest, transport scheduling, and processing plant management. The construction of these linked interests throughout supply chains alters the contours of competitive advantage and production regimes. Key changes to seed types, harvest methods, logistics, and processing plant operation in recent times all owe their genesis to this 'quality imperative'. Furthermore, as quality attributes are more comprehensively measured and documented, this contributes to an increase in the potential for the spatial substitutability of certain products, especially aseptic-packaged hot-break tomato paste. In the past, different product attributes in Europe and North America provided a non-tariff barrier mitigating the extent of trans-Continental trade in tomato products. However, while these differences still exist, they can now more readily be overcome. Moreover, such tendencies are likely to strengthen in future years.

Second, within this framework the contribution of basic processing activities (the conversion of raw tomatoes into paste or to be canned) will continue to become less significant as a source of value and profit. This cost-profit squeeze will mean intensified restructuring and heightened vulnerability of first-stage processors across the world. The clearest means to escape the cost-profit squeeze has been to reorganize industry logistics and exploit the economies of scale. These issues are driving considerable change at the present time, and are variously manifested in increased production in relatively cheap production sites (notably Xinjiang Province, China); the rise of Morning Star and demise of Tri Valley Growers in California; the restructuring of cooperatives in northern Italy; the rapid concentration of canners in Southern Italy; crisis within the French, Greek, Turkish and Spanish canning sectors, and; the threat of wholesale plant closures in Australia.

Third, the cost-profit pressure on first stage processing firms plus the quality imperative is further encouraging farm-level restructuring. The advent of mechanized harvesting in California paved the way for larger farm sizes and shifts to farm management practices in the 1960s and 1970s. During the 1980s and 1990s mechanized harvesting expanded internationally. Even in Xinjiang Province, China, where harvest wages are of the order of $5 (US) per day, mechanical harvesters have been gradually introduced from the 2000 season. Rapid reductions in tomato grower numbers are occurring in the Mediterranean region, as the restructuring of cooperatives and processors combines with the reform of European Union farm assistance. At one level, because these developments are proceeding at different rates and operate at different levels within the global framework, they highlight the importance of regional diversity within the processing tomato sector. Nevertheless, agri-industrialization and the attempt to generate scale economies is a strong undercurrent across the global industry. These trends provide profit opportunities for large, high-capitalized farms, but are also consistent with a shrinking share of value and profit within the farming component of the processing tomato supply chain.

Fourth, although tomato growers and first-stage processors face a cost-profit squeeze, components of the supply chain characterized by the deployment of intellectual property are provided with new arenas for value and profit. This includes hybrid seed development, seedling transplants, logistics and management, food machinery and technology, and branding and niche marketing. With respect to seeds and transplanting, the logic of the cost-profit squeeze and the quality imperative has encouraged greater attention to the role of plant genetics in optimising consistency and reliability. The use of hybrid variety seedling transplants has expanded rapidly to facilitate better management of plant stocks. Similarly, logistics and operational management has been critically important in optimizing the efficiencies of processing factories. Finally, with the production cost of tomato paste and canned tomatoes being reduced through international competition and productivity, there have been greater opportunities to add value through the use of these products as bulk inputs to further-processed foodstuffs. The extraction of value and profit through these means has relied heavily on knowledge-intensive intellectual property assets such as recipes, brand names and marketing devices. Alongside these developments, new markets have opened up with affluent consumers increasingly demanding organics and other specialized foodstuffs. These trends are evidenced through the recent corporate strategies of H.J. Heinz Co., which in the past decade has strengthened its tomato seeds business, made efforts to bolster the profit it extracts from its stable of brand names and trademarks, and has moved into organics, while at the same time extensively rationalizing its ownership of processing facilities.

The global scope of these developments within one commodity system highlights the important trends emerging through the process of global agri-food restructuring occurring at this historical juncture. Profits are being captured both by firms and agents able to exploit key attributes of price-competition (for example, in generating economies of scale and global sourcing), and position their outputs within particular geographical territories and market segments in ways that attract additional value. In different situations, therefore, economic actors are generating value and profit in a variety of ways.

Accordingly, globalization processes in the processing tomato industry are helping construct a more geographically varied set of commodity chain processes, which simultaneously create both opportunity and vulnerability for actors and regions within the industry. Processes of globalization are not converging towards the ascendency of a single globalization 'model'. Rather, they are creating a global environment of experimentation and change. Taken together, what we are seeing, in many varied sites across the world, is a restructuring of activities in line with perceptions of newly emerging, *globalized*, opportunities for profit, that themselves are in constant evolution.

Note

1 It needs to be noted that the comparison of company sales with GDP does not measure similar accounting concepts. GDP is a measure of value-added (that is, output minus relevant inputs), whereas sales data are a gross measure. However for the purposes of this book, the comparison serves the purpose of highlighting the increased role of transnational corporations relative to global economic output (UNCTAD, 2002, p.90).

Bibliography

Agribusiness Worldwide (1987), 'Project to Increase Thai Tomato Processing Capacity', *Agribusiness Worldwide*, April, p.13.

Allen, J. and Thompson, G. (1997), 'Think Global, then Think Again: Economic Globalization in Context', *Area*, Vol. 29(3), pp.213-27.

AMITOM (2000), *Tomato Products in the European Market*, AMITOM, Avignon (France).

Anderson, K. (1998), 'Agriculture and the WTO into the 21st century', in A. Oxley, (ed.), *Liberalising World Trade in Agriculture: Strategies for Cairns Group Countries in the WTO*, Rural Industries Research and Development Corporation, Canberra, pp.13-26.

Anonymous (1965), 'Mechanization 1942-1965', mimeographed field research notes in W.H. Friedland's files. Available from the authors.

Anonymous (1976), 'Electronic Sorter Featured at California Tomato Day', *California Farmer*, 21 February, pp.12-13.

Anti-Dumping Authority (1992), *Canned Tomatoes from Italy, Spain, Thailand and the People's Republic of China*, Report 68, Australian Government Publishing Service, Canberra.

Anti-Dumping Authority (1994), *Reconsideration of Countervailing and Dumping Duties Applicable to Imports of Canned Tomatoes from Italy*, Report 124, Australian Government Publishing Service, Canberra.

Araghi, F. and McMichael, P. (2000), 'Bringing World-History Back In: A Critique of the Postmodern Retreat in Agrarian Studies', unpublished presentation to the *World Congress of Rural Sociology*, Rio de Janeiro, August.

Ashcroft, B., Gurban, S., Holland, R., Jones, K. and Wade, S. (1990), 'Processing Tomato Cultivar Trials 1989/90', *Australian Processing Tomato Grower*, Vol. 11, pp.12-13.

Atkins, P.J. and Bowler, I. (2001), *Food in Society*, Arnold, London.

Australian Bureau of Statistics (various years), 'International Merchandise Imports, Australia', *Australian Bureau of Statistics Catalogue*, 5439.0.

Australian Customs Service (1991), *Canned Whole Tomatoes, Canned Tomato Pieces and Canned Crushed Tomatoes, in Water or Juice, from Italy, Spain, Thailand and the People's Republic of China*, Report and Preliminary Finding No. 91/20, Commonwealth of Australia, Canberra.

Ban, E. (1998), 'Grocery Shopping Survey', *Retail World* (Sydney, Australia), 12-25 October, pp.6-7.

Banaji, J. (1980), 'Summary of Selected Parts of Kautsky's "The Agrarian Question"' in H. Newby, and F.H. Buttel (eds) *The Rural Sociology of Advanced Societies*, Croom Helm, London, pp.39-82.

Bangkok Post (2001), 'Company Reports', 23 April, p.26.

Barach, J. (2000), 'Genetically Modified Organisms: Perceptions and Reality', unpublished paper presented at *IV World Congress on the Processing Tomato*, Sacramento CA, 11 June.

Barkan, J. (1978), 'Back Next Year – On Strike', *Seven Days*, October 13, p.10.

Barkin, D. and DeWalt, B. (1988), 'Sorghum and the Mexican Food Crisis', *Latin American Research Review*, Vol. 20(3), pp. 30-59.

Barndt, D. (2002), *Tangled Routes: Women, Work and Globalization on the Tomato Trail*, Garamond Press, Toronto.

Barnett, D. (1993), *China's Far West: Four Decades of Change*, Westview Press, Boulder (CO).

Becker, T. (1989), 'Asia's Most Modern Tomato Paste Factories', *Asian Development Bank Quarterly Review*, January, pp.11-13.

Becket, J.W. (1966), 'The Domestic Farm Laborer: a Study of Yolo County Tomato Pickers', *Department of Agricultural Education Research Monograph* (University of California, Davis), Vol. 2.

Becquelin, N. (2000), 'Xinjiang in the Nineties', *China Journal*, Vol. 44 pp.65-90.

Beder, S. (1997), *Global Spin: The Corporate Assault on Environmentalism*, Scribe, Melbourne.

Bell, D. and Valentine, G. (1997), *Consuming Geographies: We Are Where We Eat*, Routledge, London.

Benziger, V. (1996), 'Small Fields, Big Money: Two Successful Programs in Helping Small Farmers Make the Transition to Higher Value-added Crops', *World Development*, Vol. 24(11), pp.1681-93.

Berry, A. (2001), 'When Do Agricultural Exports Help the Rural Poor? A Political-Economy Approach', *Oxford Development Studies*, Vol. 29(2), pp.125-44.

Bessis, S. (1991), *La Faim Dans Le Monde*, Editions la Découverte, Paris.

Biswas, S. (2001), 'Kraft to Divest Lesser Canadian Names', *The Daily Deal*, 21 June.

Blackburn, M. (1979), 'New Tomato Hurts Small Growers', *New York Times*, 27 December, pp.3, 29.

Board of Investment (1990), *Thailand Investment 1991-92*, Cosmic Enterprises, Bangkok.

Board of Investment (1996), *Thailand Investment 1997-98*, Cosmic Enterprises, Bangkok.

Bonanno, A. and Constance, D. (1996), *Caught in the Net: The Global Tuna Industry, Environmentalism and the State*, University of Kansas Press, Lawrence (KS).

Boscolo, A. (2002), 'The Role of the Transplant in the Economics of Tomato Growing', unpublished paper presented to the *V World Congress on the Processing Tomato*, Istanbul, 8 June.

Bove, J. (2001), *The World is Not For Sale*, Verso, London & New York.

Bryce, I. (1999), 'Evaluating Tomato Solids per Hectare v. Tonnes per Hectare', *Australian Processing Tomato Grower*, Vol. 20, pp.20-21.

Burch, D. (1996), 'Globalized Agriculture and Agri-food Restructuring in Southeast Asia: the Thai Experience', in D. Burch, R.E. Rickson, and G. Lawrence, (eds) *Globalization and Agri-Food Restructuring: Perspectives from the Australasia Region*, Avebury, Aldershot, pp.323-44.

Burch, D. and Goss, J. (1999), 'Global Sourcing and Retail Chains: Shifting Relationships of Production of Australian Agri-foods', *Rural Sociology*, Vol. 64(2), pp.334-50.

Burch, D., Lyons, K. and Lawrence, G. (2001), 'What do we Mean by 'Green'? Consumers, Agriculture and the Food Industry', in S. Lockie, and B. Pritchard, (eds), *Consuming Foods, Sustaining Environments*, Australian Academic Press, Brisbane, pp.33-46.

Burch, D. and Pritchard, B. (1996a), 'The Uneasy Transition to Globalization: Restructuring of the Australian Processing Tomato Industry', in D. Burch, R.E. Rickson, and G. Lawrence (eds) *Globalization and Agri-food Restructuring: Perspectives from the Australasia Region*, Avebury, Aldershot, pp.107-26.

Burch, D. and Pritchard, B. (1996b), 'The Uneasy Transition to Globalization: Restructuring of the Australian Processing Tomato Industry', *Tomato News*, Vol. 8 (January), pp.23-29 and (February), pp.27-36.

Burch, D. and Pritchard, B. (2001), 'Thailand, New Zealand and the Global Processing Tomato Industry', unpublished paper delivered to *Agri-Food IX*, Agri-Food Research Network, Palmerston North (New Zealand).

Burch, D., Rickson, R.E. and Annels, R. (1992), 'The growth of agribusiness: environmental and social implications of contract farming', in G. Lawrence, F. Vanclay and B. Furze (eds) *Environment and Society: Contemporary Issues for Australia*, Macmillan, Melbourne, pp.259-77.

Burch, D. (forthcoming), *Transnational Agribusiness and Thai Agriculture: Globalization and the Political Economy of Agri-Food Restructuring*, a report to the Institute of Southeast Asian Studies, Singapore.

Business in Thailand (1991), 'Service Priorities at Champaca', *Business in Thailand*, Vol. 22(3), pp.99-100.

Buttel, F.H. (1996), 'Theoretical issues in global agri-food restructuring', in D. Burch, R.E. Rickson and G. Lawrence (eds) *Globalization and Agri-Food Restructuring: Perspectives from the Australasia Region*, Avebury, Aldershot, pp.17-44.

Buttel, F.H. (1999), 'Agricultural Biotechnology: Its Recent Evolution and Implications for Agrofood Political Economy', *Sociological Research Online*, Vol. 4.

Buttel, F., Larson, O. and Gillespie, G. (1990), *The Sociology of Agriculture*, Greenwood Press, New York.

Californian Tomato Growers Association (2001), 'Californian Tomatoes for Processing', at www.ctga.org, accessed 7 February, 2002.

Campbell, H. (1996), 'Organic Agriculture in New Zealand: Corporate Greening, Transnational Corporations and Sustainable Agriculture', in D. Burch, R.E. Rickson, and G. Lawrence (eds) *Globalization and Agri-food Restructuring: Perspectives from the Australasia Region*, Avebury, Aldershot, pp.153-69.

Campbell, H. and Coombes, B. (1999), 'New Zealand's Organic Food Exports: Current Interpretations and New Directions in Research', in D. Burch, J. Goss, and G. Lawrence, (eds) *Restructuring Global and Regional Agricultures: Transformations in Australasian Agri-Food Economies and Spaces*, Ashgate, Aldershot, pp.61-74.

Canadian Horticultural Council (1990), *Submission to the Canadian International Trade Tribunal Inquiry Concerning the Competitiveness of the Fresh and Processed Fruit and Vegetable Industry*, Canadian Horticultural Council, Toronto.

Canadian International Trade Tribunal (1991a), *An Assessment of the Relative Competitiveness of the Canadian and United States Tomato Industries: Exhibits*, Price Waterhouse, Consultants Report Ref. No. GC-90-001.

Canadian International Trade Tribunal (1991b), *An Inquiry into the Competitiveness of the Canadian Fresh and Processed Fruit and Vegetable Industry*, Ministry of Supply and Services, Ottawa.

Canned Food Information Service, (1991), *Application for Dumping and Countervailing Duties on Imported Canned Tomatoes,* Melbourne.

Castells, M. (1996), *The Rise of the Network Society*, Oxford University Press, Oxford.

Cedenco (2001), 'Cedenco', at www.Cedenco.co.nz, accessed 10 September, 2002.

Chan, Y.H. (1998), 'Army Mammoth Gears up to Take Care of Business', *South China Morning Post*, 6 September, p.6.

Clark, G. (1993), *Pensions and Corporate Restructuring in American Industry: A Crisis of Regulation*, John Hopkins University Press, Baltimore.

Classico (2002), 'Classico' at www.classico.com, accessed 14 January, 2002.

Clinton, S.K. (1998), 'Lycopene: Chemistry, Biology and Implications for Human Health and Disease', *Nutrition Reviews*, Vol. 56(21), pp.35-51.

Colusa Canning Company (2002), 'Colusa Canning', at www.colusacanningco.com, accessed 17 April.

Coote, B. and LeQuesne, C. (1996), *The Trade Trap: Poverty and the Global Commodity Markets*, Oxfam, Oxford.

Craig, S. (1994), 'Free Trade Puts Heat on Unions', *Financial Post* (Toronto), 28 May, p.15.

Del Monte (2001), '10-K Filing' Securities and Exchange Commission (US Government), Document f75859e10-k.txt.

Department of Finance, California (2000), 'County Statistics—Colusa', at www.dof.gov.ca, accessed 25 May, 2001.

Di Bella, S. (2001), 'Consumption Patterns of the European Tomato Market', *Tomato News*, Vol. 13 (January), pp.15-18.

Dicken, P. (1998), *Global Shift: Transforming the World Economy*, Paul Chapman Publishers, London.

Dixon, J. (1999), 'A Cultural Economy Model for Studying Food Systems' *Agriculture and Human Values*, Vol. 16(2), pp.151-60.

Dixon, J. (2002), *The Changing Chicken*, UNSW Press, Kensington (Sydney, Australia).

Dolinshy, D.J. (1991), *Contract Farming at Lam Nam Oon: An Operational Model for Rural Development*, Thai Development Research Institute, Bangkok.

Du Puis, M. (1993), 'Sub-national State Institutions and the Organization of Agricultural Resource Use: The Case of the Dairy Industry', *Rural Sociology*, Vol. 58(3), pp.440-60.

Duscha, S. (1976), 'Idled Hands: UFW Protests Machine's Harvest of Jobs', *Sacramento Bee*, 20 August, p.4.

Dwyer, M. (2002), 'In the shadow of the Han', *The Australian Financial Review Magazine*, July, pp.40-47.

Eckholm, E. (1999), 'Remaking a vast frontier in China's image', *The New York Times*, 17 July, p.1 and p.5.

Economist, The (2000), 'Go West, Young Han: China's Great Leap West', 23 December, p.1.

Estrada, R.T. (2001), 'Bankruptcy Filing Idles Sacramento, Calif.-Area Cannery', *Sacramento Bee*, 13 July, p.3.

EU Director General, Trade (1999), *EU Activities with the World Trade Organization: Report to the European Parliament*, European Commission, Brussels.

European Commission, (2000), *Agenda 2000: For a Stronger and Wider Union*, European Commission, Brussels.

Fagan, R. (1997), 'Local Food/Global Food: Globalization and Local Restructuring', in R. Lee and J. Wills (eds) *Geographies of Economies*, Arnold, London, pp.197-208.

Fargeix, A. (2000), 'North America – Consumption Trends', unpublished paper presented at the *IV World Congress on the Processing Tomato*, Sacramento CA, 10 June.

Fels, A. (1990), *Victorian Tomato Processing Industry: Determination by Professor A.H.M. Fels, Upon a Reference by the Honourable Barry Rowe, Minister for Agriculture and Rural Affairs*, Parliament of the State of Victoria, Melbourne.

Ferdinand, P. (1994), 'Xinjiang: Relations with China and Abroad', in D. Goodman and G. Segal (eds), *China Deconstructs: Politics, Trade and Regionalism*, Routledge, London, pp.271-85.

Ferris, S. and Sandoval, R. (1997), *The Fight in the Fields: Cesar Chavez and the Farmworkers' Movement*, Harcourt Brace & Co., New York.

Fidler R. (1998), 'Home Meal Replacements: The Changing Face of Supermarkets', *Food Australia*, Vol. 6, p.265

Financial Post, The (1994), 'Heinz Cuts 200 Jobs', *The Financial Post* (Toronto), 22 January, p.4.

Fisher, W., Lewien, D. & Horn, B. (1994), *Review of the Tomato Processing Industry Development Order*, unpublished report provided to the Minister for Agriculture, State of Victoria, Melbourne.

Fold, N. (2000), 'Globalisation, State Regulation and Industrial Upgrading of the Oilseed Industries in Malaysia and Brazil', *Singapore Journal of Tropical Geography*, Vol. 21(3), pp.263-78.

Fold, N. (2002), 'Lead Firms and Competition in 'Bi-polar' Commodity Chains: Grinders and Branders in the Global Cocoa-Chocolate Industry', *Journal of Agrarian Change*, Vol. 2(2), pp.228-47.

Food and Agriculture Organization [FAO] of the United Nations (2002), FAOSTATS Database, at www.fao.org, accessed 12 September, 2002.

Forbes (2001), 'China's 100 Richest Business People', 12 November (Special Supplement)

Fraser, J. & Oppenheim, J. (1997), 'What's New about Globalization?' *The McKinsey Quarterly*, Vol. 14, pp.168-79.

Frederick, D.A. (1990), 'Agricultural Bargaining Law: Policy in Flux', *Arkansas Law Review*, Vol. 43, pp.679-99.

Frederick, D.A. (1993), 'Legal Rights of Producers to Collectively Negotiate', *William Mitchell Law Review*, Vol. 19, pp.433-55.

Freeman, F. and Roberts, I. (1999), 'Multifunctionality: A Pretext for Protection?', *ABARE Current Issues*, Vol. 99(3).

Fresno Bee (2001), 'California Farmers Struggle to Make New Concepts Work for Them', *Fresno Bee*, 14 January, p.7.

Friedberg, S. (2001), 'On the Trail of the Global Green Bean: Methodological Considerations in Multi-site Ethnography', *Global Networks*, Vol. 1(4), pp.353-68.

Friedland, W.H. (1991), '"Engineering" Social Change in Agriculture', *University of Dayton Review*, Vol. 21(1), pp.25-42.

Friedland, W.H. (1994), 'The New Globalisation: the Case of Fresh Produce' in A. Bonanno, L. Busch, W.H. Friedland, L. Gouveia and E. Mingione (eds) *From Columbus to ConAgra: The Globalization of Agriculture and Food*, University of Kansas Press, Lawrence, KS, pp.210-31.

Friedland, W.H. (2001), 'Reprise on Commodity Systems Methodology', *International Journal of Sociology of Agriculture and Food*, Vol. 9, pp.82-103.

Friedland, W.H. and Barton, A. (1975), 'Destalking the Wily Tomato: a Case Study in Social Consequences in Californian Agricultural Research', *Department of Applied Behavioral Sciences, College of Agricultural and Environmental Sciences, University of California at Davis, Research Monograph Series*, Vol. 15.

Friedland, W.H. and Barton, A. (1976), 'Tomato Technology', *Society*, Vol. 13, not paginated.

Friedland, W.H., Barton, A. and Thomas, R. (1981), *Manufacturing Green Gold*, Cambridge University Press, Cambridge.

Friedmann, H. (1993), 'The Political Economy of Food: A Global Crisis', *New Left Review*, Vol. 197, pp.28-59.

Friedmann, H. (1994), 'Distance and Durability: Shaky Foundations of the World Food Economy', in P. McMichael (ed.) *The Global Restructuring of Agro-Food Systems*, Cornell University Press, Ithaca, pp.258-276.

Friedmann, H. and McMichael, P. (1989), 'Agriculture and the State System: The Rise and Decline of National Agricultures, 1870 to the Present', *Sociologia Ruralis* Vol. 29, pp.93-117.

Frozen Food Digest (1996), 'Total Pizza Sales Hit New High', *Frozen Food Digest*, Vol. 11(4), p.24.

Frozen Food Digest (2001), 'Processing Vegetables', *Frozen Food Digest*, Vol. 16(3) pp.12-16.

Fusheng, L. (1997), *Xinjiang Bingtuan Tuken Shubian Shi (The Xinjiang Bingtuan, Border Defence and Land Reclamation)*, Xinjiang Keti Weisheng Chubanshe, Urumqui (China).

Galarza, E. (1964), *Merchants of Labor: The Mexican Bracero Story*, McNally & Lofton, Charlotte (North Carolina).

Gandolfi (1998), 'EU Regulation and its Adverse Impact Upon Buyer/Seller Relationship', unpublished paper presented to the *III World Congress on the Processing Tomato*, Pamplona Spain, 27 May. Available from Gandolfi s.r.l., Viale Piacenza, 12/B, 43100 Parma, Italia.

Garoyan, L. and Moulton, K. (1987), *Reports on the World Processing Tomato Industry*, University of California at Berkeley Cooperative Extension Service, Berkeley, CA.

Gereffi, G. (1996), 'Global Commodity Chains: New Forms of Coordination and Control Among Nations and Firms in International Industries', *Competition and Change*, Vol. 1, pp.427-39.

Gerster, H. (1997), 'The Potential Role of Lycopene for Human Health', *Journal of the American College of Nutrition*, Vol. 16(2), pp.109-26.

Giovannucci, E., Ascherio, A., Rimm, E.B., Stampfer, M.J., Colditz, G.A., and Willett, W.C. (1995), 'Intake of Carotenoids and Retinol in Relation to Risk of Prostate Cancer', *Journal of the National Cancer Institute*, Vol. 87(23), pp.1767-1776.

Goodman, D. and Redclift, M. (1991), *Refashioning Nature*, Routledge, New York.

Goodman, D. and Watts, M. (eds) (1997), *Globalizing Food*, Routledge, New York.

Goss, J. (2002), *Fields of Inequality: The Waning of National Developmentalism and the Political Economy of Agribusiness in Siam*, unpublished Ph.D. thesis, Griffith University, Brisbane.

Gould, W.A. (1992), *Tomato Production, Processing and Technology*, CTI Publications, Baltimore.

Government of Thailand, (various years), *Statistical Reports of the Regions*, National Statistical Office, Bangkok.

Granovetter, M. and Swedberg, R. (1992), *The Sociology of Economic Life*, Westview, Boulder (CO).

Grunder, E. (2001), 'Grunder's Column', *The Record* (Stockton, CA), 15 April, p.6.

H. J. Heinz Co., (1997), 'Heinz Reorganizes for Sales and Earnings Growth and Shareholder Value', *Media Release*, 14 March.

Harrell, B. (1988), 'Nation's Vegetables Get Pampered Start: South Georgia Farmers Tend Plants Destined for Fields in the North', *Atlanta Journal*, 6 March, p.B3.

Harvey, M. (1999), 'Genetic Modification as a Bio-Socio-Economic Process: One Case of Tomato Puree', *Centre for Research on Innovation and Competition, University of Manchester, Discussion Paper*, Vol. 31.

Hay, C. and Marsh, D. (2000), *Demystifying Globalization*, Macmillan, Basingstoke.

Heasman, M. and Mellentin, J. (2001), *The Functional Foods Revolution: Healthy People, Healthy Profits?*, Earthscan, London.

Heinrichs, P. (2000), 'Who said the Bush is all Bad?', *The Sunday Age* (Melbourne), 13 February, pp.1, 13.

Heinz Canada (2002), 'Heinz Canada's Leamington Plant Announces $4.3 Million Investment', *Media Release*, 5 March.

Heinz Seeds (2001), 'Heinz Seeds', at www.heinzseeds.com, accessed 19 December 2001.

Hilmer, F. (1993), *National Competition Policy. Report of the Independent Committee of Inquiry into Competition Policy in Australia*, Australian Government Publishing Service, Canberra.

Hiscock, J. (2001), 'Most Trusted Brands', *Marketing* (London), 1 March, pp.32-33.

Hoggart, K. and Paniagua, A. (2001), 'The Restructuring of Rural Spain', *Journal of Rural Studies*, Vol. 17, pp.63-80.

Holt, R. (1965), 'Manager's Annual Report', *Californian Tomato Grower*, Vol. 8, pp.9-10.

Horn, B. (1997), 'The Processing Tomato Industry in Asia', *Tomato News*, Vol. 9 (March), pp.8-21.

Horn, B. (various years), *Annual Industry Survey*, Barry Horn Consulting, Melbourne.

Hudson, R. (1988), 'Uneven Development in Capitalist Societies: Changing Spatial Divisions of Labour, Forms of Spatial Organisation of Production and Service Provision, and their Impacts for Localities', *Transactions, Institute of British Geographers*, Vol. 13, pp.484-96.

Hulme, J., Hickey, M., Ashcroft, B. and Qassim, A. (2000), 'Processing Tomato Growers Lead the Way in Irrigation', *Australian Processing Tomato Grower*, Vol. 21, pp. 24-25.

Hume, S. (2001), 'Minding the Store', *Restaurants and Institutions*, Vol. 111(17), pp.36-51.

Hurt, R.D. (1991), *Agricultural Technology in the Twentieth Century*, Sunflower University Press, Yuma (Kansas).

Iacobelli, H. (2001), personal communication.

IGEME, (2000), 'Tomato Products', *Export Promotion Center of Turkey Information Bulletin*, at www.igeme.gov.tr, accessed 19 May, 2001.

Ilbery, B. and Kneafsey, M. (1999), 'Niche Markets and Regional Speciality Food Products in Europe: Towards a Research Agenda', *Environment and Planning A*, Vol. 31, pp.2207-2222.

IMI/KMIT (1996), *Agro-Industry for the Development of Small Farmers: A Case Study of the Royal Project Food Processing Section*, Food and Agricultural Organization, Rome.

Iritani, E (2001), 'Farmers' Fears Take Root: U.S. groups seek protection amid concerns about a glut of cheap fruits and vegetables', *The New York Times*, 8 August, pp.1, A7.

Jardine, J. (2001), 'Cooperative Buys Woodland, Calif., Tomato Processing Facility', *Modesto Bee*, California, 31 July, p.3.

Jiminez, J. (2000), 'Keynote Address – View of the Worldwide Tomato Industry' unpublished paper presented at the *IV World Congress on the Processing Tomato*, Sacramento CA, 10 June.

Judd, T. (1989), 'In Perspective: Trends in Rural Development Policy and Programs in Thailand, 1947-87', *Research Report Series*, Vol. 41, (Payap University Center for Research and Development, Chiengmai).

Kanchananaga, S. (1973), *Resources and Products of Thailand*, Siam Communications Ltd, Bangkok.

Kaptur, M. (2001), 'A new declaration of economic independence for rural America: Speech to the CLOUT Conference, Iowa State University, 16 March', at www.house.gov/kaptur/speeches/sp_010316_iowa_state_university.htm, accessed 1 July, 2002.

Kautsky, K. (1988), *The Agrarian Question*, Zwan Publications, London (first published 1899).

Ke, W. (1996), 'Xinjiang Boosts Exports Through Trade Fair', *China Daily*, 18 October, pp.5-6.

Kelly, P. (1997), 'Globalization, Power and the Politics of Scale in The Philippines', *Geoforum*, Vol. 28, pp.151-71.

Khoo, C.S. (2000), 'Overview of Tomato and Health Research', unpublished paper presented at the *IV World Congress on the Processing Tomato*, Sacramento CA, 10 June.

Kim, C.K. and Curry, J. (1993), 'Fordism, Flexible Specialization and Agri-industrial Restructuring: The case of the U.S. Broiler Industry', *Sociologia Ruralis*, Vol. 33(1), pp.61-80.

Koc, M. (1994), 'Globalization as a Discourse', in A. Bonanno, L. Busch, W.H. Friedland, L. Gouveia and E. Mingione (eds) *From Columbus to ConAgra: The Globalization of Agriculture and Food*, University of Kansas Press, Lawrence, KS, pp.265-80.

Kompass (various years), *Thailand: Food and Beverages*, Kompass Direct, Bangkok.

Kwanchai, R. (1998), 'Tricon Waives Royalty Fees For The Pizza Plc', *The Nation* (Bangkok), 1 August, p.4.

Kwang, M. (2000), 'Revolutionary Flower Blooms in Desert' *The Straits Times (Singapore)*, 3 September, pp.51-2.

La Doria, (2000), 'La Doria – selected quality', brochure published by La Doria S.p.A., Angri, Italy.

La Prise, J. (2001), personal communication.

Laogoi Research Foundation (1996), 'The World Bank and the Chinese military', at www.laogoi.org/reports/worldbnk.html, accessed 13 July, 2001.

Laramee, P. (1975), 'Problems of Small Farmers Under Contract Marketing, with Special Reference to a Case in Chiengmai Province, Thailand', *Economic Bulletin for Asia and the Pacific,* Vol. XXVI, Economic Commission for Asia and the Pacific, United Nations, New York.

Larner, J. (1999), *Marco Polo and the Discovery of the World*, Yale University Press, New Haven and London.

Lash, S. and Urry, J. (1994), *Economies of Signs and Space*, Thousand Oaks, London.

Lawrence, G. (1987), *Capitalism and the Countryside*, Pluto Press, Sydney.

Le Heron, R. (1993), *Globalized Agriculture: Political Choice*, Pergamon, Oxford

Le Heron, R. and Roche, M. (1995), 'A "Fresh" Place in Food's Space', *Area*, Vol. 27(1), pp.23-33.

Lee, C. (1992), 'Corporate Leader Sets Aim on China', *South China Morning Post*, 11 November, p.1.

Linden, T. (1999), 'California: Processing Capacity Expanding', *Tomato News*, Vol. 11 (December), pp.14-18.

Ling, Z. and Selden, M. (1993), 'Agricultural Cooperation and the Family Farm in China' *Bulletin of Concerned Asian Scholars*, Vol. 25(3), pp.3-12.

Lipton Inc. (2002), 'Five Brothers', at www.fivebrothers.com, accessed 5 September, 2001.

Little, P. and Watts, M. (eds) (1994), *Living Under Contract: Contract Farming and Agrarian Transformation in Sub-Saharan Africa*, University of Wisconsin Press, Madison.

Lockie, S. (2001), 'The Invisible Mouth', unpublished presentation to the *IX Agri-Food Research Network Conference*, Palmerston North, New Zealand, 7 December.

Lockie, S. (2002), 'Why Do People Eat Organic Food? A Multivariate Analysis of Organic Food Choice by Australian Consumers', unpublished presentation to the *XV World Congress of Sociology*, Brisbane, 11 July.

Lowe, P. Buller, H. and Ward, N. (2002), 'Setting the Next Agenda? British and French Approaches to the Second Pillar of the Common Agricultural Policy', *Journal of Rural Studies*, Vol. 18(1), pp.1-17.

Lyons, K. (2001), 'From Sandals to Suits: Green Consumers and the Institutionalisation of Organic Agriculture' in S. Lockie, and B. Pritchard (eds), *Consuming Foods, Sustaining Environments*, Australian Academic Press, Brisbane, pp.83-93.

Mabbett, J. and Carter, I. (1999), 'Contract Farming in the New Zealand Wine Industry: An Example of Real Subsumption', in D. Burch, J. Goss, and G. Lawrence (eds), *Restructuring Global and Regional Agricultures: Transformations in Australasian Agri-Food Economies and Spaces*, Ashgate, Aldershot, pp.275-88.

Majka, L.C. and Majka, T.J. (1982), *Farm Workers, Agribusiness and the State*, Temple University Press, Philadelphia.

Marketing Week (2001), 'Safeway Signs to Launch Heinz Pre-packed Sandwiches', *Marketing Week*, 6 December, p.8.

Marsden, T.K. (1997), 'Creating space for food: the distinctiveness of recent agrarian development', in D. Goodman, and M. Watts, (eds) Globalising Food, Routledge, London, pp.169-91.

Martineau, B. (2001), *First Fruit: The Creation of the Flavr Savr™ Tomato and the Birth of Biotech Food*, McGraw-Hill, Boston.

Mathews, J. (1989), *Tools of Change: New Technology and the Democratisation of Work*, Pluto Press, Sydney.

McDonald's Corporation (2001), 'Annual Report', at www.mcdonalds.com, accessed 14 January, 2002.

McGrew, A (1992), 'A Global Society?', in S. Hall, D. Held, and D. McGrew (eds), *Modernity and its Futures*, Polity Press, Oxford, pp.61-116.

McKenna, M., Roche, M. and Le Heron, R. (1999), 'An Apple a Day: Renegotiating Concepts, Revisiting Context in New Zealand's Pipfruit Industry' in D. Burch, J. Goss, and G. Lawrence, (eds), *Restructuring Global and Regional Agricultures: Transformations in Australasian Agri-Food Economies and Spaces*, Ashgate, Aldershot, pp.41-59.

McKenna-Frazier, L. (2001), 'Orestes, Ind. Tomato Processor Buys Bankrupt Packer's Brand Names', *The News Sentinel* (Fort Wayne, Indiana), 12 April, p.17.

McMichael, P. (1994), 'Introduction: Agro-food System Restructuring - Unity in Diversity', in P. McMichael (ed.), *The Global Restructuring of Agro-Food Systems*, Cornell University Press, Ithaca, pp.1-18.

McMichael, P. (1996), 'Globalization: Myths and Realities', *Rural Sociology*, Vol. 61(1), pp.25-55.

McMichael, P. (1998), 'Global Food Politics', *Monthly Review*, Vol. 50(3), pp.97-122.

McMichael, P. (1999), 'Virtual Capitalism and Agri-Food Restructuring', in D. Burch, J. Goss, and G. Lawrence, (eds), *Restructuring Global and Regional Agricultures: Transformations in Australasian Agri-Food Economies and Spaces*, Ashgate, Aldershot, pp.3-22.

McMichael, P. (2000), 'A Global Interpretation of the Rise of the East Asian Food Import Complex', *World Development*, Vol. 28(3), pp.409-24.

McMichael, P. and Myhre, D. (1991), 'Global Regulation vs. the Nation-State: Agro-food Systems and the New Politics of Capital', *Capital and Class*, Vol. 43, pp.83-105.

McMillen, D. (1981), 'Xinjiang and the Production and Construction Corps: A Han Organization in a non-Han region', *The Australian Journal of Chinese Affairs*, Vol. 1(6), pp.65-96.

McMurchy, J.C. (1990), 'Agricultural Marketing Legislation in Ontario', mimeographed report provided by the Ontario Processing Vegetable Growers.

Mingione, E. and Pugliese, E. (1994), 'Rural Subsistence, Migration, Urbanization, and the New Global Food Regime', in A. Bonanno, L. Busch, W.H. Friedland, L. Gouviea, and

E. Mingione, (eds) *From Columbus to ConAgra: The Globalization of Agriculture and Food*, University of Kansas Press, Lawrence Kansas, pp.52-68.

Mintz, S. (1985), *Sweetness and Power*, Viking-Penguin, New York.

Mintz, S. (1995), 'Food and its Relationship to Concepts of Power', in P. McMichael (ed.), *Food and Agrarian Orders in the World-Economy*, Praeger, Westport Conn, pp.3-14.

Mooney, P.H. and Majka, T.J. (1995), *Farmers' and Farm Workers' Movements*, Twayne Publishers, New York.

Morgan, D. (1980), *Merchants of Grain*, Penguin, New York.

Morning Star Packing Co. (2002a), 'The Morning Star Company', at www.morningstarco.com, accessed various times, 2002.

Morning Star Packing Co. (2002b), personal communication.

Moyano-Estrada, E., Entrena, F. and Serrano del Rosal, R. (2001), 'Federations of Co-operatives and Organized Interests in Agriculture. An Analysis of the Spanish Experience', *Sociologia Ruralis*, Vol. 41(2), pp.237-53.

Murray, M. (2001), personal communication. (Mr Murray is Colusa Country Agricultural Extension Officer.)

Nabisco (2001), *Nabisco 1903-2000*, Nabisco company brochure, Dresden (Canada).

Nation, The (1999), 'Company Reports', 15 June, p.32.

Newby, H. (1982), 'Review: Manufacturing Green Gold', *Sociology*, Vol. 16(3), pp.468-69.

O'Brien, R. (1992), *Global Financial Integration: The End of Geography*, Royal Institute for International Affairs, London.

Ohmae, K. (1995), *The End of the Nation State The Rise of Regional Economies*, Harper Collins, London.

Ontario Processing Vegetable Growers (various years), *Newsletter*, OPVG, Ontario.

Ontario Processing Vegetable Growers (2000) 'Presentation to CLOUT Conference', Iowa State University 16-17 March, unpublished.

Oxfam, (2002), *Rigged Rules and Double Standards*, Oxfam, Oxford.

Parsons, P.S. (1966), 'Costs of Mechanical Tomato Harvesting Compared to Hand Harvesting', *University of California Agricultural Extension Service, Publication* AXT-224.

Pendergrast, M. (2000), *Uncommon Grounds: The History of Coffee and How It Transformed Our World*, Basic Books, New York.

Peng, Z. (2001), personal communication.

Perez, C. (1985), 'Microelectronics, Long Waves and World Structural Change: New Perspectives for Developing Countries', *World Development*, Vol. 13(3), pp.441-63.

Piore, M. and Sabel, C. (1984), *The Second Industrial Divide: Possibilities for Prosperity*, Basic Books, New York.

Pistorius, R. and van Wijk, J. (1999), *The Exploitation of Plant Genetic Information: Political Strategies in Crop Development*, CABI Publishing, Oxon, UK.

Plummer, C. (2000), 'Modeling the US Processing Tomato Industry', *Tomato News*, Vol. 12 (January), pp.36-42.

Podbury, T. (2000), 'US and EU Agricultural Support: Who Does it Benefit?', *ABARE Current Issues*, Vol. 00(2).

Podbury, T. and Roberts, I. (1999), 'WTO Agricultural Negotiations: Important Market Access Issues', *ABARE Research Report*, Vol. 99(3).

Polkwamdee, N. (1998), 'Malee Sampran First Firm to be Relisted', *Bangkok Post*, 24 February, p.17.

Pollock, D. (2001), 'Fresno, Calif.- Area Tomato Crop Ripe, Ready for Consumption', *Fresno Bee*, 21 July, p.15.

Pritchard, B. (1993), 'Finance Capital as an Engine of Restructuring: The 1980s Merger Wave', *Journal of Australian Political Economy*, Vol. 33, pp.1-20.

Pritchard, B. (1995a), 'Uneven Globalisation: Restructuring in the Vegetable and Dairy Processing Sectors', unpublished PhD thesis, Department of Geography, University of Sydney.

Pritchard, B. (1995b), 'Foreign Ownership in Australian Food Processing: The 1995 Sale of the Pacific Dunlop Food Division', *Journal of Australian Political Economy*, Vol. 36, pp.26-47.

Pritchard, B. (1999a), 'Switzerland's Billabong? Brand Management in the Global Food System and Nestle Australia', in J. Goss, D. Burch, and G. Lawrence, (eds), *Restructuring Global and Regional Agricultures: Transformations in Australasian Agri-Food Economies and Spaces*, Avebury, Aldershot, pp.23-40.

Pritchard, B. (1999b), 'National Competition Policy in Action: The Politics of Agricultural Deregulation and the Murrumbidgee Irrigation Area Wine Grapes Marketing Board', *Rural Society*, Vol. 9(2), pp.421-43.

Pritchard, B. (1999c), 'The Regulation of Grower-Processor Relations: A Case Study from the Australian Wine Industry' *Sociologia Ruralis*, Vol. 39(2), pp.186-201.

Pritchard, B. (2000a), 'Transnational Corporate Networks and the Case of Breakfast Cereals in Asia', *Environment and Planning A*, Vol. 32(5), pp.789-804.

Pritchard, B. (2000b), 'Geographies of the Firm and Agro-food Corporations in East Asia', *Singapore Journal of Tropical Geography*, Vol. 21(3), pp.246-62.

Pritchard, B. (2001), 'Transnationality Matters: Related Party Transactions in the Australian Food Industry', *Journal of Australian Political Economy*, Vol. 48, pp.23-45.

Pritchard, B. and McManus, P. (eds) (2000), *Land of Discontent: The Dynamics of Change in Rural and Regional Australia*, University of New South Wales Press, Kensington (Sydney).

Pritchard, W. (1996), 'Shifts in Food Regimes and the Place of Producer Co-operatives: Insights from the Australian and United States Dairy Industries', *Environment and Planning A*, Vol. 28, pp.857-75.

Pritchard, W. and Fagan, R. (1999), 'Circuits of Capital and Transnational Corporate Spatial Behaviour: Nestlé in Southeast Asia', *International Journal of Sociology of Food and Agriculture*, Vol. 8, pp.3-20.

Private Sector Agricultural Trade Task Force (2002), 'Supporting Document to the Agricultural Trade Task Force Communique', *World Food Summit of the Food and Agricultural Organization of the United Nations*, 10-13 June. At www.fao.org/worldfoodsummit/sideevents/papers/WEF-en.htm, accessed 22 June 2002.

Processing Tomato Advisory Board (2001), 'PTAB data', at www.ptab.org, accessed 2 May 2001.

Quantxing, Y. and Feiyu, C. (2000), 'Now is the Time for Rapid Development of Xinjiang', *Peoples' Daily (Beijing)*, 15 November.

Rainat, J. (1991), 'Chatchai Boonyarat: Creating an Agribusiness Empire', *Business in Thailand*, Vol. 22(2), pp.10-13.

Rasmussen, W.D. (1968), 'Advances in American Agriculture: the Mechanical Tomato Harvester as a Case Study', *Technology and Culture*, Vol. 9(4), pp.531-43.

Rausser, G., Fargeix, A. and Lear-Nordby, K. (2000) 'USA: The Tomato Processing Industry Outlook' *Tomato News*, Vol. 12 (May), pp. 9-18.

Raynolds, L. (2000), 'Negotiating contract farming in the Dominican Republic', *Human Organization*, Vol. 59(4), pp.441-51.

Rendell McGuckian (2001), 'Development of the Processing Tomato Industry Research and Development Strategic Plan', *Horticulture Australia Project Reports* TM00004, Horticulture Australia, Sydney.

Rendell McGuckian, Kelliher Consulting, Barry Horn & Associates, Agriculture Victoria, (1996), *Benchmarking the Australian Processing Tomato Industry: A Report Released by the Tomato Industry Negotiating Committee*, Report available from Agriculture Victoria, Tatura, 3616 Victoria, Australia.

Reyes, S. (2001), 'Unleashing the Big Red Rocket', *Brandweek*, Vol. 42(38), pp.M6-M11.

Rickson, R.E. and Burch, D.F. (1996), 'Contract Farming in Organizational Agriculture: The Effects Upon Farmers and the Environment', in D. Burch, R.E. Rickson and G. Lawrence (eds) *Globalization and Agri-Food Restructuring: Perspectives from the Australasia Region*, Avebury, Aldershot, pp.173-202.

Ritzer, G. (1993), *The McDonaldization of Society: An Investigation into the Changing Character of Contemporary Social Life*, Pine Forge Press, New York.

Roberts, I., Podbury, T., Freeman, F. & Tulpule, V. (2001),'A Vision for Multilateral Agricultural Policy Reform', *ABARE Current Issues*, Vol. 01(2).

Rogers, P. (2001), 'Deal a Meal' *Prepared Foods*, July, pp.10-16.

Rosset, P., Rice, R. and Watts, M. (1999), 'Thailand and the World Tomato: Globalization, New Agricultural Countries (NACs) and the Agrarian Question', *International Journal of Sociology of Agriculture and Food*, Vol. 8, pp.71-94.

Rosset, P. and Vandermeer, J.H. (1986), 'The Confrontation Between Labor and Capital in the Midwest Processing Tomato Industry and the Role of the Agricultural Research and Extension establishment', *Agriculture and Human Values*, Vol. 3, pp.26-31.

Rufer, C.J. (1997), 'The Impact of In-Season Economics of Processing Tomatoes on Annual Production Stability in California', *Tomato News*, Vol. 9, (June), pp.8-13.

Rufer, C.J., Evans, M.E. and Gashaw, D. (1999), *Stability in the Processing Tomato Industry: Myths, Practices and Proposals*, The Morning Star Company, Woodland (California).

Russo Group (2000), 'Tomato Processing Results 1999 crop', unpublished memorandum, available from authors.

Russo Group (not dated), 'La Scelta Naturale', brochure published by the Russo Group, S. Antonio Abate, NA, Italy.

Sacramento Business Journal (2000), 'Carmichael Firm Aims High with Online Tomato Trading', *Sacramento Business Journal*, 29 June, p.11.

Sadler, D. (1992), *The Global Region*, Pergamon, Oxford.

Salaman, R.N. (1949), *The History and Social Influence of the Potato*, Cambridge University Press, Cambridge.

San Francisco Chronicle (1990), 'Pizza Hut tackles cheese-hating nation', 11 September, pp. C7.

Sanderson, S. (1986), 'The emergence of the 'world steer': internationalisation and foreign domination in Latin American cattle production', in F.L. Tullis, and W.L. Hollist, (eds), *Food, The State, and International Political Economy*, University of Nebraska Press, Lincoln, NB, pp.123-48.

Sasithorn, O. (2000), 'Sawasdi Upbeat on Debt Restructuring', *The Nation* (Bangkok), 21 March.

Schlosser, E. (2001), *Fast Food Nation: The Dark Side of the All-American Meal*, Houghton-Mifflin, Boston.

Schmitz, A. and Seckler, D. (1970), 'Mechanized Agriculture and Social Welfare: the Case of the Tomato Harvester', *American Journal of Agricultural Economics*, Vol. 52, pp.569-77.

Schnitt, P. (2000), 'California Agriculture Cooperatives Problems are Deeply Rooted', *Sacramento Bee*, 24 July, p.1.

Schnitt, P. (2001), 'Tomato Farmers Form Partnership to Operate Colusa, Calif. Cannery', *Sacramento Bee*, 8 March, p.6.

Schrag, P. (1978), 'Rubber Tomatoes', *Harper's*, June, pp.48-52.

Simon, H. (1959), 'Theories of decision making in economics', *American Economic Review*, Vol. 49, pp.253-83.

Sims, W.L., Zobel, M.P. and King, R.C. (1968), *Mechanized Growing and Harvesting of Processing Tomatoes*, University of California Agricultural Extension Service, Davis.

Sims, W.L., Zobel, M.P. and May, D.M. (1979), *Mechanized Growing and Harvesting of Processing Tomatoes*, University of California Division of Agricultural Sciences, Davis, Leaflet 2686.

Sjerven, J. (1991), 'Tomato Processor Brings Opportunity to Esarn', *Agribusiness Worldwide*, Nov/Dec, pp.20-21.

Skorburg, J. (2002), 'China and the World Supply of Garlic', *China Briefing Book (American Farm Bureau Federation)*, Vol. 10, at http://www.fb.com/issues/analysis/China_Briefing_Issue10.html, accessed 18 October, 2002.

Smellie, P. (1998), 'Brierley Out to Recover Loans', *The Sunday Star-Times* (Auckland), 6 December, p.B1.

Smith, A.F. (1994), *The Tomato in America*, University of South Carolina Press, Columbia, SC.

Somluck, S. (1999), 'British Firm Could Take Third Stake in UFC', *The Nation* (Bangkok), 20 November, p.11.

Sonito, (2000), 'Assemblee Generale Ordinaire, 21 Juin 2000' unpublished memorandum available from Sonito, 54, chemin de Bonaventure 84000 Avignon, France.

Sonnenfeld, D., Schotzko, T. and Jussaume, R. (1998), 'Globalization of the Washington Apple Industry: its Evolution and Impacts', *International Journal of the Sociology of Agriculture and Food*, Vol. 7, pp.151-80.

Spence, P. (2000), 'Consumer Trends and Marketing Opportunities-a Look into the Future for Tomato as a Functional Food', unpublished presentation at *IV World Congress on the Processing Tomato*, Sacramento, CA, 8 June.

Spoor, M. (2002), 'Policy Regimes and Performance of the Agricultural Sector in Latin America and the Caribbean During the Last Three Decades', *Journal of Agrarian Change*, Vol. 2(3), pp.381-400.

Stanley, K. (1994), 'Industrial and Labor Market Transformation in the US Meatpacking Industry', in P.McMichael (ed.) *The Global Restructuring of Agro-Food Systems*, Cornell University Press, Ithaca, pp.129-44.

Statistics Canada (various years), 'Finished Product Volumes – Processing Tomatoes', Government of Canada, Ottawa.

Stiglitz, J. (1985), 'Information and economic analysis: a perspective', Economics, Vol. 95 (supplement), pp.21-41.

Stitt, K. (2001), 'Marketing 1000', *Advertising Age* (Midwest edition), Vol. 72(41), p.S25.

Stork, P., Ashcroft, B. and Jerie, P. (2000), 'Improving Soluble Solids in Trickle Irrigated Processing Tomato Crops', *Australian Processing Tomato Grower*, Vol. 21, pp.20-23.

Storper, M. (1999), 'Conventions and the Genesis of Institutions', unpublished paper presented at Institute for International Studies, University of California, Berkeley, October 25.

Stride, N. (2001), 'Cedenco Buyer Eyes Expansion', *National Business Review* (Auckland), 11 May, p.3.

Stuller, J. (1997), 'As American as Pizza Pie', *The Smithsonian*, Vol. 28(3), pp.138-47.

Sumner, D.A., Rickard, B.J. and Hart, D.S. (2001), *Economic Consequences of European Union Subsidies for Processing Tomatoes*, University of California Agricultural Issues Center, Davis CA.

Taper, B. (1980), 'The Bittersweet Harvest', *Science*, November, pp.79-84.

Thailand: Department of Customs (various years), *Foreign Trade Statistics of Thailand,* Ministry of Finance, Bangkok.

Thurow, L. (1996), *The Future of Capitalism*, Allen and Unwin, Sydney.

Tomato Growers Council Victoria - Victorian Farmers Federation (1998), 'Price Negotiations', *Tomato Topics* (Shepparton, Australia), Vol. 7(2), p.7.

Tomato Industry Negotiating Committee (1994), 'Industry Code of Practice for Tomato Processors and Tomato Producers'. Available from Department of Natural Resources and Environment, Tatura, Victoria.

Tomato Land, (2002), 'The Tomato Interview: Mr Chris Rufer', *Tomato Market Report*, 17 May (circular distributed to delegates at the *V World Congress on the Processing Tomato*, Istanbul).

Tomato News (1992), 'The 1992 season', *Tomato News*, Vol. 4 (December), pp.5-16.

Tomato News (1993a), 'Balance sheet of USA foreign trade', *Tomato News*, Vol. 5 (May), pp.34-35.

Tomato News (1993b), 'USA and Chile are advancing on the Japanese paste market', *Tomato News*, Vol. 5 (May), p.38.

Tomato News (1993c), 'Southern hemisphere and tropical countries in the northern hemisphere: mediocre 92/93 season', *Tomato News*, Vol. 5 (June), pp.15-19.

Tomato News (1994a), 'The agro-industrial situation in Latin America', *Tomato News*, Vol. 6 (March), pp.13-21.

Tomato News (1994b) 'China: production development', *Tomato News*, Vol. 6 (June), p.22.

Tomato News (1994c), 'Countries with January to July harvest', *Tomato News*, Vol. 6 (December), pp.18-21.

Tomato News (1995a), 'Chilean tomato situation', *Tomato News*, Vol. 7 (January), pp.29-34.

Tomato News (1995b), '1994 exports from Chile', *Tomato News*, Vol. 7 (September), p.28.

Tomato News (1995c), 'Senegal: liberalisation of the import of tomato products', *Tomato News*, Vol. 7 (September), p.30.

Tomato News (1995d), 'Tomato processing in China', *Tomato News*, Vol. 7 (November), pp.17-18.

Tomato News (1996), 'Processing Tomatoes, 1995/96 Season', *Tomato News*, Vol. 8 (June) p.30.

Tomato News (1997), 'Special Canada Issue' Vol.9 (November), pp.9-50.

Tomato News (2000a), 'Japan: Kagome, the Tomato and Vegetable Company', *Tomato News*, Vol. 12 (May), pp. 16-18.

Tomato News (2000b), 'Twenty Top Processing Tomato Varieties in California', *Tomato News*, Vol. 12 (September), p.43.

Tomato News (2000c), 'The European Pizza Market', *Tomato News*, Vol. 12 (October), p. 27.

Tomato News (2000d), 'China: Export Statistics', *Tomato News*, Vol. 12 (December), p.41.

Tomato News (2001a), 'World Production Trends', *Tomato News*, Vol. 15 (February), pp.4-16.

Tomato News (2001b), 'Japan: Turkey Leads Paste Imports', *Tomato News*, Vol. 13 (February), p.49.

Tomato News (2001c), 'Californian and Italian Paste Prices', *Tomato News*, Vol., 13 (April), p.7.

Tomato News (2001d), 'Chile: tomato processing in 1999-2000', *Tomato News*, Vol. 13 (April), p.9.

Tomlinson, J. (1999), *Globalisation and Culture*, Polity Press, Oxford.

Truss, W. (1999), 'QUINT Ministers Agree to Urgency of WTO Agriculture Round', *Media Release AFFA99/50WT*, 5 October.

Ufkes, F. (1993), Trade Liberalization, Agro-food Politics and the Globalization of Agriculture', *Political Geography*, Vol. 12(3), pp.215-31.

Unilever (2001), 'Sustainable agriculture – tomatoes', at www.unilever.com/en/si_tomatoes.html, accessed 2 February, 2002.

United Nations Conference on Trade and Development (UNCTAD) (various years) *World Investment Report*, United Nations Publications, New York and Geneva.

United States Census Bureau, (2000), 'Model-based Income and Poverty Estimates for Colusa County in 1997' at www.census.gov/hhes/www/saipe/estimate/cty/cty06011, accessed 4 March, 2001.

United States Department of Agriculture Foreign Agricultural Service (2000), 'Italy: Tomatoes and Products', *Global Agriculture Information Network Report IT0020*.

United States Department of Agriculture Foreign Agricultural Service (2001), 'Mexico: Tomatoes and Products', *Global Agriculture Information Network Report MX0174*.

United States Department of Agriculture Foreign Agricultural Service (2002a), 'Spain: Tomatoes and Products', *Global Agriculture Information Network Report SP2017*.

United States Department of Agriculture Foreign Agricultural Service (2002b), 'Italy: Tomatoes and Products', *Global Agriculture Information Network Report IT0060*.

University of California (2000), 'Agriculture is a Growing Part of the E-commerce World', *University of California: Agriculture and Natural Resources Media Release*, 14 December.

Upton, L. (2001), 'Leap of Faith', *Insight* (television documentary, broadcast 29 March on SBS Television, Australia). Transcript at www.sbs.com.au/insight, accessed 4 April, 2001.

Uyghur Information Agency (2001), 'Bingtuan Establishes Trade Relations with More Than Sixty Countries', at www.uyghurinfo.com/views/News, accessed 13 July, 2001.

Vaitsos, C. (1973a), 'Bargaining and the Distribution of Returns in the Purchase of Technology by Developing Countries', in H. Bernstein (ed.), *Underdevelopment and Development*, Penguin, Harmondsworth., pp.315-22.

Vaitsos, C. (1973b), 'Patents Revisited: Their Function in Developing Countries', in C. Cooper (ed.) *Science, Technology and Development*, Frank Cass, London., pp.71-97.

Vallat, O. (2001), personal communication.

Victorian Electoral Commission (2001), 'State election 1999', at www.vec.vic.gov.au, accessed 22 April, 2001.

Wade, S., Hickey, M., Quadir, M., Ashcroft, B., Ashburner, R. and Watters, M. (2000), 'Cultivar Trials: 1999-2000', *Australian Processing Tomato Grower*, Vol. 1, pp.16-17.

Watson, J.L. (ed.) (1997), *Golden Arches East: McDonald's in East Asia*, Stanford University Press, Stanford, CA.

Watts, M. (1992), 'Living under contract: work, production politics, and the manufacture of discontent in a peasant society', in A. Pred, and M. Watts, (eds) *Reworking Modernity: Capitalisms and Symbolic Discontent*, Rutgers University Press, New Brunswick (NJ).

Watts, M. (1996), 'Development III: the Global Agrofood System and Late Twentieth-Century Development (or Kautsky redux)', *Progress in Human Geography*, Vol. 20(2), pp. 230-45.

Wells, M. (1996), *Strawberry Fields: Politics, Class and Work in Californian Agriculture*, Cornell University Press, Ithaca and London.

Wilkinson, J. (1997), 'Regional Integration and the Family Farm in the Mecosul Countries: New Theoretical Approaches as Supports for Alernative Strategies', in D. Goodman and

M. Watts (eds) *Globalising Food: Agrarian Questions and Global Restructuring*, Routledge, London, pp.56-78.

Winson, A. (1993), *The Intimate Commodity: Food and the Development of the Agro-Industrial Complex in Canada*, Garamond Press, Toronto.

Wolf, S., Hueth, B., and Ligon, E. (2001), 'Policing Mechanisms in Agricultural Contracts', *Rural Sociology*, Vol. 66 (3), pp. 359-81.

Woodward, D. (1982), 'A New Direction for China's State Farms', *Pacific Affairs*, Vol. 55(2), pp. 231-51.

World Processing Tomato Council (2002), 'World Production Table', at www.wptc.to, accessed 12 October, 2002.

Wrigley, N. and Lowe M.S. (eds) (1996), *Retailing, Consumption and Capital: Towards a New Retail Geography*, Harlow, Longman.

Xian, L. (2002), 'Development of the Internal Market for Tomato Products in the Republic of China with Particular Reference to the Quality Problems Encountered with the Start of Tomato Processing in China and their Solutions', unpublished presentation to the *V World Congress on the Processing Tomato*, Istanbul, 10 June.

Xinhua News Agency (2001), 'KFC Earns Two Billion Yuan Annually from China', Media Release, 28 November.

Xinjiang Tienye Co. Ltd (2001), 'Xinjiang Tienye Co. Ltd', at www.xj-tienye.com, accessed 13 July, 2001.

Xinjiang Tunhe Co. Ltd (2001), 'Tunhe tomato paste' (corporate brochure), Xinjiang Tunhe Co. Ltd, Changji.

Yeung, D. (2000), unpublished panel contribution, *IV World Congress on the Processing Tomato*, Sacramento, 8 June.

Yoa, C.T. (2001), personal communication.

Zuckerman, L. (1998), *The Potato: From the Andes in the Sixteenth Century to Fish and Chips, The Story of How a Vegetable Changed History*, Faber & Faber, Boston.

Index